Lecture Notes in Computer Science 4630

Commenced Publication in 1973
Founding and Former Series Editors:
Gerhard Goos, Juris Hartmanis, and Jan van Leeuwen

T0224446

H. Jaap van den Herik   Paolo Ciancarini
H.H.L.M. (Jeroen) Donkers (Eds.)

Computers
and Games

5th International Conference, CG 2006
Turin, Italy, May 29-31, 2006
Revised Papers

 Springer

Volume Editors

H. Jaap van den Herik
Institute for Knowledge and Agent Technology (IKAT)
MICC, Universiteit Maastricht
P.O. Box 616, 6200 MD Maastricht, The Netherlands
email: herik@micc.unimaas.nl

Paolo Ciancarini
Dipartimento di Scienze dell'Informazione
Università di Bologna
Mura Anteo Zamboni 7
40127 Bologna, Italy
email: cianca@cs.unibo.it

H.H.L.M. (Jeroen) Donkers
Institute for Knowledge and Agent Technology (IKAT)
MICC, Universiteit Maastricht
P.O. Box 616, 6200 MD Maastricht, The Netherlands
email: donkers@micc.unimaas.nl

Library of Congress Control Number: 2007936369

CR Subject Classification (1998): G, I.2.1, I.2.6, I.2.8, F.2, E.1

LNCS Sublibrary: SL 1 – Theoretical Computer Science and General Issues

ISSN   0302-9743
ISBN-10   3-540-75537-3 Springer Berlin Heidelberg New York
ISBN-13   978-3-540-75537-1 Springer Berlin Heidelberg New York

Springer is a part of Springer Science+Business Media

springer.com

© Springer-Verlag Berlin Heidelberg 2007

Typesetting: Camera-ready by author, data conversion by Scientific Publishing Services, Chennai, India
Printed on acid-free paper   SPIN: 12171316   06/3180   5 4 3 2 1 0

Preface

This book contains the papers of the Fifth Computers and Games Conference (CG 2006) held in Turin, Italy. The conference took place during May 29–31, 2006 in conjunction with the 11th Computer Olympiad and the 14th World Computer Chess Championship.

The Computers and Games conference series is a major international forum for researchers and developers interested in all aspects of artificial intelligence and computer-game playing. The Turin conference was definitively characterized by fresh ideas for a great variety of games. Earlier conferences took place in Hamamatsu, Japan (2000), Edmonton, Canada, (2002), Ramat Gan, Israel (2004), and Taipei, Taiwan (2005).

The Program Committee (PC) received 45 submissions. Each paper was initially sent to at least two referees. If conflicting views on a paper were reported, it was sent to an additional referee. Out of the 45 submissions, 2 were withdrawn before the final decisions were made. With the help of many referees (listed after the preface), the PC accepted 24 papers for presentation at the conference and publication thereafter provided that the authors submitted their contribution to a post-conference editing process. The two-step process was meant (i) to give authors the opportunity to include the results of the fruitful discussion after the lecture in their paper, and (ii) to maintain the high-quality threshold of the CG series. The authors enjoyed this procedure.

The above-mentioned set of 24 papers covers a wide range of computer games. Ten of these games are also played in practice by human players, viz., Western Chess, Chinese Chess (Xiangqi), Japanese Chess (Gunjin Shogi), Hex, Go, Lines of Action, Hearts, Skat, Lumines, and Pool. Moreover, there are three theoretical games, viz., Rush Hour, the Penny Matching Game, and One Player Can't Stop.

The games cover a wide range of research topics, including combinatorial game theory, machine learning, networked games, search, knowledge representation, and optimization.

We hope that the readers will enjoy the efforts of our researchers. Below we provide a brief characterization of the 24 contributions, in the order in which they are printed in the book.

"Computer Analysis of Chess Champions" by Matej Guid and Ivan Bratko compares the performance of World Chess Champions. The comparison is achieved with the help of a chess-playing program that analyzes games played by the World Chess Champions.

"Automated Chess Tutor" is authored by Aleksander Sadikov, Martin Možina, Matej Guid, Jana Krivec, and Ivan Bratko. The article describes a tutoring program which employs descriptions of tactical goals. Core mechanisms for the production of automated comments in terms of relevant goals are presented.

"A New Heuristic Search Algorithm for Capturing Problems in Go" by Keh-Hsun and Peigang Zhang Chen introduces a domain-specific heuristic for iterative-deepening that is based on third-order liberties. The search is applied to life-and-death problems in Go.

"An Open Boundary Safety-of-Territory Solver for the Game of Go" is written by Xiaozhen Niu and Martin Müller. It describes the SAFETY SOLVER 2.0. The program identifies open boundary problems under real game conditions, and generates moves for invading and defending such areas.

"Monte-Carlo Proof-Number Search for Computer Go" by Jahn-Takeshi Saito, Guillaume Chaslot, H. Jaap van den Herik, and Jos W.H.M. Uiterwijk introduces an enhancement to the Proof-Number Search algorithm. The new enhancement uses Monte-Carlo sampling to initialize proof and disproof numbers. The algorithm is tested on life-and-death problems in the game of Go.

"Virtual Global Search: Application to 9×9 Go" is authored by Tristan Cazenave. Virtual Global Search is a variation of Monte-Carlo search. Sample sequences are grouped according to their similarity under a variety of permutations in order to reduce the number of simulations. The algorithm is tested on 9×9 Go.

"Efficient Selectivity and Backup Operators in Monte-Carlo Tree Search" by Rémi Coulom describes the algorithmic framework underlying the Go program CRAZY STONE The framework develops a selective search tree by repeatedly applying Monte-Carlo evaluations at leaf nodes.

"Combinatorics of Go," written by John Tromp and Gunnar Farnebäck, presents a dynamic programming algorithm for computing the number of legal moves in $n \times m$ Go boards. For approximating this number on large boards, a formula is given. Moreover, the lower and upper bounds for the game-tree complexity of Go are mentioned (proof available at Tromp's Web site).

"Abstracting Knowledge from Annotated Chinese Chess game records" is a contribution by Bo-Nian Chen, Pangfang Liu, Shun-Chin Hsu, and Tsan-sheng Hsu. The authors introduce a method for automatically extracting knowledge from professional Chinese Chess game records. The article describes how the suggested procedure is successfully applied to about 2,000 game positions.

"Automatic Strategy Verification for Hex" is a joint effort by Ryan B. Hayward, Broderick Arneson, and Philip Henderson. The authors propose an AND/OR-tree annotation to describe Hex strategies concisely and an algorithm for verifying the correctness of Hex strategies. Both items are applied to Jing Yang's 7×7 center-opening strategy.

"Feature Construction for Reinforcement Learning in Hearts" is a contribution by Nathan R. Sturtevant and Adam M. White. The paper focuses on training a Hearts playing program by TD learning and stochastic linear regression. Special attention is directed to the underlying model for integrating features into the training.

"A Skat Player Based on Monte-Carlo Simulation," is by Sebastian Kupfer-schmid and Malte Helmert. The authors present a program that plays the card game Skat. The bidding engine and the card-playing engine are both based on Monte-Carlo simulation.

"A Retrogade Approximation Algorithm for One-Player Can't Stop," by James Glenn, Haw-ren Fang, and Clyde P. Kruskal, reports a retrograde approximation algorithm for solving small versions of One-Player Can't Stop based on a representation of the game as bipartite graph. The paper also gives a proof for the existence of a solution for some instances of One-Player Can't Stop.

"Improving Depth-First PN-Search: The $1 + \epsilon$ Trick" is written by Jakub Paw-lewicz and Lukasz Lew. The paper introduces an improvement for Proof-Number Search algorithms. The improvement helps to reduce the space complexity in practical applications. It mainly consists of slightly loosening the proof-number boundaries for expanding sub-trees. An application of the improved search algorithm to Lines of Action and Atari Go is presented.

"Search Versus Knowledge Revisited Again" is a contribution by Aleksander Sadikov and Ivan Bratko. The article describes the phenomenon of diminishing returns in chess and the effect of domain-dependent knowledge on diminishing returns.

"Counting the Number of Three-Player Partisan Cold Games," by Alessandro Cincotti, generalizes combinatorial game theory from two-player games to three-player games and investigates lower and upper bounds for the number of three-player games born on day n. The paper also covers the number of surreal numbers born on day n.

"LUMINES Strategies" is a contribution by Greg Aloupis, Jean Cardinal, Sébastien Collette, and Stefan Langerman. It describes how two parameters, the grid size and the frequency of block deletion, determine the solvability of instances of the computer game Lumines.

"Computing Proper Equilibria of Zero-Sum Games" is by Peter Bro Miltersen and Troels Bjerre Sørensen. A proper equilibrium of a matrix game can be found in polynomial time by solving a number of linear programs. An exemplary application is provided for the Penny Matching game.

"Comparative Study of Approximate Strategies for Playing Sum Games Based on Subgame Types," written by Cherif R.S. Andraos, Manal M. Zaky, and Salma A. Ghoneim, presents a scheme for evaluating different strategies of playing sums of left-excitable, right-excitable, or equitable combinatorial games. The presented evaluation scheme analyzes the strategies in greater detail than previously found evaluation schemes do.

"On the Symbolic Computation of the Hardest Configurations of the RUSH HOUR Game" is a paper by Sébastien Collette, Jean François Raskin, and Frédéric Servais. It describes a method for finding difficult initial configurations for the puzzle game Rush Hour. The method is based on a representation in propositional logic and requires symbolic model checking.

"Cheat-Proof Serverless Network Games" by Shunsaku Kato, Shuichi Miyazaki, Yusuke Nishimura, and Yasuo Okabe outlines an algorithm for preventing cheating in games played on peer-to-peer networks. An example application of the algorithm to Gunjin Shogi is provided.

"Monte-Carlo Methods in Pool Strategy Game Trees" is a joint effort by Will Leckie and Michael Greenspan. The paper discusses a successful application of probabilistic search techniques to the task of choosing strategies in the game of Pool.

"Optimization of a Billiard Player — Tactical Play" is a contribution by Jean-Pierre Dussault and Jean François Landry. It compares heuristics for optimizing the choice of targets after repositioning in the game of Pool.

"Gender and Cultural Differences (If Any!): South African School Children and Computer Games," written by Lizette de Wet and Theo McDonald, reports the findings of a quantitative empirical study on the computer-game-playing behavior of school children in South Africa. The results indicate that there are no major differences in the game-playing behavior dependent on gender or culture. This book would not have been produced without the help of many persons. In particular, we would like to mention the authors and the referees for their help. Moreover, the organizers of the three events in Turin (see the beginning of this preface) have contributed substantially by bringing the researchers together. We thank Tons van den Bosch, Jahn-Takeshi Saito, and Marijke Verheij for their expert assistance in making the manuscript fit for publication. They did a very good job. Without much emphasis, we recognize the work by the committee of CG 2006 as essential for this publication. Finally, the editors happily recognize the generous sponsors Fiat Group, Sao Paolo, General Electrics, and ChessBase.

March 2007

Jaap van den Herik
Paolo Ciancarini
Jeroen Donkers

Organization

Executive Committee

Editors	H. Jaap van den Herik
	Paolo Ciancarini
	H. (Jeroen) H. L. M. Donkers
Program Co-chairs	Paolo Ciancarini
	H. Jaap van den Herik

Organizing Committee

Johanna W. Hellemons (Chair)   Paolo Ciancarini
Antonia (Tons) E.M. van den Bosch   H. (Jeroen) H. L. M. Donkers
H. Jaap van den Herik

Sponsors

Institutional Sponsors	Regione Piermonte	
	Provincia di Torino	
	Cittá di Torino	
	Chessbase, Hamburg, Germany	
Main Sponsor	Fiat Group	
Official Sponsor	Sao Paolo	
	General Electrics	
Partners	Federazione Scacchistica Italiana	
	Societá Scacchistica Torinese	
	FIDE	
Official Suppliers	IKON	Blubs Viaggi
	The Place	Chessbase
	I.Net	Directa Sim
	Istituto Boella	Aquanova
	Sai Assicurazioni	

We also thank the following sponsors of the Olympics Scientific Program "Mosse d'Autore" of which the conference was part.

Compagnia di Sao Paolo
Fondazione CRT
Regione Piemonte (assessorato alla
cultura)

Program Committee

Yngvi Björnsson	Guy Haworth	Martin Müller
Bruno Bouzy	Ryan Hayward	Jacques Pitrat
Michael Buro	Jaap van den Herik	Christian Posthoff
Tristan Cazenave	Shun-Chin Hsu	Mathias Rauterberg
Guillaume Chaslot	Tsan-sheng Hsu	Jahn-Takeshi Saito
Keh-Hsun Chen	Hiroyuki Iida	Jonathan Schaeffer
Paolo Ciancarini	Graham Kendall	Pieter Spronck
Rémi Coulom	Akihiro Kishimoto	Nathan Sturtevant
Jeroen Donkers	Yoshiyuki Kotani	Jos Uiterwijk
Markus Enzenberger	Hams Kuijf	Tobias Walsh
Haw-ren Fang	Shun-Shii Lin	Jan Willemson
Aviezri Fraenkel	Ulf Lorenz	Mark Winands
Michael Greenspan	Shaul Markowitz	I-Chen Wu
Reijer Grimbergen	Alberto Martelli	Shi-Jim Yen

Referees

Ingo Althöfer	Ryan Hayward	Jacques Pitrat
Don Beal	H. Jaap van den Herik	Christian Posthoff
Yngvi Björnsson	Kai Himstedt	Jan Ramon
Bruno Bouzy	Shun-Chin Hsu	Mathias Rauterberg
Ivan Bratko	Tsan-sheng Hsu	Jeff Rollason
Andries Brouwer	Robert Hyatt	Leon Rothkrantz
Michael Buro	Hiroyuki Iida	Jahn-Takeshi Saito
Tristan Cazenave	Hermann Kaindl	Jonathan Schaeffer
Guillaume Chaslot	Kuo-Yuan Kao	Pieter Spronck
Keh-Hsun Chen	Graham Kendall	Nathan Sturtevant
Paolo Ciancarini	Akihiro Kishimoto	Gerald Tesauro
Rémi Coulom	Masashi Kiyomi	Thomas Thomsen
Jean Derks	Joghan de Koning	John Tromp
Jeroen Donkers	Richard Korf	Jos Uiterwijk
Jean-Pierre Dussault	Hans Kuijf	Paul Utgoff
Peter van Emde Boas	Jeroen Kuipers	Tobias Walsh
Markus Enzenberger	David Levy	Erik van der Werf
Haw-ren Fang	Michael Littman	Karin Wenz
Rudolf Fleischer	Richard J. Lorentz	Janet Wiles
David Fotland	Ulf Lorenz	Jan Willemson
Aviezri Fraenkel	Thomas Maarup	Mark Winands
William Fraser	Shaul Markovitch	Thomas Wolfe
Johannes Fürnkranz	Tony Marsland	I-Chen Wu
Matthew Gams	Alberto Martelli	Hiroshi Yamashita
Michael Greenspan	Martin Müller	Shi-Jim Yen
Reijer Grimbergen	Katsuhiko Nakamura	Jan van Zanten
Dap Hartmann	Dana Nau	
Guy Haworth	Kohei Noshita	

Table of Contents

Computer Analysis of Chess Champions

Matej Guid and Ivan Bratko

Artificial Intelligence Laboratory, Faculty of Computer and Information Science,
University of Ljubljana, Slovenia
{matej.guid,bratko}@fri.uni-lj.si

Abstract. Who is the best chess player of all time? Chess players are
often interested in this question that has never been answered authorita-
tively, because it requires a comparison between chess players of different
eras who never met across the board. In this contribution, we attempt
to make such a comparison. It is based on the evaluation of the games
played by the World Chess Champions in their championship matches.
The evaluation is performed by the chess-playing program CRAFTY. For
this purpose we slightly adapted CRAFTY. Our analysis takes into account the differences in players' styles to compensate the fact that calm
positional players in their typical games have less chance to commit gross
tactical errors than aggressive tactical players. Therefore, we designed a
method to assess the difficulty of positions. Some of the results of this
computer analysis might be quite surprising. Overall, the results can be
nicely interpreted by a chess expert.

1 Introduction

Who is the best chess player of all time? This is a frequently posed and interesting
question, to which there is no well founded, objective answer, because it requires
a comparison between chess players of different eras who never met across the
board. With the emergence of high-quality chess programs a possibility of such
an objective comparison arises. However, so far computers were mostly used as
a tool for statistical analysis of the players' results. Such statistical analyses
often do neither reflect the true strengths of the players, nor do they reflect
their quality of play. It is common that chess players play against opponents of
different strengths and that the quality of play changes in time. Furthermore,
in chess a single bad move can decisively influence the final outcome of a game,
even if all the rest of the moves are excellent. Therefore, the same result can be
achieved through play of completely different quality.

The most complete and resounding attempt made to determine the best chess
player in history has recently been put forward by Jeff Sonas, who has become a
leading authority in the field of statistical analysis in chess during the past years.
Sonas devised a specialized rating scheme [4], based on tournament results from
1840 to the present. The rating is calculated for each month separately, with the
player's activity taken into account. A player's rating, therefore, starts declining
when he is no longer active, which differs from the classic FIDE rating.

H.J. van den Herik et al. (Eds.): CG 2006, LNCS 4630, pp. 1–12, 2007.
© Springer-Verlag Berlin Heidelberg 2007

Having a unified system of calculating ratings represents an interesting solution to determining a "common denominator" for all chess players. However, it does not take into account that the quality of play has risen drastically in the recent decades. The first official World Champion, Steinitz, achieved his best Sonas rating, which is on a par with ratings of recent champions, in April 1876. His rating is determined from his successes in tournaments in a time when the general quality of play was well below that of today. The ratings in general reflect the players' success in competition, but not directly their quality of play.

Other estimates about who was the strongest chess player of all times, are primarily based on the analyses of their games as done by chess grandmasters; obviously these are often subjective. In his unfinished set of books *My Great Predecessors*, Gary Kasparov [3], the thirteenth World Chess Champion, analyses in detail numerous games of the best chess players in history and will most probably express his opinion regarding who was the best chess player ever. But it will be merely an *opinion*, although very appreciated in the chess world.

Our approach was different: we were interested in the chess players' quality of play regardless of the game score, which we evaluated with the help of computer analyses of individual *moves* made by each player.

2   Method

We evaluated fourteen classic-version World Champions, from the first World Chess Championship in 1886 to the present. Matches for the title of "World Chess Champion", in which players contended for or were defending the title, were selected for analysis.

Roughly, the basis for evaluation of a human's play was the difference between the position values resulting from the moves played by the human and the moves chosen as best by the chess program. This approach can be criticized on the basis that (1) sometimes there are alternative, equally strong moves, and (2) the choice between them is a matter of playing style and not merely a matter of chess strength. We will return to this issue later and provide a refinement and a justification for this approach.

Evaluation of each game started on the 12th move, without the use of an openings library, of course. This decision was based on the following careful deliberation. Not only today's chess programs poorly evaluate positions in the first phase of a game, but also analyzing games from the start would most likely favor more recent champions, due to the vast progress made in the theory of chess openings. In contrast, starting the analyses on a later move would discard too much information. The chess program CRAFTY [2], which we slightly modified for the purpose of our analyses, used. Instead of a time limit, a constant fixed search depth was applied on every move. With such an approach we achieved the following.

(1) Complex positions, which require processing larger search trees to achieve a more accurate evaluation, automatically obtain more computation time.

(2) The program could be run on different computers and still obtain the same evaluations for a given set of positions on each of the computers.

Item (2) enabled us to speed up the calculation process considerably by distributing the computation among a network of machines, and as a consequence, a greater search depth was possible. We chose to limit the search depth to 12 plies plus quiescence search. There were some speculations that a program searching 12 plies would be able to achieve a rating that is greater than that of the World Champion [1], arguably a long time ago. However, the search depth mentioned was chosen as the best alternative, since deeper search would mean a vast amount of additional computation time (more than ten full days of computation time on 36 machines with an average speed of 2.5 GHz were required to perform the analyses of all games). The limit of search was increased to 13 plies in the endgame. CRAFTY's definition of endgame was used: the endgame starts when the total combined numerical value of both white and black pieces on the board (without Pawns) is less than 15. We also changed the 'King's safety asymmetry' parameter thus achieving a shift from CRAFTY's usual defensive stance to a more neutral one where it was neither defensive nor offensive. The option to use quiescence search was left turned on to prevent horizon effects.

With each evaluated move, data was collected for different search depths (which ranged from 2 to 12), comprising (1) the best evaluated move and its evaluation , (2) the second-best evaluated move and its evaluation, (3) the move made by the human and its evaluation. We also collected data on the material state of both players from the first move on.

2.1   Average Difference Between Moves Made and Best Evaluated Moves

The basic criterion was the average difference between numerical evaluations of moves that were played by the players and numerical evaluations of moves that were suggested by computer analysis as the best possible moves.

$$\text{MeanLoss} = \frac{\sum |\text{best move evaluation} - \text{move played evaluation}|}{\text{number of moves}} \qquad (1)$$

Additional limitations were imposed upon this criterion. Moves, where both the move made and the move suggested had an evaluation outside the interval [-2, 2], were discarded and not taken into account in the calculations. The reason for this is the fact that a player with a decisive advantage often chooses not to play the best move, but rather plays a move which is still 'good enough' to lead to victory and is less risky. A similar situation arises when a player considers his[1] position to be lost, a deliberate objectively worse move may be made in such a case to give the player a higher practical chance to save the game against a fallible opponent. Such moves are, from a practical viewpoint, justified. Taking them into account would wrongly penalize players that used this legitimate approach trying (and sometimes succeeding) to obtain a desired

---

[1] For brevity we will use 'he' ('his') when 'he or she' ('his or her') is meant.

result. All positions with evaluations outside the interval specified were declared lost or won.

2.2 Blunders

Big mistakes or blunders can be quite reliably detected by a computer, even up to a high percentage of accuracy. Individual evaluations could be inaccurate, but such inaccuracies rarely prevent the machine from distinguishing blunders (made in play) from reasonable moves.

Detection of errors was similar to the aforementioned criterion. We used a measure of difference between evaluations of moves played and evaluations of moves suggested by the machine as the best ones. We label a move as a blunder when the numerical error exceeds 1.00, which is equivalent to losing a Pawn without compensation. Like before we discarded moves where both evaluations of the move made by a player and the move suggested by the machine lie outside the [-2, 2] interval, due to reasons already mentioned.

2.3 Complexity of a Position

The main deficiency of the two criteria, as detailed in the previous subsections, is in the observation that there are several types of players with specific properties, to whom the criteria do not directly apply. It is reasonable to expect that *positional players* in average commit fewer errors due to the somewhat less complex positions in which they find themselves as a result of their style of play, than *tactical players*. The latter, on average, deal with more complex positions, but are also better at handling them and use this advantage to achieve excellent results in competition.

We wanted to determine how players would perform when facing equally complex positions. In order to determine this, a comparison metric for position complexity was required.

Although there are enormous differences in the amount of search, nevertheless there are similarities regarding the way chess programs and human players conduct a search for the best possible move in a given position. They both deal with a giant search tree, with the current position as the root node of the tree, and positions that follow with all possible moves as children of the root node, and so on recursively for every node. They both search for the best continuations and doing so, they both try to discard moves that are of no importance for the evaluation of the current position. They only differ in the way they discard them. A computer is running algorithms for efficient subtree pruning whereas a human is depending mainly on his knowledge and experience. Since they are both limited in time, they cannot search to an arbitrary depth, so they eventually have to evaluate a position at one point. They both utilize partial evaluations at given depths of search. While a computer uses evaluations in a numerical form, a human player usually has in mind descriptive evaluations, such as "small advantage", "decisive advantage", "unclear position", etc. Since they may have a great impact on the evaluation, they both check all forced variations (the computer uses quiescence search for that purpose) before giving an assessment to

```
complexity := 0

FOR (depth 2 to 12)
   IF (depth > 2) {
      IF (previous_best_move NOT EQUAL current_best_move) {
         complexity += |best_move_evaluation
         - second_best_move_evaluation|
      }
   }
   previous_best_move := current_best_move
}
```

Fig. 1. An algorithm for calculating the complexity of a position

the root position. One can therefore draw many parallels between machine and human best-move search procedures, which served as a basis for assessing the complexity of positions.

The basic idea is as follows: a given position is difficult with respect to the task of accurate evaluation and finding the best move, when different "best moves", which considerably alter the evaluation of the root position, are discovered at different search depths. In such a situation, a player has to analyze more continuations and to search to a greater depth from the initial position to find moves that may greatly influence the assessment of the initial position and then eventually choose the best continuation.

As complexity metric for an individual move, we chose the sum of the absolute differences between the evaluation of the best and the second best move. It is invoked at every time that a change in evaluation occurs when the search depth is increased. A corresponding algorithm for calculating the complexity of a position is given in Fig. 1.

The difference between the evaluations of the best and the second-best move represents the significance of change in the best move when the search depth is increased. It is reasonable to assume that a position is of higher complexity, and that it is more difficult to make a decision on a move, when larger changes regarding the best move are detected when increasing search depth. Merely counting the number of changes of the best move at different search depths would give an inadequate metric, because making a good decision should not be difficult in positions where several equally good choices arise.

We used the described metric of position complexity to determine the distribution of moves played across different intervals of complexity, based on positions that players had faced themselves. This, in turn, largely defines their *style of play*. For example, Capablanca who is regarded as a calm positional player, had much less dealing with complex situations compared to Tal, who is to be regarded as a tactical player. For each player who was taken into consideration, the distribution over complexity was determined and the average error for each complexity interval was calculated (numerical scale of complexity was divided into intervals in steps of 0.1). We also calculated an average distribution of complexity of moves made for the described intervals for all players combined.

The described approach enabled us to calculate an expected average error of World Champions in a hypothetical case where they would all play equally complex positions. We calculated the errors for two cases. Firstly, for a game of average complexity, averaged among games played by all players and, secondly, for a game of average complexity, averaged among games played by a single player. The latter represents an attempt to determine how well the players would play, should they all play in the style of Capablanca, Tal, etc.

2.4   Percentage of Best Moves Played and the Difference in Best Move Evaluations

The percentage of best moves played alone does not actually describe the quality of a player as much as one might expect. In certain types of position it is much easier to find a good move than in others. Experiments showed that the percentage of best moves played is highly correlated to the difference in evaluations of the best and second-best move in a given position. The greater the difference, the better was the percentage of player's success in making the best move (see Fig. 2).

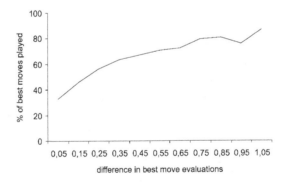

Fig. 2. Proportion of the best moves played in dependence of the difference in best-move evaluations

Such a correlation makes sense, because the bigger the difference between the best two moves, the greater the error made when selecting the wrong move. The height of the curve is amplified by the fact that we are dealing with World Champions, experts at the game of chess. Analysis of weaker players would give a curve of lesser height.

By analyzing the correlation between (1) the percentage of best moves played and (2) the difference in best two moves' evaluations, we derive information about the quality of each individual player. It turned out that curves for individual players differ significantly. This behavior served as a basis for creating a criterion, used to infer information on the quality of individual players.

For each player we calculated the distribution of moves across separate intervals of the difference in evaluations of two best moves (where the step was 0.1). We also calculated an average distribution for all players combined. Given this average distribution, we then determined the expected percentage of the best moves played for each individual player. Due to reasons already mentioned, we did not count clearly lost or won positions in this statistics.

2.5 Material

The purpose of calculating the average material quantity, that is the sum of the numerically expressed values of all pieces on board, was not to determine the quality of play, but to collect additional information on a player's style of play. We mainly tried to observe a player's inclination to simplify positions.

2.6 Credibility of Crafty as an Analysis Tool

It is important to determine whether CRAFTY represents a valid analysis tool for evaluating World Champions. Chess programs of the present time are still being regarded as somewhat weaker than the best human chess players. It is very likely that CRAFTY is weaker than at least some of the World Champions who were taken into consideration.

There are many arguments in favor of computer programs being an appropriate tool for evaluating chess players: (1) they use numerical values as evaluations, (2) they adhere to the same rules all the time, and (3) they are therefore more consistent than human observers. In particular, they are very good at evaluating tactical positions, where a great deal of computation is required.

Modified CRAFTY that was used in our work, has a great advantage when compared to standard chess programs. By limiting and fixing the search depth we achieved automatic adaptation of time used to the complexity of a given position. Another important fact is that we were able to analyze a relatively large sample of 1,397 games, containing over 37,000 positions. As a direct consequence, occasional errors made in the evaluating of positions do only marginally affect the final, averaged results (see Fig. 3).

To assess how trustworthy CRAFTY is as our assumed golden standard, we checked the correlation between our calculated error rates made in the games and the actual outcomes of these games. As stated before, in our opinion game results do not always reflect the actual quality of play and therefore the statistical analysis of game outcomes alone is not sufficient to compare World Champions. Because of this, we did not expect absolute correlation, but for CRAFTY's credibility a significant level of correlation should be detected nonetheless. We determined the correlation between the difference in measured average errors made by opposing players in a given game and the outcome of that game. Calculated Spearman correlation was found to be $\rho = 0.89$ (with significance level $p < 0.0001$).

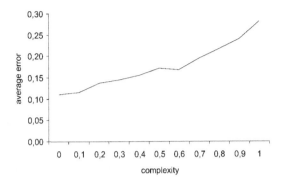

Fig. 3. Graph of errors made by players at different levels of complexity clearly indicates the validity of the chosen measure of complexity of positions. The players made little errors in simple positions, and the error rate increased with increasing complexity.

3   Results

Below we list the fourteen World Champions by five different criteria (Subsections 3.1 to 3.5). Finally, in Subsection 3.6, we provide an overview of the players' tendency to exchange pieces.

3.1   The Basic Criterion

The basic criterion for evaluating World Champions was the average difference between moves played and best evaluated moves by computer analysis.

   According to this analysis (see Fig. 4), the winner was the third World Champion, José Raúl Capablanca. We expected positional players to perform better by this criterion than tactical players. Capablanca is widely renowned to be a pure positional player. In compliance with this observation, Steinitz, who lived in an era of tactical 'romantic chess', took clearly last place.

3.2   The Blunder-Rate Measurement

The results of blunder-rate measurement are similar (see Fig. 5). We note the excellent result by Petrosian, who is widely renowned as a player who almost never blundered. Gary Kasparov [3] describes Capablanca with the following words: "He contrived to win the most important tournaments and matches, going undefeated for years (of all the champions he lost the fewest games)." and "his style, one of the purest, most crystal-clear in the entire history of chess, astonishes one with his logic."

3.3   The Complexity Measurement

Capablanca is renowned for playing a 'simple' chess and avoiding complications, while it is common that Steinitz and Tal faced many 'wild' positions in their

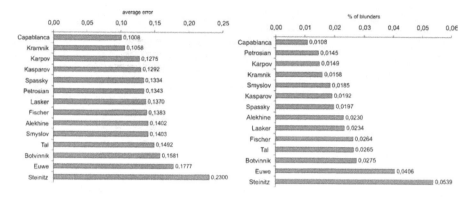

Fig. 4. Average difference between moves played and best evaluated moves (loss per move)

Fig. 5. Blunder rate

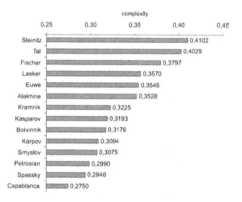

Fig. 6. Average position complexity

games. The results of the complexity measurement (see Fig. 6) clearly coincide with this common opinion.

3.4 The Player's Style

Figure 7 demonstrates that Capablanca indeed had much less dealings with complex positions compared to Tal. Distribution of moves in different intervals regarding complexity is closely related with a player's *style*. Calculated players' expected errors with a variety of such distributions was another criterion. The winner was the fourteenth World Champion Vladimir Kramnik. Kramnik also had the best performance of all the matches; the average error in his match against Kasparov (London, 2000) was only 0.0903. It is interesting to notice that Kasparov would outperform Karpov, providing they both played in Tal's style.

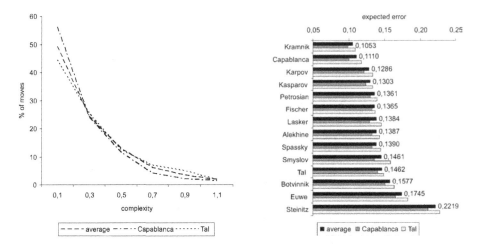

Fig. 7. Expected errors when playing in different styles

3.5   The Expected Number of Best Moves Played

A fifth criterion was the expected number of best moves played providing that all players dealt with positions with equal difference between the best two moves, as was described in the previous section. It represents another attempt to bring the champions to a common denominator (see Fig. 8). Kramnik, Fischer, and Alekhine had the highest percentage of best moves played, but also the above-mentioned difference was high. In contrast, Capablanca, who was right next regarding the percentage of the best move played, on average dealt with the smallest difference between the best two moves. The winner by this criterion was once again Capablanca. He and Kramnik again clearly outperformed the others.

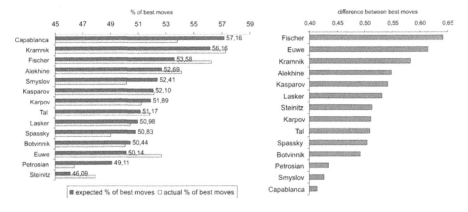

Fig. 8. Percentage of the best move played and the difference between the best two moves

3.6 The Tendency to Exchange Pieces

The graphs in Fig. 9 show the players' tendencies to exchange pieces. Among the players who stand out from the others, Kramnik obviously dealt with less material on board. The opposite could be said for Steinitz, Spassky, and Petrosian.

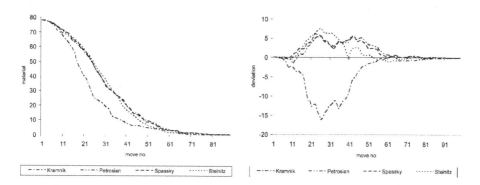

Fig. 9. Material during the game and players' deviations regarding it

4   Conclusion and Future Work

We applied the slightly modified chess program CRAFTY as tool for computer analysis of games played by World Chess Champions aiming at an objective comparison of chess players of different eras. Generally, the results of our computer analysis can be nicely interpreted by a chess expert. Some of the results might appear quite surprising and may thus be considered also as an interesting contribution to the field of chess. Capablanca's outstanding score in terms of mean value loss will probably appear to many as such an interesting finding, although it probably should not come as a complete surprise. As we did in the study, this result should be interpreted in the light of the comparatively low complexity of positions in Capablanca's games which is quite in line with the known assessments in the chess literature of his style. For example, Kasparov [3] when commenting Capablanca's games speculates that Capablanca occasionally did not even bother to calculate deep tactical variations. The Cuban simply preferred to play moves that were clear and positionally so strongly justified that calculation of variations was simply not necessary.

Our approach assumes that CRAFTY's evaluation, based on search limited to 12 ply plus quiescence, is sufficiently accurate to be used as the golden standard. It seems indeed that this worked fine in our analysis. Even if CRAFTY's evaluations are not always perfect, for our analysis they just need to be sufficiently accurate on average since small occasional errors cancel out through statistical averaging.

Still, as one idea for future work, it would be nice to obtain some more firm, quantitatively supported evidence about the evaluation error of CRAFTY with respect to some sort of ideal evaluation.

A related question is whether using more recent chess programs that in tournaments perform better than CRAFTY would make a significant difference if applied instead of CRAFTY. This question is difficult to answer directly. Since by simply plugging another program into the analysis system instead of CRAFTY, these other programs would have to be modified for the analysis similarly to CRAFTY. It would require source code of these programs that was not available. An indirect way of tentatively answering this question is however possible by evaluating these strong chess programs by our method using CRAFTY. High scores of these programs evaluated by CRAFTY would indicate that CRAFTY competently appreciates the strength of these programs, and that thus using these programs to evaluate human players instead of CRAFTY would be likely to produce similar results. To retain the style of human play, we chose to use for this experiment games played between these top programs against top human players. The results of the evaluation, presented in Table 1, give some indication that using other very strong chess programs instead of CRAFTY would probably not affect the results significantly.

Table 1. Evaluation of strong chess programs by CRAFTY

Program	Mean loss per move	Games	Opponent	Place	Year
DEEP BLUE	0.0757	6	Kasparov	New York	1997
DEEP FRITZ	0.0617	8	Kramnik	Bahrain	2002
DEEP JUNIOR	0.0865	6	Kasparov	New York	2003
FRITZ X3D	0.0904	4	Kasparov	New York	2003
HYDRA	0.0743	6	Adams	London	2005

As the mean evaluation loss per move is obviously not sufficient to assess a player's strength, we also took into account the average difficulty of positions encountered in the player's games. This made it possible to compare players of different playing styles. Our measure of position complexity seems to have produced sensible results. These results are qualitatively much in line to the observation of how an expert chess commentator would describe the players in this study in terms of their playing style. As another line of future work, it would be interesting to explore by means of a psychological study, how well our complexity measure reflects the true cognitive difficulty of a chess position.

References

1. Hsu, F., Anantharaman, T., Campbell, M., Nowatzyk, A.: A grandmaster chess machine. Scientific American 263(4), 44–50 (1990)
2. Hyatt, R.: The Crafty ftp site (2006), ftp://ftp.cis.uab.edu/pub/hyatt/
3. Kasparov, G.: My Great Predecessors, Parts 1–5. Everyman Chess, London, (2003–2006). ISBN 1857444043
4. Sonas, J.: Chessmetrics (2005), http://www.chessmetrics.com

Automated Chess Tutor

Aleksander Sadikov, Martin Možina, Matej Guid,
Jana Krivec, and Ivan Bratko

Artificial Intelligence Laboratory, Faculty of Computer and Information Science,
University of Ljubljana, Slovenia
{aleksander.sadikov,martin.mozina,matej.guid,bratko}@fri.uni-lj.si

Abstract. While recently the strength of chess-playing programs has grown immensely, their capability of explaining in human understandable terms why some moves are good or bad has enjoyed little attention. Progress towards programs with an ability to provide intelligent commentary on chess games, either played by a program or by a human, has been negligible in comparison with the progress concerning playing strength. The typical style of a program's "comments" (in terms of the best variations and their numerical scores) is of little use to a human who wants to learn important concepts behind the variations.

In this paper, we present some core mechanisms for automated commenting in terms of relevant goals to be achieved or preserved in a given position. By combining these mechanisms with an actual chess engine we were able to transform this engine into a chess tutor/annotator that is capable of generating rather intelligent commentary. The main advantages of our work over related approaches are: (a) it has the ability to act as a tutor for the whole game of chess, and (b) it has a relatively solid chess understanding and is thus able to adequately comment on positional aspects.

1 Introduction

In the last five years the strength of computer chess programs has grown immensely. They have now reached and most likely surpassed the level of the human World Champion. These programs (some of them are free to use) run on ordinary personal computers, thus enabling practically everyone to have a world-champion caliber player at home to play with whenever they wish to. This is a commodity that no one imagined possible a decade or two ago.

However, there is one big difference between having a strong computer program and having a human grandmaster at home. The latter can be asked questions, you can learn from him or her[1], you can analyze games together and he or she can explain complex concepts to you. But why cannot a computer program do that? A program is tactically impeccable and in play it exhibits an amount of strategic knowledge comparable to a solid grandmaster. In 2003, Garry Kasparov, the highest rated chess player in the world, said that computers are over

---

[1] In the remainder of this paper we use 'he' and 'his' whenever 'he or she' and 'his or her' are meant.

H.J. van den Herik et al. (Eds.): CG 2006, LNCS 4630, pp. 13–25, 2007.

2800 ELO in tactics and about 2500 ELO in strategy. Since then, they improved in both categories. So why cannot they give us lessons?

The answer lies in the output the programs give. They usually output nothing more than a principal variation and the numerical evaluation of several best moves — in other words, some moves and numbers are given. Typically, people are not good in comprehending numbers, but in this case we grew accustomed to interpreting that a value of -3 would mean losing a Knight or some other equivalent of three Pawns. Yet, we are puzzled by that cheeky evaluation of -0.39; what does that mean? That we are 0.39 Pawns behind in material? What is 0.39 Pawns? The point we are trying to make is that there exists a communication barrier between humans and chess programs. We clearly do not speak the same language. While we can understand the tactical lines with a forced win or a loss of material (even though it would not hurt hearing it in words like "you lose a Knight"), we may have difficulties understanding the quiet, strategic lines.

This is a pity. On the one hand we have at our disposal an oracle of knowledge, yet on the other hand we cannot understand what it is saying for the most part. Imagine that you were given to understand what that tricky -0.39 really stands for. For example, "You own a bishop pair, and have a safer King, but it does not compensate for the two isolated Pawns on the queenside and the strong Knight's outpost on d3. Your position is somewhat (-0.39) worse." This would be something that one could learn from. There is a myriad of games between strong players played or relayed on the internet. With a chess program that could provide comments that we understand, we would be able to learn while watching those games.

This is the long-term goal of our research. The current paper presents an initial, though perhaps most important, step towards this goal. We took Robert Hyatt's CRAFTY [4], a strong freeware chess program, as our underlying chess engine and on top of it built an annotating/tutoring module. The chess program is thus turned into an intelligent chess tutor.

2   Related Work

The idea of the automatic construction of commentary is not new. Its importance was realized long ago. Around 1980, Donald Michie was probably the first to propose research in this direction. The ICGA (then still ICCA) Journal started an annual competition for the best chess annotation program — named after the late Professor Herschberg. The first winner of the competition was CHESSMASTER 4000 for the year 1993 [1]. The program commented practically every move with some prose, but it was mostly very simplistic. The next year, Jeff Mallet's INNOVATION, an amateur program, won as the only entrance. It gave interesting comments, but was hampered by a relatively poor chess engine. Then, three years in a row (with no award given for 1997) different versions of the Dutch program FRITZ won the prize. It had a very strong engine and gave good analysis — but its commentary lacked prose. This handicap slightly improved from version to version, yet the version that was the winner of 1998 — it was the last winner — still lacks a great deal in this respect. It has a good selection of (1) when to comment, (2) which subvariations to give (not easy problems at all), and (3) how long the variations should be. A very fine point is that it also points out the tactical pitfalls the players avoided. Yet, it still explains only tactical nuances and ignores what we discussed in the introduction — why is a position −0.39 in strategic terms.

Still, in the mid 1990s, the spirits were high and the progress made even prompted David Levy and Tony Marsland [5], members of the competition jury, to write: "If progress in this field continues to be made at the same rate we would expect that, by the end of the century, there will be annotation programs available which can offer a very useful service to chess-players of every strength." Surprisingly, though, the 1998 FRITZ was the last entrance in the competition to this day.

On the other side of the field, Seidel [7] made an attempt to create endgame programs that could annotate their play. He tackled elementary endgames of the lone-king type such as KRK and KBBK. At the core of the system is a general rule for generating moves for the stronger side that (1) determines the threats by the weaker side, (2) generates possible actions, and (3) eliminates those moves that do not parry the threats. Threats and actions are very specific for various endgames and rely heavily on chess theory. They represent the chess knowledge of the system and are defined manually. The annotation mechanism follows simply from the move-generating scheme by commenting which actions and threats are encountered. The chess knowledge in this form is both a strong and a weak point — strong, because it enables specific annotations, and weak because a vast amount of manually entered knowledge is required even for very simple endgames. For example, there are six different types of moves (actions) for just a part of the KBBK endgame. This severely limits the extendability of the scheme to more complex endgames, not to mention the whole game. A problem with the whole game, apart from just the raw amount of knowledge needed, is the lack of strict rules — something Seidel's system requires.

Herbeck and Barth [3] took a different and in some ways perhaps a more conventional approach. They combined alpha-beta search with knowledge in the form of rules. Their domains were also chess endgames (up to four pieces), though the knowledge was somewhat more involved than in Seidel's [7] case. Herbeck and Barth's algorithm performed a standard search and then followed the principal variation until a position was encountered in which a rule from the knowledge base stated the outcome (or a bound on the outcome) of the game. This rule (or a set of them) is offered to the user as an explanation. The algorithm is also capable of giving comments (rules) for alternative variations. An example of the rules in KPKP is "White wins, because the white Pawn promotes first with check and Black is in a bad position". As in Seidel's case the rule base can be viewed both as a strong and a weak point of the scheme — strong because the annotations are quite good, and weak because (a) the rules are manually crafted, and (b) there are no such exact rules for more complex endgames, let alone for the middlegame. The latter prevents the scheme to be extended beyond relatively simple endgames.

We should also briefly mention here Gadwal et al.'s work on tutoring King and Bishop and two Pawns on the same file versus lone King endgame [2]. As with the other studies mentioned the main problem with it is the lack of extendability beyond the simple endgames.

3 The Tutoring System

Our main underlying idea is to use the chess engine's evaluation function's features to describe the changes in the position when a move is made. These elementary features can later be combined to form higher-level concepts understandable to humans. In this manner we bridge the communication barrier between machines and humans.

Our tutoring system consists of three components: (1) the chess engine, bringing chess knowledge and search to the table, (2) the commenting module for generating raw comments from the engine's evaluations, and (3) a simple expert system for refining the raw comments into (an even more) human understandable language.

The first component, the chess engine, serves to (1a) calculate variations, (1b) analyze positions, and (1c) evaluate possible moves. It is essentially an unmodified version of CRAFTY (apart for some routines that transfer calculated variations, moves, and detailed evaluations to our commenting module). The depth of search is either fixed or time-controlled; higher depths provide better variations and thus better (in terms of playing strength) commentary. As for the greater part the other two components are relatively independent of the engine, any other chess engine could be used in CRAFTY's place. In fact, we believe it would be quite advantageous to use an engine that champions knowledge in favor of search.

The other two components, which form the backbone of the tutoring system, will be described in Sections 3.1 and 3.2.

3.1 The Commenting Module

At first glance, moves can be divided into two categories: good moves and bad moves. Yet, when delving deeper into the matter, one can see that most moves in games, pitting two players with opposing interests against each other, are a sort of tradeoff of positive and negative characteristics of the position. With most moves you gain something and you lose something.

These characteristics and their tradeoffs are what our commenting module is calculating and analyzing. For any given move, it shows what characteristics of the position have changed and on the basis of this and the change in score, as given by the engine, the tutor can elaborate what is the essence of the given move or variation. The general merit of the move, however, is obtained by simply comparing its score with the scores of other possible moves in the given position.

Most chess engines, CRAFTY being no different, employ an evaluation function in the form of a weighted sum of the position's features. These features, along with their associated weights, are actually the position's characteristics on which our commenting module is operating. The weights are important too, because they define the relative importance of the features (characteristics).

Commenting Good Characteristics. In general, the tutoring system has two possibilities to generate a comment why a certain move is good: (a) the move achieves progress towards some goal, or (b) the move eliminates some weakness or deficiency of the current position. Let us take a look at both options in more detail.

The basic idea behind the first option is that making headway involves achieving progress towards goals, eventually accomplishing them. The goals in our schema are simply the evaluation function's features. We believe this is a natural way to state straightforward, comprehensible goals to be achieved. Later we show how the expert system can combine several straightforward goals into a more structured one thus increasing the expressive power of the tutoring system.

Figure 1a illustrates the setting for this idea. First, a search procedure is employed to obtain a principal variation starting with the move to be commented upon. The final position in the principal variation represents the goal position — this is the position that one can reach from the initial position with the move under investigation. This position might be viewed as envisioned by the player when he made the move. After we have obtained this envisioned position, we calculate which features of the evaluation function have changed and by how much they changed when comparing this position with the starting position. If one looks at the evaluation function's estimation of a position as a vector of values this operation is a simple difference between the corresponding position vectors.

The positive characteristics (or rather positively changing characteristics) achieved are those that have positive values in the resulting vector of differences (here we assume that we comment from White's perspective; otherwise it is just the opposite). In the raw output, each such characteristic represents a separate comment of what the move under investigation aims to achieve. Basically, at this stage, the commentary is a list of positive characteristics the move (or rather the principal variation starting with the move under investigation) aims to achieve.

For example, if the following two characteristics were singled out as the ones that changed positively:

WHITE_KNIGHTS_CENTRALIZATION
WHITE_KING_SAFETY

then the raw commentary would be "The move aims to centralize the Knight and to improve King's safety".

It should be noted that both, the starting position and the envisioned position, must be quiescent. The starting position should be quiescent, because there is

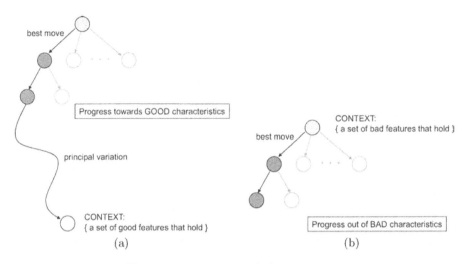

Fig. 1. Commenting good characteristics

no point in commenting in the middle of a tactical sequence; such a sequence should be viewed as a whole. The envisioned position should be quiescent for obvious reasons — the evaluation based on a non-quiescent position is completely unreliable and thus of little use.

Let us now take a look at the other possibility why a move can be good — namely, because it eliminates some weakness or deficiency of the current position. This situation is illustrated in Fig. 1b. However, computationally this possibility is implemented exactly as the first one. We note that good characteristics for the opponent are bad for us, and are represented with negative numbers in the position's vector. For example, one such negative characteristic (position's deficiency) could be:

BLACK_BISHOP_PAIR

and in this case the generated raw commentary would be "The move aims to eliminate the Black's bishop pair".

Commenting Bad Characteristics. The tutoring system has three possibilities to generate a comment why a certain move is bad: (a) the move creates a weakness or deficiency in the position, (b) the move spoils some good characteristic of the current position, or (c) the move is compared to a better (the best) possible move and the differences are pointed out. Possibilities (a) and (b) are quite similar to the two possibilities we encountered earlier when discussing how commenting of good aspects of a move is accomplished. Therefore we first briefly debate these two options and then take a look at option (c).

Possibility (a) mirrors possibility (a) for generating comments for good moves. The difference is that there we attempted to achieve some positive goals, while here the move achieves some negative goal(s). Similarly, possibility (b) mirrors possibility (b) for generating comments for good moves. Here, the difference is that instead of removing some weakness or deficiency of the current position, the move removes some positive characteristic of the current position.

From the computational point of view, the only difference is that we are now looking at the evaluation features that are negative in the vector of differences between the starting position and the envisioned position. The rest is the same. For example, if the following features were flagged as changed for the worse (negatively changing characteristics):

EVALUATE_PAWNS
BLACK_ROOK_BEHIND_PASSED_PAWN

the raw commentary generated would be "The move allows the opponent to improve the pawn structure and to get a Rook behind a passed Pawn".

However, there is a further difficulty when commenting really bad moves — be it bad in positional or tactical sense. The nature of minimax search is such that it does not search for a sort of "bad envisioned position", but rather allows the mistake (as it is forced upon it by the user) and then searches for best play for both sides from that point on. So, in essence, the envisioned position at the end of the principal variation returned by the search may not necessarily reflect the real weakness of the bad move (which is what we would like to comment upon), because this weakness was perhaps traded for some other weakness later in the variation.

Let us illustrate this by an example. Assume that White made a mistake by moving a Pawn and that as a consequence Black gained a strong outpost for his Knight. Later in the principal variation, stemming from this initial pawn move, however, Black exchanged this strong Knight for White's Bishop and so eliminated White's strong bishop pair and doubled White's Pawns. The tutoring system, comparing the starting position with the position at the end of the principal variation, would comment that "The move allows the opponent to eliminate your bishop pair and to weaken your pawn structure". While in principle this is true, it may be more in the spirit of tutoring to say instead "The move allows the opponent to gain a strong knight outpost". The initial comment can prove too abstract to the user. Or, after all, the user can choose not to follow the principal variation at all.

The difficulty we described is actually a special case of a more general problem — namely, how long should the principal variation be and where in it we should decide to comment. This is a cognitive problem and amongst other things it depends on the chess strength of the user whom the system attempts to tutor. In some cases only a single move should be commented, in other cases the whole variation, and in a third group of cases a part of the variation. This problem indicates our future work.

The idea behind possibility (c) is different from the concepts we discussed so far. Instead of observing what the move achieved, positive or negative, we observe what the move did *not* achieve although it could have. Therefore, we compare the move with the best move that is available. In essence, we generate comments for the move played and for the best move that could have been played, and attempt to compare their respective merits. This possibility is still highly experimental and is mentioned here just to give the reader the idea how we plan to extend our tutoring system's capabilities.

3.2   The Expert System for Commentary Refinement

In principle, the raw comments generated by the commenting module are in itself sufficient to produce relatively intelligent tutoring and annotations. However, it is both easy and rather beneficial to refine these comments to obtain quite human-like commentary. This is the purpose of the expert system for commentary refinement.

It is useful to refine the raw comments for several reasons: (a) to remove the unnecessary complementary comments, (b) to explain further or better the meaning of some features in the given context, and (c) to combine some basic features into higher-level, structured features that are more understandable to humans, thus increasing the expressive power of the system. Below we take a more detailed look at these options.

The structure of CRAFTY's evaluation function (other engines are probably quite similar in this respect) is such that it separately looks for the same features for White and for Black [4]. For example, it looks for White's Rook on an open file and for Black's Rook on an open file. These are two different features. Yet, they can be looked upon as a pair of complementary features. If, in a given variation, the commenting module realized that both these features changed — meaning that both players got a Rook on the open file — it would comment on both of them with something like "The move aims to get a Rook on open file. The move allows the opponent to get a Rook on open file". The expert system removes such comments provided that the complementary features are worth about the same amount (by directly comparing their respective values in the vector of differences)[2].

The features sometimes have a different meaning depending on the context as defined by other features. The expert system detects which meaning is appropriate in a given situation and modifies the raw comment accordingly. Let us explain this by an example. Assume that the commenting module flagged the feature WHITE_BISHOP_PLUS_PAWNS_ON_COLOR as a characteristic that changed positively for White. This feature is designed to modify (reduce) the value of a Bishop if there are own Pawns on the squares of the same color that this Bishop moves on. There are two very different cases when this feature can change positively. In the first case, we improve the Bishop by moving some Pawns thus creating more space for the Bishop. In this case the raw comment "The move aims to improve Bishop plus Pawns with respect to color" is slightly

---

[2] The complementary features are not always worth the same, for example, White's bishop pair can be better than Black's bishop pair. That is why the expert system has to take their worth into account when attempting to remove comments based on complementary features.

awkward but still understandable. However, in the second case this (bad) Bishop is simply exchanged. Now the raw comment is completely out of place; though in essence correct if one knows the hidden meaning. By looking at other features (in this case whether the Bishop is exchanged) the expert system knows what commentary should be given; in this case "Bad Bishop was exchanged".

The third role of the expert system is to combine several elementary features into higher-level, structured features that represent concepts understandable to humans. Let us take a look at one example rule in the expert system that does so:

```
if (BLACK_BISHOPS = 2) then
    if (BLACK_BISHOPS_POSITION + BLACK_BISHOPS_MOBILITY +
        BLACK_BISHOPS_BLOCK_CENTER + BLACK_BISHOP_PAIR +
        BLACK_BISHOP_KING_SAFETY <= -65) then
            comment("Black has an active bishop pair.")
```

This rule combines six elementary features to derive whether Black has an active bishop pair (higher-level concept). As we can see the bishop pair can be active for various reasons. It can simply be that the Bishops have great mobility. Or it can be that the Bishops guard the King well and at the same time are well positioned. The raw comments for every elementary aspect of the position are less instructive in this case than a single high-level comment. Even more, not all elementary features need to change positively for the same side. In this case the raw comments might even be confusing. The high-level comment "Black has an active bishop pair" is usually preferable to commenting separately on the good positions of the Bishops and how the Bishops defend the King well. Moreover, the commentary can easily be extended to explain why the bishop pair is active — simply state which parts of the rule contribute (most) to the score.

Our expert system consists of about 25 rules. The rules were constructed by chess experts[3] who also set the threshold values in the rules (cf. −65 in the above example rule). The expert system can easily be extended with new rules or by refining the existing ones. Further refining and in particular fine-tuning of the thresholds can perhaps also be accomplished by machine learning methods.

4 Some Tutoring Examples

In the previous sections we explained the ideas behind our tutoring/annotating mechanism. Now we shall demonstrate how they work in practice on some positions taken from grandmaster games. The positions are taken from John Nunn's tutoring book [6].

Let our tutoring system work on the position from the game Kasparov versus Shirov shown in Fig. 2a. It, or rather CRAFTY, evaluates the position after 16. ... Nc5 as slightly worse for Black (+0.43). In its place it recommends the move

---

[3] The chess expertise was provided by FM Matej Guid and WIM Jana Krivec who are both rated over 2300 ELO.

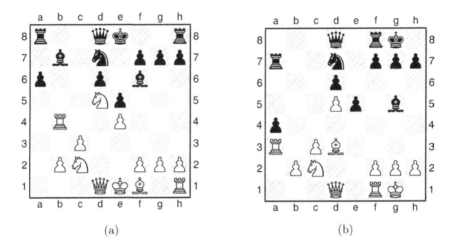

Fig. 2. Kasparov versus Shirov, Horgen, 1994

Table 1. Vector of differences

Feature (positive change)	Score	Feature (negative change)	Score
black bishop pair	20	king tropism	−51
black back rank	12	evaluate pawns	−27
black bishop plus pawns on color	12	white knights outposts	−15
white bishops mobility	12	white back rank	−12
white bishops position	10	white knights centralization	−6
black bishops mobility	6		

16. ... Ra7, which, it suggests, would lead to an equal position (−0.01) after 17. Bd3 O-O 18. O-O a5 19. Rb3 Bxd5 20. exd5 a4 21. Ra3 Bg5. The position after 21. ... Bg5 is our envisioned position and is shown in Fig. 2b.

Table 1 shows the vector of differences for the 16. ... Ra7 variation. Let the tutor comment from Black's perspective. The positive changes for Black are listed on the right side of the table, and negative changes for Black on the left side of the table. The raw commentary would thus be "The move 16. ... Ra7 aims to improve king tropism, improve pawn structure, eliminate opponent's knight outposts, weaken opponent's back rank, and reduce opponent's knights centralization. The move allows the opponent to eliminate your bishop pair, weaken your back rank, weaken your Bishop plus Pawns on same color, increase the Bishops' mobility, improve the Bishop's position, decrease your Bishops' mobility". As we can see, the raw commentary is directly translated from the vector of differences. Let us now take a look how the expert system for commentary refinement improves the raw comments.

First of all, the back rank weakness comments are eliminated as they are worth the same (for their respective sides) and are complementary. Next, from material features it is observed that Black lost the bishop pair. However, truly interesting are the rules below.

```
if (WHITE_ACTIVITY_CHANGE + BLACK_ACTIVITY_CHANGE <= -10) then
        comment("Black has improved the activity of pieces")
if (WHITE_BISHOPS = 1 and WHITE_BISHOPS_MOBILITY >= 12 and
    WHITE_BISHOPS_POSITION >= 6) then
        comment("White has improved the bishop")
if (abs(KING_TROPISM + EVALUATE_KING_SAFETY) > 60) then
        comment("White has the initiative against black's king")
if (EVALUATE_PAWNS <= -10)
        comment("Black has improved the pawn structure")
if (WHITE_KNIGHTS_CENTRALIZATION + WHITE_KNIGHTS_OUTPOSTS <= -14)
        then comment("White no longer has a strong knight")
```

These rules achieve the following. The first one combines two elementary features to detect that Black has improved the activity of pieces. The second rule again combines three elementary features to observe that White's Bishop has improved. The third rule actually removes the king tropism raw comment, because it should be combined with King's safety and the required threshold for the initiative against the King was not reached. The fourth rule is simple (and should be refined with further work) and observes that Black improved the pawn structure. The last rule combines two elementary features to note that White lost (exchanged) a strong Knight. After applying these rules the refined commentary is: "Black has improved the pawn structure, White no longer has a strong Knight, and Black has improved the activity of his pieces. On the other hand: Black no longer has the advantage of a bishop pair, White has improved the Bishop." We can see that the refined commentary, using higher-level concepts, is much better than the original raw commentary.

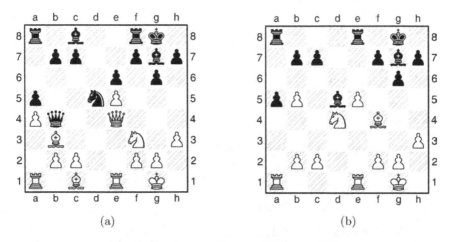

(a)   (b)

Fig. 3. Short versus Timman, Tilburg, 1991

A second position is from the famous game Short versus Timman. In the position of the diagram in Fig. 3a Short played 17. Bc4 and GM John Nunn in his book agrees with him [6]: "White is not deflected by the prospect of winning a Pawn with 17.Bxd5 exd5 18.Qxd5 Be6, when Black would have reasonable drawing chances in view of his good development and active Bishops." The tutor, however, prefers 17. Bxd5 exd5 18. Qxd5 Be6 19. Qb5 Qxb5 20. axb5 Rfe8 21. Nd4 Bd5 22. Bf4 (position in Fig. 3b) and backs it up with: "White has won a Pawn. Black has gained a bishop pair, Black has improved the Bishops". In essence, the tutor agrees that Black has reasonable compensation for the Pawn.

5   Conclusions and Future Work

The work presented in this paper is intended as a founding step towards turning a chess engine into a competent chess tutor and annotator. Even at this early stage of development, the core mechanism exhibited that it is able to generate rather intelligent commentary as demonstrated by the examples shown. The main advantage of our approach over the related proposals is that it could be, and consequently was, applied to the complete domain of chess. The advantage of our program over various annotating programs is that while our program is able to understand and comment on tactical positions, it also exhibits a solid understanding[4] of positional intricacies of positions. Hence, it is able to provide adequate commentary on positional aspects as well.

An interesting side-effect of our approach is that it can potentially prove useful to chess engine programmers as they can now practically see how their programs "think".

There is an array of cognitive issues still to be solved to arrive at really good tutoring system, e.g., (1) when to comment and when not to, (2) which move (the best or some similar but better) to compare the move to, (3) the most suitable depth of lookahead, to name but a few. These are obvious tasks for future work.

The system's knowledge is easily extendable, either by employing a more knowledge-oriented chess engine or by enriching the expert system with new concepts. Also, the functionality of the system can be extended by adding various functions, e.g., implementing the ability to ask the tutor questions, but this is more a matter of engineering than of scientific work.

References

1. The Board of the ICCA: The Best Annotation Award for 1993. ICCA Journal 17(2), 106–108 (1994)
2. Gadwal, D., Greer, J.E., McCalla, G.I.: Tutoring Bishop-Pawn Endgames: An Experiment in using Knowledge-Based Chess as a Domain for Intelligent Tutoring. Applied Intelligence 3(3), 207–224 (1993)

---

[4] Of course, positional understanding of the tutor is limited with the level of positional understanding as exhibited by today's chess programs — e.g., long-term planning is still absent.

3. Herbeck, H., Barth, W.: An Explanation Tool for Chess Endgames Based on Rules. ICCA Journal 19(2), 75–82 (1996)
4. Hyatt, R.: The CRAFTY ftp site. (2006), `ftp://ftp.cis.uab.edu/pub/hyatt/`
5. Levy, D., Marsland, T.: The ICCA Best Annotation Award for 1995. ICCA Journal 19(2), 135–136 (1996)
6. Nunn, J.: Understanding Chess Move by Move. Gambit Publications Limited (2001)
7. Seidel, R.: Self-annotating Elementary Endgames. ICCA Journal 17(2), 51–62 (1994)

A New Heuristic Search Algorithm for Capturing Problems in Go

Keh-Hsun Chen and Peigang Zhang

Department of Computer Science,
University of North Carolina at Charlotte, NC, USA
{chen,pzhang1}@uncc.edu

Abstract. We propose a highly selective heuristic search algorithm for capturing problems in Go. This iterative deepening search works on the crucial chain in which the prey block is located. The algorithm starts using three order liberties of the chain as the basis of the position evaluation, the value is then adjusted by the presence of few liberty-surrounding opponent blocks. The algorithm solved most capturing problems in Kano's four volumes of graded Go problems. Moreover, it is fast enough to be used by Go programs in real time.

1 Introduction

Whether a block of stones in a Go board configuration can be captured by the opposite side is a fundamentally important knowledge item for a Go program. Without any knowledge of the block capture ability, other tactical problems in Go, such as life/death and connection cannot be accurately resolved. The global positional judgement could be totally wrong due to misjudgment of the capture ability of some blocks. Move decisions could become big blunders if made without recognizing the capture ability of some key stones [4]. Most Go programs spend over 90% of their processing time in doing capturing search of blocks of stones on the board. The quality of their capturing search routine is a key factor in the strength of a Go program.

In this paper, we shall describe a highly selective heuristic capturing search algorithm based on the classical α-β game-tree-search paradigm. The algorithm, called HuPrey (from Hunter-Prey), is a significantly improved version of the capturing algorithm used in Go Intellect in the past. Experimental results show that its performance compares favorably to other Go capturing algorithms [1,2,11].

Section 2 describes ladder capturing, which is a simple form of capturing; its routine is used extensively in our heuristic search capturing algorithm. We discuss the capturing target in Sect. 3 and its evaluation in Sect. 4; the generation of candidate moves in Sect. 5. The search paradigm and enhancements are examined in Sect. 6 and 7, respectively. The special situations of seki and ko are considered in Sect. 8. The experimental results are presented in Sect. 9. The paper is concluded with suggested future work in Sect. 10.

H.J. van den Herik et al. (Eds.): CG 2006, LNCS 4630, pp. 26–36, 2007.

2   Ladder

We shall call the capturing target block the prey (block). The player who attacks the prey is the Hunter and the player who defends the prey block is called Prey. The simplest form of capturing is so-called ladder capturing. In this case, the prey block should have fewer than 3 liberties. Once the prey gets 3 or more liberties, the ladder chase fails. In a game, the Hunter will continue to play atari moves. The Prey will always extend its only liberty or capture adjacent opponent's blocks with just one liberty if they exist. The move sequence in a ladder chase usually forms a zigzag pattern like a step ladder. This is a key routine that every Go-playing program must have.

Keeping track of the prey's liberties and those of the adjacent blocks under atari is the key of an efficient ladder algorithm. We note that the average branching factor of a ladder search is near 1; so, the ladder procedure is extremely fast. A normal diagonal ladder from one side of the board to the other side takes less than one millisecond on a modern computer.

A capturing search routine frequently needs to know (1) in its move generation and (2) in its position evaluation whether a block can be captured by a ladder. Obviously, for the *general* capturing problem, we cannot just consider the prey's liberties and those of the adjacent blocks in atari as candidate moves like in a ladder search.

3   Crucial Block and Crucial Chain

To generate the candidate moves adequately and to evaluate positions appropriately, one has to consider the whole chain, called the crucial chain, to which the prey block (also called crucial block) belongs. Figure 1 shows a simple example, in which just considering liberties and second liberties of the prey block (marked by a triangle) is obviously inadequate in finding the correct capture/escape move. The key move 'a' is not a liberty or secondary liberty of the crucial block marked by the triangle. The whole crucial chain, to which the prey belongs, needs to be considered.

A chain is a set of blocks of same color, which cannot be separated by the opponent [3]. We recognize two blocks being in the same chain if they obey either one of C1 to C3.

C1 They share two or more common liberties.

C2 They share one protected common liberty. By a 'protected empty point', we mean either it is illegal for the opponent to play there or if the opponent plays there, the played stone can be captured by a ladder (and the opponent cannot reoccupy the point through ko or snapback).

C3 They are adjacent to a common opponent's dead block.

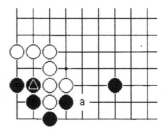

Fig. 1. Why the crucial chain should be connected

In practice, we can compute the transitive closure of the crucial block under C1, C2, and C3 to obtain the crucial chain. We should note that C1 to C3 do not completely define connectivity; they are just simple heuristics used by the capturing search routine in recognizing chains.

4   Position Evaluation

We define the empty points vertically or horizontally adjacent to a block as its liberties. A block is captured when it loses all its liberties. The first-order liberties are the ordinary liberties as defined above. We define a second-order liberty as a liberty of the augmented block when one additional stone of the block's color is added to the block (this stone may join additional blocks), which is not a first-order liberty. Similarly, we define a third-order liberty as a liberty of the augmented block when two additional stones of the block's color are added to the block (these stones may join additional blocks), which is not a first-order or second-order liberty. The first-order, second-order, and third-order liberties of the crucial chain can be computed via a breadth-first search. Let n_1, n_2, and n_3 be the total numbers of the first-order, second-order, and third-order liberties respectively. Assume Prey is to play. The basic evaluation is

$$V = n_1 \times 16 + n_2 \times 8 + n_3 \times 4 \,. \tag{1}$$

We consider a first-order liberty twice as valuable as a second-order liberty and four times as valuable as a third-order liberty. If there is an adjacent Hunter block which can be captured by a ladder, we add 50 to V. If Prey has a block that can be captured by a ladder, we subtract 25 (since Prey plays next, the block could be saved).

When the crucial chain has two eyes, then it does not matter how many liberties it has, it is completely safe. Hence, in addition to liberties we should also check for solid eyes of the crucial chain. We go through each of the first-order liberties of the crucial chain to determine whether it is a solid eye. An eye is solid if (1) a liberty is surrounded only by stones of Prey and borders plus at least 3 of the 4 diagonal board points are secure for liberties away from edges, or (2) both diagonals are secure for liberties on a board edge, or (3) the diagonal is secure for liberties at board corners.

By a diagonal being secure, we mean it is occupied by a Prey's stone or it is a protected empty point. When we find a chain having two solid eyes, the evaluation value V will be reassigned to a large positive constant representing *definite safe*. Of course, when the prey is captured, the evaluation will be a negative constant with a large absolute value representing capture. We use the numbers of liberties of surrounding opponent chains to adjust the evaluation. We add 15 to the evaluation for each surrounding opponent chain with 2 or fewer effective liberties.

When it is Hunter to play, then we reverse the sign of evaluation V except when the prey block can be captured by a ladder, a very large positive constant representing successful capture will be assigned to V. This evaluation is used in

the node evaluation as well as in the candidate move generating and ordering in
the search tree.

5   Candidate Move Generation

Below we shall discuss the generation of candidate moves. The relevant blocks
include:

(A) the adjacent blocks of blocks in the crucial chain with:
   (a1) max(3, #liberties of crucial block) or fewer liberties if Hunter is to play;
   (a2) max(3, #liberties of crucial block) + 1 or fewer liberties if Prey is to
        play;
(B) the adjacent blocks of the relevant hunter blocks, not in the crucial chain
    and with 3 or fewer liberties.

The blocks listed under (A) will be called the relevant hunter blocks, the blocks
listed under (B) will be called the relevant prey blocks.

We first take up to 4 first-order and second-order liberties at which the move
would produce the best evaluations (see Section 4). Almost all Prey's moves that
can be captured by a ladder or that fill its own solid eyes are eliminated. Hunter's
sacrifice moves are allowed. We then first add 2 or 3 selected liberties of the
relevant hunter blocks and subsequently add 2 selected liberties of the relevant
prey blocks. The algorithm selects the best liberties by checking the surroundings
of the liberties. It favors the liberties of the blocks with fewer liberties; liberties
could further increase liberties or could connect to other blocks.

If a move can be captured by a ladder then we add the first move of the ladder
capturing sequence into the candidate move set. This procedure introduces moves
with an indirect approach. All the capturing moves, which can capture a block
in a crucial chain or a relevant block or a block adjacent to a relevant prey block,
are promoted or inserted to the front of the move list during the move ordering.
When the number of candidate moves is two or one, Pass move is added as a
candidate.

The power of HuPrey comes from the compact yet generally adequate candidate move set. The following three examples (see Figs. 2 to 4) illustrate the
point.

Figure 2 shows candidate moves of Hunter (Black) in an attempt to capture
the marked block. Candidate moves 1 to 4 are all from first-order and second-
order liberties of the crucial block which is marked by a triangle. The relevant
hunter blocks contain only the adjacent single stone block. It is Hunter's turn to
play, the adjacent hunter blocks with more liberties than the prey are not considered. But the only relevant hunter block did not contribute any new candidate
moves in this case. The relevant prey blocks are empty. Move 1 was promoted
to become first candidate, since it is a ladder capture move (of the single stone
adjacent hunter block). Candidate move 3 is the correct first move to capture
the prey (followed by W2 andB4).

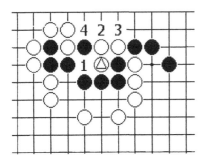

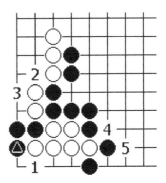

Fig. 2. Candidate moves for Black for capturing the marked white block. The correct black first move is at 3.

Fig. 3. A compact set of candidate moves from the crucial chain, relevant hunter blocks, and relevant prey blocks contains the winning first move at location 2

The next example (Fig. 3) is for Black to play to save the marked prey. Move 1 is the only move from the first-order liberties and second-order liberties of the prey, the other liberty moves can be captured by ladders, and are thus suppressed. Candidate moves 2 and 3 are liberties of the relevant hunter blocks and candidates 4 and 5 are liberties from the relevant prey blocks. Candidate move 2 is the key first move to save the prey.

In Fig. 4, Black to play, the crucial chain has two blocks, the marked one and the sin-

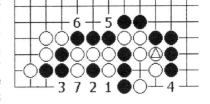

Fig. 4. Candidate moves for Black to capture the marked white block. The correct black first move is at 7 (followed by W2, B4).

gle white stone at the edge. They produce only one candidate move 4, a secondary liberty of the prey. Candidates 1, 2, and 3 are all ladder capturing moves, they have been promoted to the front. The two three-in-a-line black blocks form the relevant hunter blocks. They produce candidates 5 and 6. The relevant prey block is the n shape white five-stone block in the middle, which produces candidate move 7 - the winning first move.

6   Iterative Deepening with Hash Table

We use iterative deepening with a minimum of depth 2, a maximum of depth 20, and a depth increment of 2. The search terminates when the definitive result is obtained, or the maximum depth is reached, or the time limit is over 50% consumed. When Hunter is the first player, a definitive result means the crucial block (1) is removed from the board or (2) has two eyes or (3) is in seki. When

Prey is the first player, a definitive result means that the absolute value of the evaluation reaches a value above the search cut value (150).

A hash table is used to store the search results from transposition and from previous iterations. Zobrist's hashing method [13] is used to produce hash codes. Subtree search results with sufficient search depth are stored. At each node of the search tree, the hash table is searched to find the previously found best next move, which will be tried first.

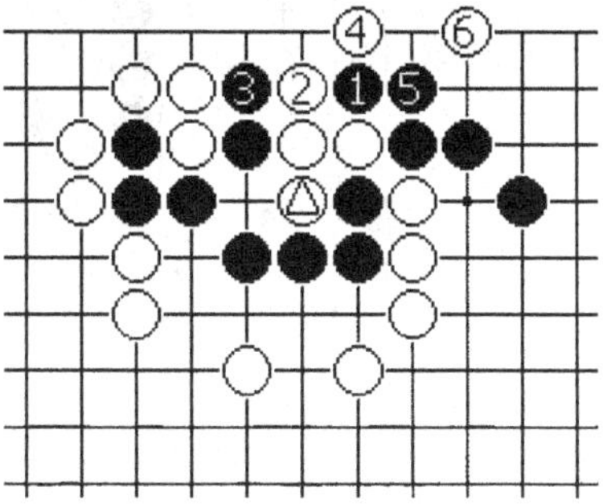

Forward pruning is used in a conservative way: only when the evaluation of the current node is outside the α-β window by the search cut, Δ value (50) or more, a forward pruning is performed and no further node expansion will be done under the node.

Figure 5 shows the solution sequence found by the heuristic capturing search

Fig. 5. Search outcome of the problem in Fig. 2. Note that at this point the algorithm knows the prey can be captured by a ladder, it returns success.

HuPrey on the problem in Fig. 2. It took 0.422 second searching to 6 plies. When Hunter finds that the prey block can be captured by a ladder, the search terminates. The moves in the ladder capturing sequence are not counted in the search depth.

7 Try Opponent's Best Refute Move

GoTools [12] tries the opponent's best refute move as the next candidate move in life/death search with great success. So, we borrowed this idea in HuPrey. It explores the opponent's refute move next regardless whether the move is in the current candidate list or not. This technique speeds up the search significantly. Since when a refute move does not help, it slows the search by a small fraction; when a refute move helps identifying a key sequence, it usually cuts the search time drastically. On the capturing problems in Kano [7], this technique speeds up the search by about 20% on average (see Table 1).

There is an additional merit of this technique. When the set of selected candidate moves does not include the winning key move, the technique may introduce the missing move in the search tree. In Fig. 6, the left-hand-side diagram shows the initial candidate moves for White. The right-hand-side diagram shows the successful capturing sequence found by the augmented HuPrey. Looking closely on the left-hand-side diagram, we notice the "best" four first-order and second-order liberties of the crucial chain are candidates 3, 4, 5, and 6 — the winning first move was not even there!

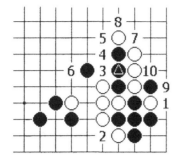 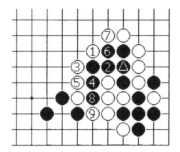

Fig. 6. The left diagram shows the initial candidate moves of Hunter (White). The right diagram shows the winning capturing sequence produced by HuPREY. The winning first move was not generated by the move generator. The technique of trying the opponent's best refute move introduced the winning move to the search tree.

8   Seki and Ko

When there are two consecutive passes, the prey is considered safe — either it has made two eyes or it has created a seki situation. In this case a large value will be returned by the evaluation function instead of first counting the three order liberties.

We use a simple approach to deal with ko. We assume the first player for the capturing problem can win kos up to three times and the other player cannot win any ko. If the capturing relied on winning kos, the value returned would not be as high as capturing without requiring to win a ko. This ko treatment is modified from Kierulf [9].

9   Experimental Results

We tested HuPREY on capturing problems in Kano's *Graded Go Problems for Beginners* four-book series [5,6,7,8]. Book 1 contains trivial problems. Book 2 requires some Go knowledge. Book 3 contains problems interesting to average players. Book 4 contains problems challenging to advanced amateur players. The books contain all types of Go problems: life/death, connection, capturing, and opening. We test our algorithm exclusively on capturing problems. There are a total of 180 capturing problems in the four books.

Book 1 (34 problems) – problems 1→22, 61→64, 179→182, 193→196.
Book 2 (34 problems) – problems 31→36, 130→141, 219→222, 316→327.
Book 3 (61 problems) – problems 2, 3, 7, 8, 11→13,15, 16, 20, 21, 24→26,
          30, 31, 36, 38→43, 45, 95, 98, 105→108, 112→115, 118, 123, 124,
          127, 128, 132, 157→160, 163, 278→293.
Book 4 (51 problems) – problems 1, 2, 11, 18, 19, 23, 27→41, 103→115,
          135→138, 168, 184, 257→263, 387→390.

Table 1. Performance of HuPrey on Kano's graded Go capturing problems. The columns contain the following item for each book (from left to right column): the average time in seconds per problem, the average number of nodes per problem, the average depth per problem, the number of solved problems, the number of unsolved problems.

	ave. time	ave. nodes	ave. depth	num. solved	num. unsolved
Book 1	0.007	27	2.00	34	0
Book 2	0.830	4,912	6.76	33	1
Book 3	0.750	4,304	6.29	55	6
Book 4	8.100	35,933	5.97	40	11

We found that HuPrey can solve all the capturing problems in Books 1 and 2, 90% of the capturing problems in Book 3, and 80% of the capturing problems in Book 4, which outperforms all known testing results so far [1,2].

HuPrey is implemented in the latest version of Go Intellect. We run our tests on a 2.8 GHz Pentium 4. Table 1 summarizes the average testing results per solved problem.

For HuPrey, Book 2 capturing problems are slightly more time consuming than Book 3 capturing problems. Problem 28 of Book 2 can be solved by HuPrey in its stated setting in 165 seconds generating over 300,000 nodes. In order not to obscure the average statistics, we count it as unsolved in Table 1. This problem is not really a hard problem. After widening the candidate move set by one, our algorithm solved it in less than 0.01 second.

Every capturing problem in Book 1 was solved by HuPrey in less than 0.1 second. Figs. 7, 8, and 9 show the time allowance vs. percentage of problems solved for capturing problems in Kano Books 2, 3, and 4, respectively [6,7,8]. In each figure, the X-axis represents log(time allowance in seconds), the Y-axis represents the percentage of problems solved under the time constraint.

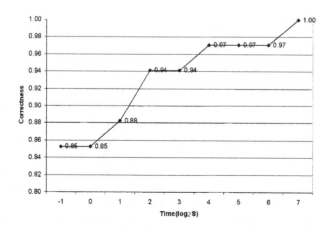

Fig. 7. The algorithm's time and solvability trade-off on the capturing problems in Kano Book 2

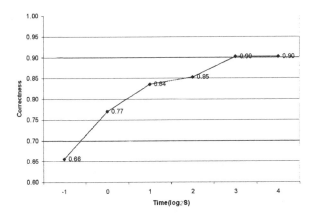

Fig. 8. The algorithm's time and solvability trade-off on the capturing problems in Kano Book 3

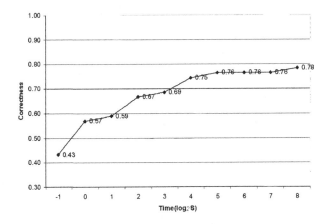

Fig. 9. The algorithm's time and solvability trade-off on the capturing problems in Kano Book 4

Figure 9 shows that about 43% of the capturing problems in Book 4 can be solved in less than 0.5 seconds. If we double the time limit to one second, about 57% of the problems can be solved. Double the time again to 2 seconds, 59% the problems can be solved. With a 4-second time limit, 67% of the problems can be solved. After passing the 75% result by this procedure, doubling the time produces very little improvements on the number of problems that can be solved.

Typically, a Go program will allocate 1 second or less for a capturing tactic search. Longer than one second time allocation for a capturing problem can be made only during the opponent time in a tournament setting. In practice, Hunter is set to be the first player. When the search finds that the prey can be captured, a second search with Prey going first is done to determine whether the prey can escape. Solved problems are stored in a solution hash table to be retrieved many times until the surrounding changes, which will be reflected in the hash code. This solution hash table is not the hash table used by the capturing search, which is cleared for each new problem before the start of the capturing search.

10   Future Work

Since the candidate move generation of HuPrey is highly selective, inevitably it will miss some key moves once in a while and if these key moves are not produced by the opponent's best refute then the algorithm may spend unnecessary long time in searching for a solution - it may even claim a wrong conclusion. Incorporating some suitable widening scheme into the iterative deepening framework is likely to fix the problem and to produce a more powerful capturing search algorithm. We shall investigate such a mixed widening/deepening algorithm in the near future.

Proof number search has been used very successfully on the tactical problems with fully enclosed boundaries [10]. Capturing problems are in general quite open, requiring dynamic determination of the candidate move region. Whether PN+ search can work nearly as well for open-region capturing problems is worth investigating.

Acknowledgments

We would like to thank Martin Müller for providing SGF files of the graded Go problems in Kano Books 1, 2, and 3 [5,6,7]. As a result, we only needed to enter the capturing problems of Book 4 by hand [8]. This saved us much time.

References

1. Cazenave, T.: Abstract Proof Search. In: Marsland, T., Frank, I. (eds.) CG 2001. LNCS, vol. 2063, pp. 39–54. Springer, Heidelberg (2002)
2. Cazenave, T.: Iterative Widening. In: Nebel, B. (ed.) Proceedings of IJCAI-01, vol. 1, pp. 523–528 (2001)
3. Chen, K.: Group Identification in Computer Go. In: Levy, D., Beal, D. (eds.) Heuristic Programming in Artificial Intelligence, pp. 195–210. Ellis Horwood, Chichester (1989)
4. Chen, K.: Computer Go: Knowledge, Search, and Move Decision. ICGA Journal 24(4), 203–215 (2001)
5. Kano, Y.: Graded Go Problems For Beginners. In: Introductory Problems, vol. 1. Kiseido Publishing Company (1985)
6. Kano, Y.: Graded Go Problems For Beginners. In: Elementary Problems, vol. 2. Kiseido Publishing Company. ISBN 1985
7. Kano, Y.: Graded Go Problems For Beginners. In: Intermediate Problems, vol. 3. Kiseido Publishing Company (1987) ISBN 4-906574-48-3
8. Kano, Y.: Graded Go Problems For Beginners. In: Advanced Problems, vol. 4. Kiseido Publishing Company (1990)

9. Kierulf, A.: Smart Game Board: A Workbench for Game Playing Programs, with Go and Othello as Case Studies. PhD thesis, ETH Zurich (1990)
10. Kishimoto, A.: Correct and Efficient Search Algorithms in the Presence of Repetitions, PhD thesis, University of Alberta (2005)
11. Thomsen, T.: Lambda-Search in Game Trees - with Application to Go. ICGA Journal 23(4), 203–217 (2000)
12. Wolf, T.: Forward Pruning and Other Heuristic Search Techniques in Tsume Go. Information Sciences 122(1), 59–76 (2000)
13. Zobrist, A.L.: A New Hashing Method with Application for Game Playing, Techn. Rep. No. 88, Univ. of Wisconsin, Madison, 1970. Republished in 1990, ICGA Journal, 13(2):69–73 (1970)

An Open Boundary Safety-of-Territory Solver for the Game of Go

Xiaozhen Niu and Martin Müller

Department of Computing Science,
University of Alberta, Edmonton, Canada
{xiaozhen,mmueller}@cs.ualberta.ca

Abstract. This paper presents SAFETY SOLVER 2.0, a safety-of-territory solver for the game of Go that can solve problems in areas with open boundaries. Previous work on assessing safety of territory has concentrated on regions that are completely surrounded by stones of one player. SAFETY SOLVER 2.0 can identify open boundary problems under real game conditions, and generate moves for invading or defending such areas. Several search enhancements improve the solver's performance. The experimental results demonstrate that the solver can find good moves in small to medium-size open boundary areas.

1 Introduction

Since the final score of the game of Go is determined by who can create more territory, estimating the safety of territories is a major component of a Go program. In previous work [7,8], the search-based safety-of-territory solver SAFETY SOLVER 1.0 was used to determine the safety status of a given region. However, the program operated under several restrictions. First, the region had to be completely enclosed. Second, the search was strictly local and did not utilize any external liberties of the boundary blocks.

In real games, most territories do not become fully enclosed until the late endgame. During most of the game, they have open boundaries. The surrounding conditions of a region, such as the number of external liberties of boundary blocks, are also very important. Ignoring them can lead to overly pessimistic safety estimates.

SAFETY SOLVER 2.0 is an improved safety-of-territory solver for open boundary areas. This program can generate moves for either invading or defending such areas. Additional searches provide answers to a series of related safety questions such as whether who plays first matters, whether the number of external liberties changes the local safety status, and whether changing the winner of ko fights affects safety.

The effectiveness of SAFETY SOLVER 2.0 depends crucially on providing it with useful input. A heuristic board-partitioning technique based on the concept of *zones* as implemented in the Go program EXPLORER [6] is used to identify open boundary areas in a full-board position.

H.J. van den Herik et al. (Eds.): CG 2006, LNCS 4630, pp. 37–49, 2007.

1.1 Safety of an Open Boundary Area

In a game of Go, as more and more stones are played, the board is gradually partitioned into relatively small areas. These areas can be completely surrounded by stones of one player, but more often they are not. Current Go programs use heuristic rules or pattern matching to generate moves to invade or defend open boundary areas. This approach is hit-and-miss, and chances to play better moves often exist. Figure 1 shows a typical example from our test set.

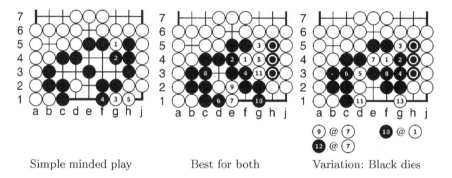

Simple minded play   Best for both   Variation: Black dies

Fig. 1. An open boundary example

Assume that White, the attacker, plays first. In this position most Go programs will play the simple endgame moves shown on the left of Figure 1. These moves lead to a 10 point Black territory. Eventually, Black must add one more stone inside to prevent double-atari at **f4**. However, **f4** is a much better first move for White. The diagram in the middle of Figure 1 shows a best move sequence. After this sequence, in order to make the whole group alive Black has to give up the block marked by ◉ The rightmost diagram in Figure 1 demonstrates that if Black resists White **f4** with **g4**, the whole group dies. In this example although the size of Black's open area is small (only 10), White's best attacking play gains large benefit compared to the simple-minded play.

The safety status of an open boundary area is strongly related to its external context such as the number of external liberties of boundary blocks, possible outside connections, and the number and size of ko threats for each player. In the above example, if Black's boundary blocks had more external liberties, then the whole area would be safe territory. In many cases, the number of external ko threats also affects the local safety status. Therefore, an exact open boundary safety-of-territory solver needs to take all relevant context information into account.

The major difference between a life-and-death solver and a safety-of-territory solver is that a life-and-death solver does not consider territory. As long as a group is alive, the three alternatives of living with a large area, living with two small eyes, and forming a seki are all equally correct. In contrast, the goal of an open boundary safety-of-territory solver is to maximize (or minimize the opponent's) safe territory. SAFETY SOLVER 2.0 achieves that in a flexible way. It can switch between different search goals, including a life-and-death solver. Subsection 2.2 describes these options in detail.

1.2   Related Work

Many approaches to territory evaluation in Go have been proposed in the literature. Static methods include Benson's algorithm for *unconditionally alive blocks* [1], Van der Werf's extension of Müller's static rules for safety under alternating play, and Vilà and Cazenave's static classification rules for regions up to a size of 7 points [11]. Search-based methods include Müller's local search for identifying the safety of regions by alternating play, and SAFETY SOLVER 1.0 as described in [7,8]. All these methods can handle only fully enclosed problems.

Several approaches for dealing with open boundary areas have been discussed or developed for life-and-death solvers. Solvers such as Kishimoto's tsume-Go solver [3] can deal with cases where the defender's boundary is open. However, the whole search region must be fully enclosed by safe attacker stones. Wolf [12] discusses practical issues about extending his program GoTools to open problems. Heuristic static eye-space analysis introduced by Chen and Chen [2] can also handle open problems.

1.3   Contributions

The main contributions of this paper are as follows.

- SAFETY SOLVER 2.0, a new improved solver for open boundary safety of territory problems.
- Re-search techniques for different search goals in the same position.
- Forward pruning techniques to safely reduce the search space and improve the performance of the solver.
- A heuristic method, using the Go program EXPLORER, to identify open boundary problems in real game positions.

The structure of this paper is as follows. Section 2 describes the structure and processing steps of the safety solver. Section 3 describes heuristic board partitioning for identifying open boundary problems. Section 4 introduces multiple searches for solving different goals. Section 5 introduces two forward pruning techniques. Section 6 provides experimental results. Conclusions and discussions about future work are given in the final section.

2   Open Boundary Safety-of-Territory Solver

2.1   Safety Solver

The safety solver described in this paper extends previous ones in [5,7,8]. It has been integrated into the Go program EXPLORER [5]. The solver uses the df-pn(r) search algorithm [3,4] with domain-specific heuristic functions for initializing proof and disproof numbers. SAFETY SOLVER 2.0 contains the following new features.

- Search goals customized by different parameters.
- Re-searches to provide solutions for different search goals.
- Preliminary integration into the Go program EXPLORER.

2.2   Input Parameters for the Search

Defining search goals for open boundary areas is more complex than for enclosed areas. In SAFETY SOLVER 2.0, the following input parameters specify a search goal for an open area. Figure 2 provides an example for two of these parameters.

- A set of points, called the area.
- The color of the defender and attacker.
- The first player. The search result is from this player's point of view.
- Two different search goals: prove *boundary safe* or *territory safe*? For the boundary safe goal, all boundary blocks surrounding an area must be proven safe. This goal is equivalent to a life-and-death search. For the territory safe goal, all interior points of the area need to be proven safe as well. The attacker may neither live inside the area, nor reduce the area from the outside.

  For example in Figure 2, White is the attacker and plays first at **a2**. If the goal is boundary safe, Black can play **c1** to make eyes and does not need to consider whether the territory is safe, as shown in the left of Figure 2. However if the goal is territory safe, then Black has to disconnect White by playing at **a3**. The move sequence on the right of Figure 2 indicates that eventually White has the choice to form a seki or kill all black blocks by ko. The black zone is boundary safe but not territory safe.
- *Seki recognition. Seki* is a position where neither player can capture the opponent's stones, so coexistence is the best result. When seki recognition is switched on, the solver distinguishes between the three results - win, loss and seki - by using the re-search techniques of [7]. Otherwise, the only possible outcomes are a win or a loss for the first player.
- Using external liberties. An *external liberty* of a boundary block lies outside the search area. Such liberties can affect the safety status of an area. For example, in Figure 2 the black block ◉ has no external liberties and block ▲ has one at **f1**. If either of them had one more external liberty, the area would be safe territory. When external liberties are not considered by the solver, it performs a quicker search to decide whether the area is locally safe. If external liberties are used, then the solver generates such external moves for the attacker and the defender as well, leading to a larger search space. Subsection 5.1 contains details on extended move generation for external liberties.
- The *ko winner* can also affect the safety status. For example, after the move sequence shown in the right of Figure 2, given enough ko threats White can capture all Black blocks. If Black is the ko winner, then Black can make the territory safe by winning a "thousand-year ko".

SAFETY SOLVER 2.0 concentrates on proving whether an area is safe or not *locally*. It does not consider whether the defender's boundary blocks can connect

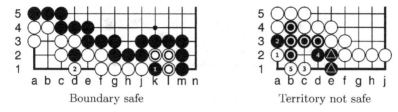

Boundary safe   Territory not safe

Fig. 2. Safety status changes under different goals

to other outside safe blocks from a full-board view. By default, the search goal of the solver is set to prove territory safe. To ensure that the local result is correct, the solver must handle seki and count external liberties. However, no ko winner needs to be set initially, which relaxes the assumption in SAFETY SOLVER 1.0 that the attacker wins all ko's.

2.3   Integration with EXPLORER: First Steps

SAFETY SOLVER 1.0 as described in [7,8] was implemented within the framework of the Go program EXPLORER but not called during game play. In SAFETY SOLVER 2.0, attacking and defending moves generated by the safety solver are passed to EXPLORER's global move generator. The solver uses the standard goal setting as described in Section 2.2. The heuristic value of these moves is set as follows.

- If the current search goal is boundary safe, then the best attacking or defending move is given the defender's minimum block value.
- If the search goal is territory safe, then the best attacking or defending move receives a value proportional to the number of interior points of the area.

3   Board Partitioning for Identifying Open Boundary Problems

The Go program EXPLORER uses heuristic territory evaluation to partition the board into *zones* [6]. On a board, empty points in gaps between stones of the same color can function as *dividers* or *potential dividers*. Dividers are small gaps that are sufficiently narrow so that the opponent can always be stopped from connecting through. Potential dividers are larger gaps that can be converted into dividers by a single move. Zones are computed by using dividers, potential dividers, and heuristic knowledge. They are surrounded by stones, dividers, and potential dividers. Figure 3 shows two black zones. The boundary of the left zone consists of three stones and four dividers marked by A. The right zone's boundary contains six stones and two dividers marked by B. Both zones have 15 interior points.

Search areas for SAFETY SOLVER 2.0 are computed by processing all zones. The first step selects candidate zones for search. Zones that are too large, with

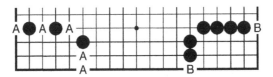

Fig. 3. Two examples of open boundary zones

over 15 empty interior points are ignored since such large areas can rarely be solved in reasonable time by the current system. Zones that are too open are also not suitable. In the current implementation, if more than one third of a zone boundary consists of dividers, it is not searched. The left zone in Figure 3 is such an example. The solver will only search the right zone.

As in SAFETY SOLVER 1.0 [8], *related zones* need to be merged. Two zones are related if they share one or more common boundary blocks. The merging algorithm in SAFETY SOLVER 2.0 extends the one for fully enclosed zones by dealing with dividers. The left side of Figure 4 shows two related black zones. Their interior points are marked by A and B. *Inner dividers* which are adjacent only to related zones are added to the merged zone. In the example, **d3** is a divider between A and B and is added to the merged zone M shown on the right side of the figure.

As a final step, the opponent's connectible points are removed from the areas to avoid trivial attacker "wins". For example, in the right side of Figure 4, **h1** is originally part of the merged area M. However, the attacker can directly play there and is connected to the outside through **j1**.

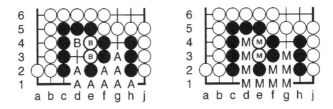

Fig. 4. Merging related zones

4   Multiple Searches for Related Goals

Since df-pn(r) only resolves boolean questions, the re-search technique described in [7] is used to distinguish between wins, seki, and losses. However, in order to provide solutions for different search goals, further searches with different goal settings may be required. One important case is switching the first player. In a given board position, assume that Black plays first. The zone processing step discussed in the previous section provides the solver with a list of black and white areas as inputs. For white areas, performing seki re-searches when Black plays first as the attacker can distinguish between a black win, black loss and a seki. However, for black zones, a second search with a different first player may be required. The detailed steps are as follows.

1. Set the first player to White and perform the first seki re-search to determine the outcome. If the result is a loss, White cannot do anything in this area and the black area is safe. In this case further search is unnecessary.
2. If the outcome is win or seki, the zone is unsafe when White plays first. A second seki re-search is performed to check whether the zone is safe if Black plays first. If the second outcome is a win, Black can successfully defend the zone. If the second outcome is seki, then the zone is unsafe, and the best result for Black is to form a seki. If the outcome is a loss, the black zone is not safe.

Further searches can be used to determine when external liberties affect the safety status of an area. Figure 5 on the left shows a black bent-four area with no outside liberties. If White plays at **b1** the result is ko. Black playing first can defend the territory. The same example with three external liberties is shown on the right of Figure 5. If White tries **b1** again, Black can live with **B:a1, W:a2, B:c2** because of the external liberties. In this scenario the question is, if White starts taking external liberties such as **c3**, when does Black need to respond? To answer such questions, a search taking external liberties into account is required. The detailed steps, explained using the example from Figure 5 are as follows.

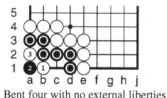
Bent four with no external liberties

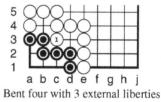
Bent four with 3 external liberties

Fig. 5. External liberties affect zone's safety status

- Step 1: After **W: c3** it is Black's turn. The first search is local within Black's area and pessimistically assumes no external liberties. If the outcome is a win, the area is safe without using any external liberties. No further search is necessary.
- If the outcome is seki or loss for Black, then a second search using the real external liberties is performed. In the left of Figure 5, the first search returns a loss. The second search establishes that with two external liberties the Black area is safe. In the future, after White plays **b4** or **d3**, another search will show that Black needs to defend the territory to avoid ko.

5   Forward Pruning Techniques

For an enclosed area, all legal moves inside the area are generated for both players. In an open boundary area, many external moves must also be generated. This leads to a larger search space, with the increase depending on how many external moves have been added. In order to improve the performance, two forward pruning techniques for defender moves are used in SAFETY SOLVER 2.0.

5.1 External Moves

The first technique is used for generating external moves. Figure 6 shows an open boundary example. First, the program generates all legal moves on interior points. In this example, all 12 interior points will be generated for either attacker or defender.

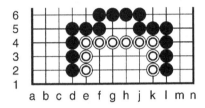

Fig. 6. External move generation for an open boundary area

Second, the generator needs to identify all external liberties for all boundary blocks of the zone. In this example, the white boundary block has 4 external liberties at **e1, g5, h5, k1**. Among them **e1** and **k1** are dividers of the area, and **g5, h5** are pure external liberties not related to dividers. Attacker plays on these external liberties can affect the safety of the area. However, defender plays on these liberties can be pruned since they can only decrease the liberties of defender blocks, and can neither be used to form an eye nor to create a seki. SAFETY SOLVER 2.0 focuses on proving the local safety status, and ignores possible outside connections or counterattacks on outside attacker stones. Therefore, pure external liberties, moves such as **g5, h5**, can be pruned for the defender.

Third, the program generates moves around dividers. In this example, if White plays **d1** Black can block by playing at **c1**. Moves such as **d1, l1** are called *layer one moves* and moves **c1, m1** are called *layer two moves*. The defender must generate layer one moves but not layer two moves because playing directly in layer two moves will not help to prove the inside area safety (the connection problem is ignored). Therefore, layer two moves are pruned for the defender. For the attacker, moves in both layers around dividers will be generated. In this example, the program generates 16 moves for the defender. For the attacker, since layer one and dividers are all empty points, layer two moves are pruned, giving 18 attacker moves including **g5** and **h5**.

5.2 Inner Eyes

During search, the defender might be able to form eyes inside the area. In this paper, only one-point eyes inside the area are called *inner eyes*. Inner eyes are

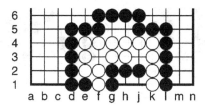

Fig. 7. Pruning inner eyes for the defender

very helpful for the evaluation function to evaluate the safety status. Since the defender should not fill them, such defender moves are pruned. For example, in Figure 7 the move **f3** can be pruned for the defender. Since the attacker playing at **f3** is illegal, **f3** will also be pruned for the attacker. In this example, five moves are generated for the attacker (**g5, h1, h5, j1, j3**) and only three for the defender (**h1, j1, j3**).

6   Experimental Results

Three test sets were created for three experiments. Set 1 contains a total of 120 test positions, including 60 original open boundary problems and 60 modified problems that are versions of the original problems with some external liberties added. Most original positions are taken from the classic *Guan Zi Pu* [9]. The remaining problems are collected from several resources, including [10] and positions created by the authors. Set 2 contains 20 positions from computer-computer games. Since SAFETY SOLVER 1.0 cannot handle open boundary problems, it cannot solve any positions from Set 1 and 2. Set 3 contains 100 final positions of 19 × 19 games played by amateur and professional players. All three test sets are available at: http://games.cs.ualberta.ca/go/open/.

Experiments were performed on a Pentium IV/2.4GHz with 1024 Mb memory. The time limit is 200 seconds per position in experiment one and two, and 10 seconds per local area in experiment three.

6.1   Experiment One: Correctness Test

The purpose of this experiment is to test the correctness of the solver. The difficulty of positions varies from easy to hard. The standard search setup described in the end of Subsection 2.2 is used for all these positions. The solver correctly proves that the territories of all 60 original problems are not safe, and generates White's best invading or reducing move. In the 60 modified problems, the solver proves that the territories are safe due to the external liberties added.

Figure 8 shows four Black open boundary positions from the 60 original problems in Test set 1. White as the attacker plays first.

The two black zones on the top are easy problems. On the left, after White's best invading move at **a18**, Black can choose to either let White connect back through **a17**, or form a seki as shown in the Figure. Either way the territory is

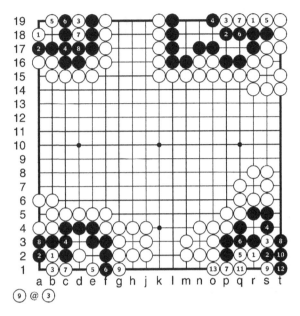

Fig. 8. Four open positions in Test set 1

not safe. The solver finds the best invading move **a18** in 0.01 seconds. On the right, the solver finds the best move **r19** in 0.1 seconds.

The problem at the bottom left corner is moderately difficult. It takes 51 seconds for the solver to find the best invading move **b2**. White can form a seki inside Black's zone. The problem at the bottom right corner is hard. The solver finds the best invading move **r2** in 179 seconds. Figure 8 shows strongest move sequences for each problem.

6.2 Experiment Two: Game Play Test

The purpose of this test set is to test whether EXPLORER enhanced by the safety solver is able to play the correct defending or invading move in 20 real game positions. Before integrating the safety solver, EXPLORER failed all these test cases. The standard setup described at the end of Subsection 2.2 is used. The time limit is 200 seconds per position. The results show that current EXPLORER correctly identifies the problems and generates best first moves for all of them.

Figure 9 shows an example in Test set 2. Before using the safety solver, EXPLORER as Black played **k1** to capture three white stones marked by ⊙. White played **d2** to capture one black stone and finished the play in this local area with a safe territory of 9 points. After integrating the safety solver, EXPLORER finds the move **d1**, which leads to a seki in the corner. Black is still able to capture the three White stones ⊙.

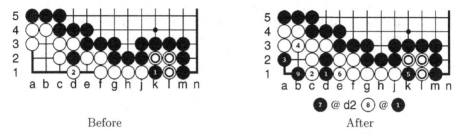

Before   After

Fig. 9. An example from Test set 2

6.3 Experiment Three: Comparison of Solvers

This experiment compares the performance of four solvers BENSON, STATIC, SAFETY SOLVER 1.0, and SAFETY SOLVER 2.0 on 100 completed games. The proven safe points are computed by each program starting from the end of the games, then backwards every 10 moves. Figure 10 shows the average number of proven safe points by solvers. At the end of the game, SAFETY SOLVER 2.0 proves 202 points on average, while SAFETY SOLVER 1.0, STATIC and BENSON prove 137, 85, and 46 points respectively. SAFETY SOLVER 2.0 outperforms SAFETY SOLVER 1.0 by a large margin, proving 47% more points. More importantly, the solver is useful much earlier in the game. Even 80 moves before the end of games, SAFETY SOLVER 2.0 can prove 42 points safe on average, while the other solvers can hardly prove any.

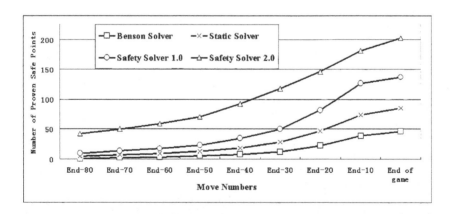

Fig. 10. Comparison of four solvers on 100 games

7 Conclusions and Future Work

SAFETY SOLVER 2.0 is an improved safety solver that provides exact evaluations for the safety status of open boundary areas. By using re-search techniques, it can compute different search goals in the same position. In addition, it has been used in Go program EXPLORER to identify and solve open boundary problems in real game positions. Although experimental results are very encouraging, there are still numerous ideas on how to improve the performance.

One major limitation of the current safety solver is the *size of an open area*. The search space for an open area is often much larger than for an enclosed area, depending on the number of dividers and external liberties. An open area with 15 interior points can easily have over 20 legal moves for each player. In practice the current size limit for the solver is around 15 interior points. A second problem is time control. Under tournament time conditions, it is not feasible to spend too much time on one local area. In addition, searching every unsafe zone using the same amount of time is also not wise. Concentrating the effort on areas where good invading or defending moves are most likely to exist would greatly improve the effectiveness of the solver in game play. Developing a flexible time control scheme that uses heuristics to select suitable problems to solve is an interesting topic for future research.

Three further ideas for improvements are as follows.

1. Store a subtree of search results of local problems in a hash table, then look up solutions when they are needed.
2. Use the most promising move from the search as the best try, even if a search runs out of time.
3. Integrate other tactical solvers into the safety solver. For example, if an open area has been proved as locally unsafe, it should be checked whether its boundary blocks can be connected to other outside blocks.

References

1. Benson, D.B.: Life in the Game of Go. Information Sciences, 10, 17–29 (1976). Reprinted In: Levy, D. (ed.) Computer Games, vol. II, pp. 203–213. Springer, New York (1988)
2. Chen, K., Chen, Z.: Static Analysis of Life and Death in the Game of Go. Information Science 121, 113–134 (1999)
3. Kishimoto, A.: Correct and Efficient Search Algorithms in the Presence of Repetitions. PhD thesis, Department of Computing Science, University of Alberta (2005)
4. Kishimoto, A., Müller, M.: Df-pn in Go: Application to the One-eye Problem. In: van den Herik, H.J., Iida, H., Heinz, E.A. (eds.) 10th Advances in Computer Games (ACG10), Many Games, Many Challenges, pp. 125–141. Kluwer Academic Publishers, Dordrecht (2004)
5. Müller, M.: Computer Go as a Sum of Local Games: An Application of Combinatorial Game Theory. PhD thesis, ETH Zürich, Diss. ETH Nr. 11.006 (1995)
6. Müller, M.: Counting the Score: Position Evaluation in Computer Go. ICGA Journal 25(4), 219–228 (2002)
7. Niu, X., Kishimoto, A., Müller, M.: Recognizing Seki in Computer Go. In: van den Herik, H.J., Hsu, S.-C., Hsu, T.-s., Donkers, H.H.L.M. (eds.) CG 2005. LNCS, vol. 4250, pp. 88–103. Springer, Heidelberg (2006)
8. Niu, X., Müller, M.: An Improved Safety Solver for Computer Go. In: van den Herik, H.J., Björnsson, Y., Netanyahu, N.S. (eds.) CG 2004. LNCS, vol. 3846, pp. 97–112. Springer, Heidelberg (2006)

9. Tao, S.Y.: Guan Zi Pu. 1689 (Reprinted) In: Jiang, M., Jiang, Z. (eds.) Wei Qi Ji Qiao Da Quan. Shu Rong Qi Yi Press, Cheng Du, China (1996)
10. van der Werf, E.: AI Techniques for the Game of Go. PhD thesis, Maastricht University (2005)
11. Vilà, R., Cazenave, T.: When One Eye is Sufficient: a Static Classification. In: van den Herik, H.J., Iida, H., Heinz, E.A. (eds.) 10th Advances in Computer Games (ACG10), Many Games, Many Challenges, pp. 109–124. Kluwer, Dordrecht (2004)
12. Wolf, T.: About Problems in Generalizing a Tsumego Program to Open Positions. In: proceedings of the Game Programming Workshop in Hakone/Japan, pp. 20–26 (1996)

Monte-Carlo Proof-Number Search for Computer Go

Jahn-Takeshi Saito, Guillaume Chaslot,
Jos W.H.M. Uiterwijk, and H. Jaap van den Herik

MICC-IKAT,
Maastricht University, Maastricht, The Netherlands
{j.saito,g.chaslot,uiterwijk,herik}@micc.unimaas.nl

Abstract. In the last decade, proof-number search and Monte-Carlo methods have successfully been applied to the combinatorial-games domain. Proof-number search is a reliable algorithm. It requires a well defined goal to prove. This can be seen as a disadvantage. In contrast to proof-number search, Monte-Carlo evaluation is a flexible stochastic evaluation for game-tree search. In order to improve the efficiency of proof-number search, we introduce a new algorithm, Monte-Carlo Proof-Number search. It enhances proof-number search by adding the flexible Monte-Carlo evaluation. We present the new algorithm and evaluate it on a sub-problem of Go, the Life-and-Death problem. The results show a clear improvement in time efficiency and memory usage: the test problems are solved two times faster and four times less nodes are expanded on average. Future work will assess possibilities to extend this method to other enhanced Proof-Number techniques.

1 Introduction

Since 1994, proof-number search (PNS) has enjoyed wide acceptance in the combinatorial-games domain (cf. Sect. 1.1). The advantage of PNS is its effectiveness and speed. Given enough time and space, it finds a proof or disproof of a binary goal in a game tree. Its disadvantage lies in the requirement for a testable binary goal. In contrast, Monte-Carlo evaluations of combinatorial game positions, introduced in 1993 (cf. Sect. 1.2), offer the advantage of flexibility. A Monte-Carlo evaluation may change its goal because it looks for scores.

This article describes a new algorithm, Monte-Carlo Proof-Number search. It successfully exploits the flexibility of Monte-Carlo evaluation and so it improves the efficiency of the reliable proof-number search.

The remainder of this section outlines the two established techniques, presents the motivation for the new algorithm, and situates the work in the context of proof-number search and Monte-Carlo evaluation. Section 2 offers a detailed description of the new algorithm. Section 3 describes an experiment applying the algorithm to the Life-and-Death problem in Go. Section 4 presents experimental results. Section 5 discusses the findings. Section 6 provides a conclusion and an outlook on future research.

H.J. van den Herik et al. (Eds.): CG 2006, LNCS 4630, pp. 50–61, 2007.

1.1 Proof-Number Search

Proof-number search (PNS) mechanisms are nowadays accepted as standard means of game-tree search. They were introduced by [1] and since then developed into a whole family of search algorithms (e.g., [5,10,15]) with applications to many combinatorial games such as Shogi [12], the one-eye problem in Go [9], and checkers [11]. PNS is a heuristic search applied to game trees to prove or disprove a goal. The status of the goal has to be determined as reachable or non-reachable.[1] The core idea of standard PNS is to order branches efficiently. To that end, the expansion mechanism prefers nodes which require the least number of estimated further expansions for proving or disproving the goal. To achieve such an ordering, the algorithm performs a best-first search strategy based on two figures in each node. One of the numbers is the *proof number*. It maintains the minimal number of successor nodes that require expansion to prove the goal. Analogously, the *disproof number* represents the minimal number of successors necessary to disprove the goal. Leaf nodes are evaluated, interior nodes receive their respective proof and disproof numbers by bottom-up back-propagation. The heuristic exploits the AND/OR tree characteristic that proving the goal requires only to prove the single best OR branch.

1.2 Monte-Carlo Evaluation

Extensive investigation has been conducted on the Monte-Carlo method within the field of Go. The first occurrence of Monte-Carlo (MC) evaluation for the game of Go emerged more than a decade ago [6]. Since then, MC evaluation has been attracting attention in the field of computer Go, particularly in recent years [2,3,4,7,13]. The core idea of the MC approach is to evaluate a game position statistically. This is achieved by playing a number of randomly generated games. The results are stored and then mapped by statistical evaluation to a single value, e.g., by calculating the mean of the scores of all games. In this paper, we call the mapping that achieves the statistical evaluation on the games' values the *accumulative function*. Each random game we call a *simulated game*. A *simulated move* is a move played in a simulated game. A defining characteristic of the MC evaluation lies in its scarce requirement for domain-specific knowledge as the amount of knowledge necessary for playing random games and for scoring finished games suffices.

1.3 Integrating MC and PNS

For integrating MC evaluation and PNS two approaches are possible. First, a global MC move-selection framework might take advantage of local PNS. In this case a top-level co-ordination mechanism needs to determine when to shift from MC evaluation to local PNS. Second, joining MC evaluation with PNS for solving local tactical search may be advantageous. In this article, we address only the latter approach.

---

[1] One variation of the algorithm also deals with the *unknown* status [8, Section 2.1.8].

2   MC-PNS

Below, we propose a new algorithm, Monte-Carlo Proof-Number Search (MC-PNS). It extends PNS' best-first heuristic by adding an MC evaluation. The aim is to achieve a better ordering of moves and thereby to omit investigating branches which are not promising according to the MC evaluation.

2.1   Algorithm

MC-PNS performs a best-first search in an AND/OR game tree. The search aims at proving or disproving a binary goal, i.e., a goal that can be reached by player MAX or be refuted by player MIN under optimal play by both sides. Each node n in the tree contains two real-valued numbers called the proof number ($pn(n)$) and the disproof number ($dn(n)$), respectively.

MC-PNS performs a two-step cycle fully identical to that of PNS. For reasons of readability we summarize the two-step cycle here. In the first step, the best-first strategy requires the algorithm to traverse down the tree starting at the root guided by the smallest proof or disproof number until leaves are reached and expanded. The second step of the cycle begins after that expansion of the leaf. Now, the newly assigned proof and disproof numbers are propagated back to the root updating the proof and disproof number in each node throughout the path back to the root. The cycle is complete as soon as the root has been reached and its values are updated. The cycle is repeated until the termination criterion is met. The criterion is satisfied exactly if either the root's proof number is 0 and the disproof number is infinity, or vice versa. In the first case, the goal is proven. In the latter instance it is refuted. Still, there is a difference between PNS and MC-PNS. The next paragraph outlines the details of the algorithm more formally and shows the small differences.[2]

Let l be a leaf node. If l is a node proving the goal then $pn(l) = 0$ and $dn(l) = \infty$ holds. If l is a node disproving the goal then $pn(l) = \infty$ and $dn(l) = 0$ holds. If l does not immediately prove or disprove a goal $pn(l) = pmc(l)$ and $dn(l) = dmc(l)$, where pmc and dmc are the Monte-Carlo based evaluation functions mapping a node to a target domain. This domain could be $(0, 1)$, with pmc reflecting a node's estimated probability to reach the goal, and dmc reflecting a node's expected probability not to reach a goal. The value 0 is excluded to avoid incorrectly setting $pn = 0$ or $dn = 0$ which could exclude the node from further exploration falsely.

The pmc and dmc numbers for a position with N simulated games are gained by calculating the evaluation function $eval_N : \{0, ..., N\} \to (0, 1)$. The function $eval_N$ depends on the number N_+ of simulated games in which the goal was reached and the number N_- of simulated games in which the goal was not reached. Since $N = N_+ + N_-$, $eval_N$ depends on N. We let $eval_N(0) = \epsilon$ for

---

[2] The two-step cycle can be optimised by not fully back-propagating the values when this is not required. Thereby, the cost of traversal can be reduced. The implementation used in the experiments described here does not include this optimisation.

some small positive real number $\epsilon < 1$ and $eval_N(N_+) = N_+/(N+1)$ for $N_+ \neq 0$. We set $dmc = eval_N(N_+)$ and $pmc = 1 - dmc$.

Starting at the leaf node expanded last, pn and dn values are back-propagated by PNS' back-propagation rule (Fig. 1). Note that MC-PNS assigns a new ordering to the branches.

Rules for AND nodes:

$$pn(n) = \sum_{s \in successor(n)} pn(s)$$

$$dn(n) = \min_{s \in successor(n)} (pn(s))$$

Rules for OR nodes:

$$pn(n) = \min_{s \in successor(N)} (pn(s))$$

$$dn(n) = \sum_{s \in successor(N)} dn(s)$$

Fig. 1. Rules for updating proof and disproof numbers, respectively

2.2 Controlling Parameters

MC-PNS imposes an evaluation cost for each expansion of a node. The algorithm aims at achieving a better move ordering as a trade-off. Two extreme approaches can be distinguished: (1) MC-PNS spends hardly any time on evaluation, and (2) MC-PNS takes plenty time on evaluation. Below, three control parameters are introduced which will enable a trade-off between these extremes.

Number of MC evaluations per node. The precision of the MC evaluation can be determined by the number of simulated games at each evaluated node. Henceforth, this number will be denoted by N.

Look-ahead for MC evaluation. The look-ahead (la) is the maximal length of a simulated game.

Launching level of the MC evaluation. Given a limited look-ahead, it is not useful to waste time on evaluating close to the root of the search-tree. The nodes close to the root must be expected to be expanded anyway. The launching level ($depth$) is the minimal required level of the tree at which nodes are evaluated.

3 Experimental Application of MC-PNS to Go

This section describes an experimental comparison of PNS and the MC-PNS variations introduced above (Subsection 2.2). The experiment is conducted by applying the algorithms to a test set of Life-and-Death problems. Subsection 3.1 outlines the experiment's application domain, Go. Subsection 3.2 poses research questions to be tested. Subsection 3.3 provides an account of the experimental setup.

3.1 Application to the Go Domain

We applied MC-PNS for solving instances of the Life-and-Death problem which is a frequently occurring sub-problem in Go games. Search for solving the Life-and-Death problem has been addressed before (e.g., [16]). The general Life-and-Death problem consists of a locally bounded game position with a target group of stones. Black moves first and has to determine the group's status as either alive, dead, ko, or seki. For the current investigation, however, the problem is reduced to a binary classification of the target group to either alive or dead.

In order to fit the algorithm to the Go domain, the goal to prove, i.e., the status of a group of stones as either alive or dead, is checked by a simple status-detection function. This function is called by each simulated move to determine whether to play further moves or stop expanding. Each simulated game halts after either the goal has been met or a certain previously determined number of moves has been played.

3.2 Research Questions

The experiment is conducted in order to test the viability of the MC-PNS algorithm and its variations created by the parameter settings. The experiment should answer which, if any, of the MC-PNS settings can be of practical use or may lead to enhanced algorithms. To account for the tested expectations concisely, we formulate a list of five operational research questions which should be answered by the experiment. The questions are ordered descendingly by the expected likelihood of being answered positively. The first question is therefore the weakest and the last one the strongest.

1. Can assigning a heuristic weighting in the expansion of PNS achieve a more efficient move ordering?
2. Can a MC-PNS variation achieve such an ordering?
3. Do the number of evaluations per node, the look-ahead, and the launching level trade off between the run-time characteristics of PNS and MC-PNS?
4. Do they thereby contribute to establishing a trade-off for practical work?
5. Does MC-PNS work more efficiently with respect to time and space complexity than PNS?

3.3 Setup

In order to describe the experimental setup, this subsection specifies the Life-and-Death test set, algorithmic details, and the test procedure.

Life-and-Death test set. The test set consists of 30 Life-and-Death problems. The game problems have a beginner and intermediate level (10 Kyu to 1 Dan) and are taken from a set publicly available at GoBase [14]. All test problems can be proven to be advantageous for player Black. For each case there exists a best first move. The test cases were annotated with marks for playable intersections and marks for the groups subject to the alive-or-dead classification. The number of intersections I becoming playable during search is determined by the initial empty intersections and by intersections which are initially occupied but become playable because the occupying stones are captured. For the test cases this indicator I of the search space varied from 8 to 20. The factorial of I provides a rough lower bound for the number of nodes in the fully expanded search tree. Thereby, it provides an estimate for the size of the search space.

Algorithmic and implementational details. PNS and various parameter settings for MC-PNS were implemented in a C++ framework. All experiments were conducted on a Linux workstation with AMD Opteron architecture and 2.8 GHz clock rate. 32 GB of working memory were available. No special pattern matcher or other feature detector was implemented to enhance the MC evaluation or detect the Life-and-Death status of a position. Instead, tree search and MC evaluation are carried out in a brute-force manner in the implementation. In the experiment, Zobrist hashing [17] was implemented to store proof and disproof numbers estimated by MC evaluation. In order to save memory the game boards are not stored in memory for any leaf. A complete game is played for each cycle. The proof tree is stored completely in memory.

The MC evaluation used was based on our Go program MANGO. The program's MC engine provides a speed of up to 5,000 games of 19×19 Go of an average length of 120 moves per second on the hardware specified above. The MC engine was not specially designed for the local Life-and-Death task. Its speed and memory consumption must therefore be expected to perform sub-optimally. The MC evaluation was introduced in Subsection 2.1 as $eval_N$.

Test procedure. Three independent parameters and three dependent variables describe the experiment. The independent parameters control the amount and manner of MC evaluation during search. The dependent variables measure the time and memory resources required to solve a problem by a specified configuration of the algorithm. The configuration is synonymously called parameter setting.

The independent parameters available for testing are: (1) the number of simulated games per evaluated node ($N \in \mathbb{N}$), (2) the look ahead ($la \in \mathbb{N}$), and (3) the launching depth level for MC evaluations ($depth \in \mathbb{N}$).

There are three dependent test variables measured: (1) the time spent for solving a problem measured in milliseconds ($time \in Time = \mathbb{R}$), (2) the number of nodes expanded during search ($nodes \in Nodes = \mathbb{N}$), and (3) a move ($move \in Moves = \{0, 1, ..., 361\}$) proven to reach the goal.[3] The set $Moves$ represents the range of possible intersections of the 19×19 Go board together with an additional null move (0). The null move is returned if no answer can be found in the time provided (see below). Because no other dynamic-memory cost is imposed, $nodes$ suffices for calculating the amount of memory occupied. The memory consumption of a single node in the search tree is 288 bytes.

For the experiment, each configuration consisted of a triple of preset parameters $\in (N, la, depth)$. The following parameter ranges were applied: $N \in$

---

[3] For three test problems, some proof systems found a forcing move first, i.e., there were two best first moves.

$\{3, 5, 10, 20\}$, $la \in \{3, 5, 10\}$, $depth \in \{I, \frac{1}{2}I, \frac{3}{4}I\}$. The $depth$ parameter requires additional explanation. In the experiment's implementation, the number of initially empty intersections I is employed to calculate a heuristical $depth$ value. Thus $I, \frac{1}{2}I$, and $\frac{3}{4}I$ represent functions dependent on the specific instance of I given by each test case. We will call these choices of I the starting strategies and refer to them as 1, 2, and 3 for $I, \frac{1}{2}I$, and $\frac{3}{4}I$, respectively.

The outcome of an experiment is a triple of measured variables $\in Time \times Nodes \times Moves$. Each experimental record consists of a configuration, a problem it is applied to, and an outcome. An experiment delivers a set of such records. In order to account for the randomness of the MC evaluations and potential inaccuracy for measuring the small time spans well below a 10th of a second, PNS and each configuration of MC-PNS was applied to each test case 20 times resulting in 20 records.

4   Results

This section outlines the results of the experiment. First, a statistical measure for comparing the algorithms is introduced, then the experimental results for different configurations are presented and described in detail. The section concludes by summarizing the results in six propositions.

The experiment consumed about 21 hours and produced 22,200 records. All solutions by PNS as well as by MC-PNS configurations were correct.

Aggregates for each combination of test cases and configurations are considered in order to enable a comparison of different parameter settings. For a parameter setting $p = (N, la, depth)$ and a test case ϑ the average result is the triple $(t, s, m) \in Time \times Nodes \times Moves$. For this triple t, s, and m are averaged over all 20 records with the parameter p applied to the test case ϑ. $t_{(p,\vartheta)}$ is the average time and $s_{(p,\vartheta)}$ the average number of nodes expanded for solving ϑ. In order to make a parameter setting p comparable, its average result is compared to average result of PNS. We define the gain of time as $gain_{time}(p) = \frac{1}{30} \sum_{\vartheta=1}^{30} t_{(pns,\vartheta)}/t_{(p,\vartheta)}$. (Here, $t_{(pns,\vartheta)}$ is the average time consumed by PNS to solve test case ϑ.) The gain of space is defined analogously. The positive real numbers $gain_{time}$ and $gain_{space}$ express the average gain of a parameter setting for all thirty test cases. Each gain value is a factor relative to the performance of the PNS benchmark.

In the experiment the configuration using 3 MC evaluations per node, a lookahead of 10 moves, and starting strategy 3, is found to be the fastest configuration ($p_{fast} = (3, 10, 3)$). The $gain_{time}(p_{fast}) = 2.05$ indicating that it is about twice as fast as the PNS benchmark (see Table 1, left and the definition of the gain of speed and gain of space). The $gain_{space}(p_{fast}) = 4.26$. Thus p_{fast} expands less than a quarter of nodes which PNS expands. The parameter setting $p_{narrow} = (20, 10, 3)$ is expanding the smallest number of nodes. It requires less than a fifth of the expansions compared to the benchmark on average on the test set. It is slightly slower than PNS in spite of that (see Table 1, right).

Table 1. Time and node consumption of PNS and various configurations of MC-PNS relative to PNS. Each table presents the ten best ranked settings of the 36 parameter settings. Left: ordered by a factor representing the gain of speed. Right: ordered by a factor representing the decrease of nodes.

Rank	N	la	depth	$gain_{time}$	$gain_{space}$	Rank	N	la	depth	$gain_{space}$	$gain_{time}$
1	3	10	3	2.05	4.26	1	20	10	3	5.31	0.95
2	5	10	3	1.93	4.54	2	20	10	1	5.23	0.96
3	3	10	2	1.87	3.90	3	10	10	3	4.99	1.42
4	5	10	1	1.80	4.59	4	10	10	1	4.95	1.36
5	3	10	1	1.75	4.30	5	5	10	1	4.59	1.80
6	5	10	2	1.74	4.11	6	20	10	2	4.56	0.87
7	3	5	3	1.56	2.70	7	5	10	3	4.54	1.93
8	3	5	1	1.51	2.70	8	3	10	1	4.30	1.75
9	10	10	3	1.42	4.99	9	10	10	2	4.28	1.26
10	10	10	1	1.36	4.95	10	3	10	3	4.26	2.05

Parameter settings with large N and large look-ahead require the least number of nodes to prove or disprove the goal. Parameter settings with small N but large look-ahead perform fastest.

So far, this subsection focused on outlining the results relevant for characterizing the set consisting of PNS and MC-PNS variations. The remainder of this subsection describes the results required for comparing p_{fast}, p_{narrow}, and PNS in greater detail. For this purpose, the data hidden in the aggregates of test cases are unfolded. This is achieved by comparing the average time and space performance for each test case. Figures 2 and 3 illustrate this comparison.

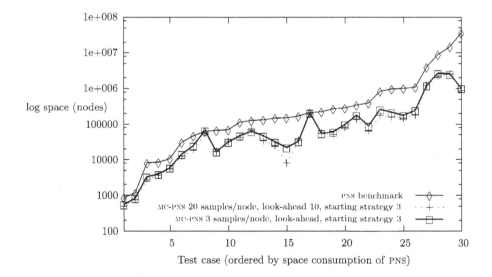

Fig. 2. Space complexity of various configurations

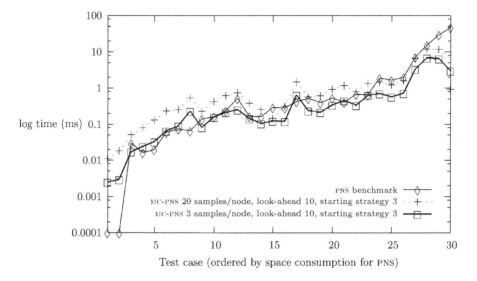

Fig. 3. Time complexity of various configurations

A comparison of the space behaviour (Figure 2) shows that PNS requires the highest number of node expansions to prove or disprove a goal irrespectively of the test case. Its memory requirements are roughly the same as those of p_{fast} and p_{narrow} except for two test cases. PNS requires much more space than the two MC-PNS variations on virtually all other test cases. p_{fast} and p_{narrow} show a similar behaviour in memory consumption with p_{narrow} performing slightly more efficient.

A comparison of the time behaviour (Figure 3) shows a pattern containing more variety. Overall, p_{narrow} is characterized by the least time efficient results: in 22 of the 30 test cases it is the slowest of the three compared proof systems. PNS performs slowest on the remaining 8 test problems. But PNS also performs as single fastest solver in 6 cases and as fast as p_{fast} in 6 cases. p_{fast} shows the most efficient time consumption in 25 cases including the 6 cases in which it is as fast as PNS.

The results show that the performance depends on the complexity inherent to the tested problem. PNS finds its proofs faster than the other problem solvers on simple problems, i.e., problems requiring fewer than 5,000 nodes to be solved. It outperforms its competitors only once in the 20 more complex tasks while achieving this five times in the 10 least complex problems. The two MC-PNS variations perform comparably faster on the 20 most complex tasks. The experimental outcome shows that the speed advantage of p_{fast} relative to PNS grows with the complexity of the tested problem.

The main results of this section can be summarized in six propositions.

(1) (3,10,3) is the fastest parameter configuration. On average it performs two times faster than PNS on the test set and expands less than a quarter of the nodes.

(2) The configuration (20,10,3) is the MC-PNS configuration with the least node expansions on average. It expands only a fifth of the number of nodes expanded by PNS.

(3) p_{fast} and p_{narrow} reliably expand considerably fewer nodes than MC-PNS.

(4) p_{fast} performs reliably faster than MC-PNS.

(5) The advantage of time performance of p_{fast} relative to PNS even grows with the complexity of the problem.

(6) PNS performs better than p_{fast} on problems with small complexity.

5 Discussion

This section discusses the results outlined in the previous section. First, explanations for the results experimentally gained are offered. A critical remark on the suitability of the setup concludes this section.

On average p_{fast} solves problems twice as fast as PNS. This is coherent with a general tendency observed. As outlined above, settings with small N (i.e., few simulated games) and large la (i.e., far look-ahead) are generally the fastest. The speed of a MC-PNS depends mainly on two items: (1) the amount of nodes it expands, and (2) the speed needed to evaluate a node. These two items are mutually dependent. The number of expansions decreases with the intensity of evaluation because the heuristic is more reliable and prunes more nodes. This is reflected by the experimental finding that nodes with thorough evaluation (large N and large la) reach their goals with few expansions (see Table 1). But more intensive evaluations require more time. Therefore, an optimization problem has to be solved to compensate for the intensity of the evaluation. The optimum trades off between the number of nodes visited and the evaluation time for each single node. This optimum is found to be $p_{fast} = (3, 10, 3)$. Extensively evaluating each node, as pursued by p_{narrow}, devotes too much time on each single evaluation. The strategy of omitting a heuristic evaluation entirely, as embodied by PNS, is cheap on each node but fails to prune the tree sufficiently on a global scale.

Thus the N and la parameters can be said to control the intensity of each evaluation successfully. The *depth* parameter has a minor impact on controlling this intensity. More importantly, only few MC evaluations per node can yield a reasonable heuristic for the MC-PNS framework.

The design of the PNS framework requires running up and down the branch leading to the currently expanded node (see Sect. 2.1). The deeper the tree grows the higher are the costs for this traversal. The cost grows linearly with the tree's depth as more nodes are expanded. Additionally, the traversal takes longer. Therefore, pruning the tree will have a much stronger impact on larger trees. The point at which pruning pays off is reached at a search tree complexity equivalent to a size which requires about 5,000 nodes to be solved by PNS. Any problem beyond this threshold of complexity must be expected to be solved faster by p_{fast}. Two thirds of the test cases employed lie beyond that threshold. One might argue that this composition of the set is arbitrary. But there are two strong reasons for assuming that the inferences made about the quality of p_{fast} are valid in general.

First, the test cases chosen are rather easy. One may therefore expect that real-life problems are harder than the cases presented. Thus in practice p_{fast} should be more relevant. Second, the absolute time saved by p_{fast} is much larger for complex problems. For instance, PNS requires 47.7 sec to solve the most complex problem whereas p_{fast} solves the problem in less than 6 sec (eight times faster). The absolute time saved is crucial for real-life applications, e.g., in a Go program. Thus we may conclude that it is valid to generalize our finding that p_{fast} is performing faster than PNS beyond our test set.

6   Conclusion and Outlook

We introduced a new algorithm, MC-PNS, based on MC evaluation within a PNS framework. An experimental application of the new algorithm and several variations to the Life-and-Death sub-problem of Go were described; moreover, its interpretation was presented. It was demonstrated experimentally that given the right setting of parameters MC-PNS will outperform PNS. For such a configuration MC-PNS will be two times faster than PNS on average and use less than a quarter of the node expansions. Section 5 presented strong evidence for assuming that this result will be generalized beyond the test cases observed.

Thus, we may conclude that all research questions posed in Subsection 3.2 can be answered positively on the basis of the experimental findings.

Future work will be required to assess the possibility of successfully extending the MC-PNS approach to the practically more relevant Depth-first Proof-Number Search (DF-PN, [10]). Its characteristics are slightly different but we believe that efficiency of DF-PN can be improved with our approach, too. Furthermore, detailed tuning of the algorithm and including more domain knowledge by applying patterns remain to be tested.

Acknowledgments

We would like to thank Mark Winands, Ulaş Türkmen, Benjamin Torben-Nielsen, Steven de Jong, and Jeroen Donkers for their support. The comments by the anonymous referees are gratefully acknowledged. In particular, we would like to thank one referee for his or her constructive criticism which led to crucial improvements of the implementational work. Without this feedback the current results would not have been possible. The work is financed by the Dutch Organization for Scientific Research in the framework of the project GO FOR GO, grant number 612.066.409.

References

1. Allis, L.V., van der Meulen, M., van den Herik, H.J.: Proof-Number Search. Artificial Intelligence 66, 91–124 (1994)
2. Bouzy, B., Helmstetter, B.: Monte Carlo Developments. In: van den Herik, H.J., Iida, H., Heinz, E.A. (eds.) 10th Advances in Computer Games (ACG10), Many Games, Many Challenges, pp. 159–174. Kluwer Academic Publishers, Dordrecht (2004)

3. Bouzy, B.: Associating Shallow and Selective Global Tree Search with Monte Carlo for 9×9 Go. In: van den Herik, H.J., Björnsson, Y., Netanyahu, N.S. (eds.) CG 2004. LNCS, vol. 3846, pp. 76–80. Springer, Heidelberg (2006)
4. Bouzy, B.: History and Territory Heuristics for Monte-Carlo Go. In: Chen, K., et al. (eds.) Joint Conference on Information Sciences JCIS 2005, p. 4 (2005)
5. Breuker, D.M.: Memory versus Search. PhD thesis, Maastricht University (1998)
6. Brügmann, B.: Monte Carlo Go. White paper (1993)
7. Cazenave, T., Helmstetter, B.: Search for Transitive Connection. Information Sciences 132(1), 93–103 (2004)
8. Kishimoto, A.: Correct and Efficient Search Algorithms in the Presence of Repetitions. PhD thesis, University of Alberta (2005)
9. Kishimoto, A., Müller, M.: DF-PN in Go: Application to the One-Eye Problem. In: van den Herik, H.J., Iida, H., Heinz, E.A. (eds.) 10th Advances in Computer Games (ACG10), Many Games, Many Challenges, pp. 125–141. Kluwer Academic Publishers, Dordrecht (2003)
10. Nagai, A.: Df-pn Algorithm for Searching AND/OR Trees and Its Applications. PhD thesis, University of Tokio (2002)
11. Schaeffer, J., Björnsson, Y., Burch, N., Kishimoto, A., Muller, M., Lake, R., Lu, P., Sutphen, S.: Solving Checkers. In: International Joint Conference on Artificial Intelligence (IJCAI), pp. 292–297 (2005)
12. Seo, M., Iida, H., Uiterwijk, J.W.H.M.: The PN*-Search Algorithm: Application to Tsume-Shogi. Artificial Intelligence 129(1-2), 253–277 (2001)
13. Sheppard, B.: Efficient Control of Selective Simulations. ICGA Journal 27(3), 67–80 (2005)
14. van der Steen, J.: GoBase.org website (2006), http://www.gobase.org
15. Winands, M.H.M., Uiterwijk, J.W.H.M., van den Herik, H.J.: An Effective Two-Level Proof-Number Search Algorithm. Theoretical Computer Science 313(3), 511–525 (2004)
16. Wolf, Th.: Forward Pruning and Other Heuristic Search Techniques in Tsume Go. Information Sciences 122(1), 59–76 (2000)
17. Zobrist, A.L.: A New Hashing Method with Application for Game Playing. ICCA Journal 13(2), 69–73 (1990)

Virtual Global Search: Application to 9×9 Go

Tristan Cazenave

LIASD, Dept. Informatique,
Université Paris 8, Saint-Denis, France
`cazenave@ai.univ-paris8.fr`

Abstract. In games, Monte-Carlo simulations can be used as an evaluation function for Alpha-Beta search. Assuming w is the width of the search tree, d its depth, and g the number of simulations at each leaf, then the total number of simulations is at least $g \times (2 \times w^{\frac{d}{2}})$. In games where moves permute, we propose to replace this algorithm by a new algorithm, *Virtual Global Search*, that only needs $g \times 2^d$ simulations for a similar number of games per leaf. The algorithm is also applicable to games where moves often but not always permute, such as Go. We specify the application for 9×9 Go.

1 Introduction

Monte-Carlo methods can be used to evaluate moves and states in perfect information games. They can be combined with Alpha-Beta search, using Monte-Carlo simulations at the leaves of the search tree to evaluate positions [1]. In the remaining of the paper, w is the width, and d the depth of the global search, and a global move is a move that can be tried in a global search. The time complexity of the combination is at least proportional to $g \times (2 \times w^{\frac{d}{2}})$, and the space complexity is linear in d.

In some games such as Hex, moves always permute. The final position of a game is the same when two moves by the same color were switched during the game. It is not true in other games such as Go. However, it is often true in Go, and this property can be used efficiently to combine Alpha-Beta search with Monte-Carlo evaluation at the leaves. The algorithm we propose, named *Virtual Global Search*(VGS), has a time complexity proportional to $g \times 2^d$, and a space complexity proportional to w^d. In games where moves always permute, VGS gives results close to the normal best moves within the complexity frameworks mentioned above. We have found that VGS is also interesting in games such as Go, where moves often permute.

The course of the article is as follows. Section 2 describes related work. Section 3 exposes Standard Global Search (SGS). Section 4 presents Virtual Global Search. Section 5 estimates the complexities of the two search algorithms. Section 6 provides details on experimental results. Section 7 outlines future work. Section 8 completes with a conclusion.

H.J. van den Herik et al. (Eds.): CG 2006, LNCS 4630, pp. 62–71, 2007.

2   Related Work

Below we briefly discuss related works. They are on Monte Carlo and games (2.1), Standard Monte-Carlo Go (2.2), Monte-Carlo Go enhancements (2.3), and Global search in Go (2.4).

2.1   Monte Carlo and Games

Monte-Carlo methods have been used in many games. In Bridge, GIB uses Monte Carlo to compute statistics on solved double dummy deals [12]. In Poker, POKI uses selective sampling and a simulation-based betting strategy for the game of Texas Hold'em [2]. In Scrabble, MAVEN controls selective simulations [15]. Moreover, Monte Carlo has also been applied to Phantom Go by randomly putting opponent stones before each random game [6], and to probabilistic combinatorial games [16].

2.2   Standard Monte-Carlo Go

The first Go program based on Monte Carlo techniques is GOBBLE [5]. It uses simulated annealing on a list of moves. The list is sorted by the mean score of the games in which the move under investigation has been played first. Moves in the list are switched with their neighbor with a probability dependent on the temperature. The moves are tried in the games in the order of the list. At the end, the temperature is set to zero for a small number of games. After all games have been played, the value of a move is the average score of the games in which it has been played as a first move. GOBBLE[5]has a good global sense but lacks tactical knowledge. For example, it often plays useless atari, or tries to save captured strings.

2.3   Monte-Carlo Go Enhancements

An enhancement of GOBBLE [5] is to combine its Monte-Carlo techniques with Go knowledge. INDIGO has been using Go knowledge to select a small number of moves that are later evaluated with the Monte-Carlo method [3]. A second use of Go knowledge is to bias the selection of moves during the random games using patterns and rules [3,8]. A third enhancement is to compute statistics on unsettled tactical goals instead of only computing statistics on moves [7].

2.4   Global Search in Go

The combination of global search with Monte-Carlo Go has been studied by Bouzy [4] for 9×9 Go. His algorithm associates progressive pruning with Alpha-Beta search to discriminate moves in a global-search tree with a Monte-Carlo evaluation at the leaves.

CRAZY STONE [11] uses a back-up operator and biased move exploration in combination with a Monte-Carlo evaluation at the leaves. It finished first in the 9×9 Go tournament in the 2006 Computer Olympiad [10].

The analysis of decision errors during selective tree search has been studied by Chen [9].

3   Standard Global Search

In this section we first present how global moves are selected (3.1), and then how they are used for Standard Global Search (3.2).

3.1   Selection of Moves

Global search in Go is highly selective due to the large number of possible moves. In order to select the global moves to be considered in the search, we use a program that combines tactical search and Monte-Carlo search [7], called TSMC. It gives a static evaluation to each move. In this paper, TSMC only uses connection search in association with Monte-Carlo simulations. The w moves that have the best static evaluation according to TSMC are selected as global moves. They are sorted according to their static evaluation. An alternative is tresholding, i.e., to select the global moves that have a static evaluation above a fixed percentage of the static evaluation of the best move, so far.

It is important to note that only the locations of the selected global moves are used for the search and not the moves themselves. The w locations are used in the global search to generate possible moves for both colors.

3.2   Standard Global Search

Standard Global Search (SGS) is a global search with Monte-Carlo evaluation at the leaves. Performing Monte-Carlo simulations at the leaves of the tree in order to evaluate them is a natural idea when combining Monte-Carlo sampling and game tree search [1].

4   Virtual Global Search

The selection of moves for Virtual Global Search (VGS) is the same as for Standard Global Search. The difference with SGS is in the second phase (the search). In this section, we first specify, how the permutation of moves is transformed into sequences of moves (4.1). Then we explain how sequences are evaluated at the end of the random games (4.2). Eventually, we explain how the global search tree is developed (4.3).

4.1   Permutation of Moves

The main idea underlying the algorithm is that a move in a random game has roughly the same influence when it is played (1) at the beginning, (2) in the course of a random game, or (3) at the end. GOBBLE [5] uses a similar idea when evaluating the moves independently of the depth where they occurred in the random games. In GOBBLE, a move is evaluated using all the games where it has been played as first move on an intersection, not taking into account whether it has been played at the beginning or at the end of a game.

We extend this idea to sequences of moves. Therefore, we assume that a sequence of moves has roughly the same value even when (in a random game) the moves of the sequence are played in any order. This leads to count a random game by sequence (instead of by move), i.e., when all the moves of the sequence have been played on their intersection during the random game. (Please note that the program only takes into account the first move played on an intersection, it does not take into account moves that are played on an intersection where a string has been captured earlier during the random game).

4.2   Random Games with Evaluation of the Sequences at the End

If we consider that a sequence is valid when its moves have been played in any order during a random game, TSMC may store at the end of the game the score of the game and associate it to the sequence.The length of the sequences considered is d, and it is assumed that the same selection of w moves at the root node is used for all positions.

TSMC has a structure which records, for every possible sequence of selected moves, the mean score of the games. We note again that the sequence has been played in any order. The size of the memory used to record mean scores is proportional to w^d as it is approximately the number of possible sequences.

A sequence of moves is associated to an index. Moreover, each global move is associated to an index in the sorted array of selected global moves. TSMC allocates b bits for representing the index of a move. If d is the maximum depth allowed, a sequence is coded with $b \times d$ bits. The move at depth i in the sequence is coded starting at the bit number $(i - 1) \times b$.

To each sequence a structure is associated. This structure records (1) the number of games where the sequence has been played in any order, and (2) the cumulated scores of the games where the sequence has been played. These two kinds of data are used at the end of the simulations to compute the mean score of the games where the sequence has been played in any order. The size of the array of structures is $2^{b \times d}$ entries, each entry has the size of two integers.

When a random game is completed, TSMC develops a search tree using the selected global moves. We call this search tree a *virtual search tree* since it does not really play moves during the expansion of the tree, but only updates the index of the sequence, for each global move chosen (it also forbids to choose a move that is already played in the sequence).

In the virtual search tree, a global move of a given color is played only if it has the same color as the first move on the intersection, in the random game. On average, the move in the random game is of the same color half of the time. Out of the w possible global moves, only $\frac{w}{2}$ have the required color in the random game on average. At the leaves of the search tree, TSMC updates the score of the sequence that has been played until the leaf: it increments the number of times the sequence has been played, and it adds the score of the game to the cumulated score. Given that TSMC tries $\frac{w}{2}$ moves at each node, and that it searches to depth d, the number of leaves of the search tree is roughly $(\frac{w}{2})^d$ (a little less in fact since moves cannot be played twice on the same intersection, and sometimes even less than $\frac{w}{2}$ moves have been played with a color).

4.3   Search After All the Random Games Are Completed

Virtual Global Search is a global search with a virtual pre-computed evaluation at the leaves. It is performed after all the random games have been played, and after all possible sequences have been associated to a mean score.

An Alpha-Beta search is used to develop the Virtual Global Search tree. Moves are not played, actually, but instead the index of the current sequence is updated at each move. The evaluation at the leaves is pre-computed; it consists in returning the mean of the random games where the current sequence has been detected. It only costs an access to the array of structures at the index of the current sequence. Developing the search tree of Virtual Global Search takes little time in comparison to the time used for the random games.

5   Estimated Complexities

In this section, we give estimations of the complexity of SGS (5.1) and of VGS (5.2).Then we give a comparison of the sibling leaves (5.3).

5.1   Complexity of Standard Global Search

Let g be the number of random games played at each leaf of the SGS tree in order to evaluate the leaf. The Alpha-Beta algorithm, with an optimal move ordering, has roughly $2 \times w^{\frac{d}{2}}$ leaves [14]. Therefore, the total number of random games played is $g \times (2 \times w^{\frac{d}{2}})$. The space complexity of the Alpha-Beta search is linear in the depth of the search.

5.2   Complexity of Virtual Global Search

There are a little less than w^d possible global sequences. At the end of each random game, approximately $(\frac{w}{2})^d$ sequences are updated. Let g_1 be the number of random games necessary to have a mean computed with g random games for each sequence. We have:

$$g_1 = \frac{g \times w^d}{(\frac{w}{2})^d} = g \times 2^d. \tag{1}$$

Therefore, TSMC has to play $g \times 2^d$ random games, for the Virtual Global Search, in order to have an equivalent of the Standard Global Search with g random games at each leaf.

The space complexity of VGS is w^d, as there are w^d possible sequences and the program has to update statistics on each detected sequence at the end of each random game.

5.3   Comparison of Sibling Leaves

Two leaves that have the same parent share half of their random games. To prove this statement we consider all the random games where the move that links the parent and one of the leaves has been played. Half of these random games also contain the move that links the parent and the other leaf. Therefore, in order to compare two leaves based on g different games as in Standard Global Search, $g \times 2^{d+1}$ random games are necessary instead of $g \times 2^d$ random games.

6   Experimental Results

Experiments were performed on a Pentium 4 3.0 GHz with 1GB of RAM. Table 1 gives the time used by different algorithms to generate the first move of a game. It is a good approximation of the mean time used per move during a game. From lines 1 and 2, we can see that for a similar precision, VGS with width 8 and at depth 3 takes 0.3 seconds when SGS with a similar setting takes 15.8 seconds.

Table 1. Comparison of times for the first move

Algorithm	w	d	Games	Time
SGS	8	3	$g = 100$	15.8s
VGS	8	3	$g_1 = 800$	0.3s
SGS	16	3	$g = 100$	118.2s
VGS	16	3	$g_1 = 800$	0.3s
VGS	81	3	$g_1 = 800$	0.6s

An interesting result is given in the last line, where VGS only takes 0.6 seconds for a full width depth-3 search, and 800 games played in total for all sequences. This number is equivalent to 100 random games at each leaf according to the description given in Subsection 5.2.

Lines 2 and 4 have the same time since the time used for the tree search is negligible compared to the time used for the random games, and the number of random games needed is not related to the width of the tree.

Table 2 compares the two different algorithms. Each line of the table resumes the result of 100 games between the two programs on 9×9 boards (50 games with Black, and 50 with White). The first column gives the name of the algorithm for the max player. The second column gives the maximum number of global moves allowed for Max. The third column gives the maximum global depth for Max. The fourth column gives the total number of random games played for VGS. The fifth column gives the minimum percentage of the best-move static evaluation required to select a global move: a move is selected if its static evaluation is greater than the static evaluation of the best move adjusted by a given percentage. The sixth column gives the average time used for VGS (including the random games). The next columns give similar information for the min player. The last two columns give the average score of the 100 games for Max, and the number of games won by Max out of the 100 games.

For example, the first line of Table 2 shows that VGS, with width eight, depth three, two thousand random games, all moves allowed, takes half a second per move and loses against SGS with similar settings. This experiment shows that

Table 2. Comparison of algorithms

Max	w	d	g_1	%	Time	Min	w	d	g	Time	Result	Won
VGS	8	3	2,000	0%	0.5s	SGS	8	3	250	11.5s	−3.3	42
VGS	8	3	8,000	0%	2.1s	SGS	8	3	100	7.2s	5.6	66
VGS	8	3	8,000	50%	2.2s	SGS	8	3	100	7.0s	7.2	75
VGS	16	3	8,000	50%	1.8s	SGS	8	3	100	6.4s	9.6	70
VGS	8	5	32,000	0%	13.1s	SGS	8	3	100	7.6s	10	73

with equivalent precisions on the evaluation (here the number of games per leaf for VGS is $\frac{2000}{2^3} = 250$, the same as for SGS), the virtual global search takes 23 times less time for an average loss of 3.3 points per game.

The next lines test different options for VGS against a fixed version of SGS (100 games per leaf, width 8, depth 3). A depth-3 VGS, with at most 16 global moves that have a static evaluation which is at least half the best static evaluation, and 8,000 games takes less than 2 seconds per move and wins by almost 10 points against a SGS that takes more than 6 seconds per move. In these experiments the number of games used to select the moves to search is the same as the number of games per leaf.

In the next experiments, we have decorrelated these two numbers. Table 3 gives some results of 100-game matches against GNUGO 3.6. The first column is the algorithm used for the max player, the second column the maximum width of the search tree, the third column the depth, the fourth column (Pre) is the number of games played before the search in order to select the moves to try, the fifth column is the number of games for each leaf of the search tree, then comes the minimum percentage of the best move used to select moves (sixth column), the average time of the search per move (seventh column), the mean result of the 100 games against GNUGO 3.6 (eight column), the associated standard deviation (nineth column), and the number of won games (tenth column).

The best number of won games is 31 for VGS with $g_1 = 80,000$ ($g = 10,000$ and $d = 3$). However, the best mean is −11.1 for SGS with $g = 1,000$ and $d = 3$, but it only wins 21 games. The two results for depth 1 are close to a standard Monte-Carlo evaluation without global search. The results show that

Table 3. Results against GNUGO 3.6

Max	w	d	Pre	g	%	Time	Mean	σ	Won
VGS	8	3	100	100	80%	0.4s	−34.4	27.6	4
VGS	8	3	1,000	1,000	80%	3.7s	−26.6	27.7	10
VGS	8	1	1,000	4,000	80%	3.7s	−17.7	28.6	16
VGS	8	3	16,000	2,000	80%	4.7s	−16.1	23.1	17
VGS	8	3	1,000	10,000	80%	37.4s	−14.4	28.5	31
SGS	8	3	100	100	80%	3.3s	−23.9	22.3	10
SGS	8	1	1,000	4,000	80%	4.4s	−17.3	24.7	16
SGS	8	3	1,000	1,000	80%	23.6s	−11.1	23.9	21

more accuracy (4,000 games instead of 1,000) may be more important in some cases than more depth when comparing lines two and three of the table.

A result of this table is that for the same number of games per leaf and for a width-8 and depth-3 search, standard search is better than virtual search.

Table 4 gives some results of 100 game matches with width-16 global search against GNUGO 3.6. In these experiments, the global search is given more importance since the number of locations investigated is 16 and those locations that are not highly evaluated by the static evaluation are searched.

Table 4. Results of width 16 against GNUGO 3.6

Max	w	d	Pre	g	%	Time	Mean	σ	Won
VGS	16	3	1,000	2,000	0%	7.6s	−12.2	25.4	29
VGS	16	3	1,000	10,000	0%	23.7s	−16.1	26.1	23
VGS	16	1	1,000	8,000	0%	4.8s	−23.0	32.4	18
SGS	16	3	1,000	2,000	0%	515.7s	−15.0	23.9	19

The results of Table 4 show that VGS outperforms SGS for width 16 and depth 3. It plays moves in 7.6 seconds instead of 515.7 seconds, scores −12.2 points instead of −15.0, and wins 29 games instead of 19.

7   Future Work

In Go, permutation of moves does not always lead to the same position. For example, in Fig. 1, White 1 followed by Black 2 does not give the same position as Black 2 followed by White 1. In other games, such as Hex, moves always permute. Using VGS in such games is appropriate.

Concerning Go, we list five potential improvements below. A first improvement of our current program would be to differentiate between permuting and non-permuting moves. There are two implementation possibilities. First, two moves at the same location and of the same color may be considered different if one captures a string, and the other not. This would enable to detect problems such as the non permutation of Fig. 1. The idea is linked with some recent work on single agent search [13]. Second,the order of the moves in the random game should be matched with the order of the moves in the sequence to evaluate.

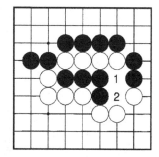

Fig. 1. The order of moves can be important

A second improvement is in selection of moves. Currently the moves are chosen according to their static evaluation at the root, not taking into account moves that are answers to other moves. Three straightforward implementations are: (1) investigate whether the move played is a "forced"' answer, (2) improve the evaluation at the leaf; (3) add Go knowledge such as in INDIGO [3,8].

A third improvement is to combine global search with tactical goals. This means searching in a global tree of goals instead of a global tree of moves, and evaluating a sequence of goals instead of a sequence of moves [7].

A fourth improvement is to combine Virtual Global Search with progressive pruning [4]. We keep investigating the global search tree while playing the random games, and stop as soon as one move is clearly superior to the others.

A fifth improvement may be to combine the virtual search with (1) new backup operators and (2) biased move exploration, such as in CRAZY STONE [10].

8   Conclusion

We presented a new algorithm VGS that combined Alpha-Beta search with Monte-Carlo simulations. It takes $g \times 2^d$ simulations and w^d memory instead of more than $g \times (2 \times w^{\frac{d}{2}})$ simulations and linear memory in d for the usual combination of Alpha-Beta and Monte-Carlo simulations (SGS). In games where moves permute VGS gives better results than SGS. In 9×9 Go, it also gives good results even through the moves do not always permute. Hence, we may conclude that VGS is a viable idea to elaborate upon even in games where the moves not always permute.

References

1. Abramson, B.: Expected-Outcome: a General Model of Static Evaluation. IEEE Transactions on PAMI 12(2), 182–193 (1990)
2. Billings, D., Davidson, A., Schaeffer, J., Szafron, D.: The Challenge of Poker. Artificial Intelligence 134(1-2), 210–240 (2002)
3. Bouzy, B.: Associating Domain-Dependent Knowledge and Monte Carlo Approaches Within a Go Program. Information Sciences 175(4), 247–257 (2005)
4. Bouzy, B.: Associating Shallow and Selective Global Tree Search with Monte Carlo for 9x9 Go. In: van den Herik, H.J., Björnsson, Y., Netanyahu, N.S. (eds.) CG 2004. LNCS, vol. 3846, pp. 67–80. Springer, Heidelberg (2006)
5. Brügmann, B.: Monte Carlo Go (1993),ftp://ftp-igs.joyjoy.net/go/computer/mcgo.tex.z
6. Cazenave, T.: A Phantom Go Program. In: van den Herik, H.J., Hsu, S.-C., Hsu, T.-s., Donkers, H.H.L.M. (eds.) CG 2005. LNCS, vol. 4250, pp. 120–126. Springer, Heidelberg (2006)
7. Cazenave, T., Helmstetter, B.: Combining Tactical Search and Monte-Carlo in the Game of Go. In: CIG'05, pp. 171–175 (2005)
8. Chaslot, G.: Apprentissage par Renforcement dans une Architecture de Go Monte Carlo. Mémoire de DEA, Ecole Centrale de Lille (September 2005)
9. Chen, K.: A Study of Decision Error in Selective Game Tree Search. Information Science 135(3-4), 177–186 (2001)
10. Coulom, R., Chen, K.: Crazy Stone wins 9×9 Go Tournament. Note. ICGA Journal 29(2), 92 (2006)
11. Coulom, R.: Efficient Selectivity and Backup Operators in Monte-Carlo Tree Search. In: van den Herik, H.J., Ciancarini, P., Donkers, H.H.L.M. (eds.) 5th Computers and Games Conference (CG 2006). LNCS, vol. 4630, pp. 73–84. Springer, Heidelberg (2007)

12. Ginsberg, M.L.: GIB: Steps Toward an Expert-level Bridge-playing Program. In: IJCAI-99, pp. 584–589, Stockholm, Sweden (1999)
13. Helmstetter, B., Cazenave, T.: Incremental Transpositions. In: van den Herik, H.J., Björnsson, Y., Netanyahu, N.S. (eds.) CG 2004. LNCS, vol. 3846, pp. 220–231. Springer, Heidelberg (2006)
14. Knuth, D.E., Moore, R.W.: An Analysis of Alpha-Beta Pruning. Artificial Intelligence 6(4), 293–326 (1975)
15. Sheppard, B.: Efficient Control of Selective Simulations. ICGA Journal 27(2), 67–80 (2004)
16. Zhao, L., Müller, M.: Solving Probabilistic Combinatorial Games. In: van den Herik, H.J., Hsu, S.-C., Hsu, T.-s., Donkers, H(J.) H.L.M. (eds.) CG 2005. LNCS, vol. 4250, pp. 225–238. Springer, Heidelberg (2006)

Efficient Selectivity and Backup Operators in Monte-Carlo Tree Search

Rémi Coulom

CNRS-LIFL, INRIA-SequeL,
Université Charles de Gaulle, Lille, France
`remi.coulom@univ-lille3.fr`

Abstract. A Monte-Carlo evaluation consists in estimating a position by averaging the outcome of several random continuations. The method can serve as an evaluation function at the leaves of a min-max tree. This paper presents a new framework to combine tree search with Monte-Carlo evaluation, that does not separate between a min-max phase and a Monte-Carlo phase. Instead of backing-up the min-max value close to the root, and the average value at some depth, a more general backup operator is defined that progressively changes from averaging to min-max as the number of simulations grows. This approach provides a fine-grained control of the tree growth, at the level of individual simulations, and allows efficient selectivity. The resulting algorithm was implemented in a 9×9 Go-playing program, CRAZY STONE, that won the 10th KGS computer-Go tournament.

1 Introduction

When writing a program to play a two-person zero-sum game with perfect information, the traditional approach consists in combining alpha-beta search with a heuristic position evaluator [20]. The heuristic evaluator is based on domain-specific knowledge, and provides values at the leaves of the search tree. This technique has been very successful for games such as chess, draughts, checkers, and Othello.

Although the traditional approach has worked well for many games, it has failed for the game of Go. Experienced human Go players still easily outplay the best programs. So, the game of Go remains an open challenge for artificial-intelligence research [8].

Among the main difficulties in writing a Go-playing program is the creation of an accurate static position evaluator [8,15]. When played on a 9×9 grid, the complexity of the game of Go, in terms of the number of legal positions, is inferior to the complexity of the game of chess [2,27]; the number of legal moves per position is similar. Nevertheless, chess-programming techniques fail to produce an artificial 9×9 Go player stronger than experienced humans. One reason is that tree search cannot be easily stopped at quiet positions, as it is done in chess. Even when no capture is available, most of the positions in the game of Go are very dynamic.

H.J. van den Herik et al. (Eds.): CG 2006, LNCS 4630, pp. 72–83, 2007.

A useful alternative to static evaluation that fits the dynamic nature of Go positions is Monte-Carlo evaluation. It consists in averaging the outcome of several continuations. Monte-Carlo evaluation is an usual technique in games with randomness or partial observability [5,14,17,23,26], but can also be applied to deterministic games, by choosing actions at random until a terminal state is reached [1,9,10].

The accuracy of Monte-Carlo evaluation can be improved with tree search. Juillé [18] proposed a selective Monte-Carlo algorithm for single-agent deterministic problems, and applied it successfully to grammar induction, sorting-network optimization, and a solitaire game. Bouzy [6] also applied a similar method to 9 × 9 Go. The algorithms of Juillé and Bouzy let grow a tree grow by iterative deepening, and prune it by keeping only the best-looking moves after each iteration. A problem with these selective methods is that they may prune a good move because of evaluation inaccuracies. Other algorithms with better asymptotic properties (given enough time and memory, they will find an optimal action) have been proposed in the formalism of Markov decision processes [12,19,22].

This paper presents a new algorithm for combining Monte-Carlo evaluation with tree search. Its basic structure is described in Section 2. Its selectivity and backup operators are presented in the Sections 3 and 4, respectively. In Section 5, game results are discussed. Section 6 summarizes the contributions of this research, and gives directions for future developments.

2   Algorithm Structure

The structure of our algorithm consists in iteratively running random simulations from the root position. This produces a tree made of several random games. The tree is stored in memory. At each node of the tree, the number of random games that passed through this node is counted, as well as the sum of the values of these games, and the sum of the squares of the values. In CRAZY STONE, the value of a simulation is the score of the game.

Our approach is similar to the algorithm of Chang, Fu, and Marcus [12], and provides some advantages over Bouzy's method [6]. First, the algorithm is anytime: each simulation brings additional information that is immediately backed up to the root, which is convenient for time management (Bouzy's algorithm only provides information at each deepening iteration). Also, the framework allows algorithms with proved convergence to the optimal move, because selectivity can be controlled at the level of individual simulations, and does not require that complete branches of the tree be cut off.

In practice, not all the nodes are stored. Storing the whole tree would waste too much time and memory. Only nodes close to the root are memorized. This is done by applying the following two rules.

- Start with only one node at the root.
- Whenever a random game goes through a node that has been visited once, create a new node at the next move, if it does not already exist.

As the number of games grows, the probability distribution for selecting a move at random is altered. In nodes that have been visited less than the number of points of the goban (this threshold has been empirically determined as a good compromise), moves are selected at random according to heuristics described in Appendix A. Beyond this number of visits, the node is called an *internal* node, and moves that have a higher value tend to be selected more often, as described in Section 3. This way, the search tree is grown in a best-first manner.

3   Selectivity

In order not to lose time exploring useless parts of the search tree, it is important to carefully allocate simulations at every node. Moves that look best should be searched more deeply, and bad moves should be searched less.

3.1   Background

Over the years, a large variety of selectivity algorithms have been proposed in the framework of Monte-Carlo evaluation. Most of them rely on the central-limit theorem, that states that the mean of N independent realizations of a random variable with mean μ and variance σ^2 approaches a normal distribution with mean μ and variance σ^2/N. When trying to compare the expected values of many random variables, this theorem allows to compute a probability that the expected value of one variable is larger than the expected value of another variable.

Bouzy [7,9] used this principle to propose progressive pruning. Progressive pruning cuts off moves whose probability of being best according to the distribution of the central-limit theorem falls below some threshold. Moves that are cut off are never searched again. This method provides a very significant acceleration.

Progressive pruning can save many simulations, but it is very dangerous in the framework of tree search. When doing tree search, the central-limit theorem does not apply, because the outcomes of random simulations are not identically distributed: as the search tree grows, move probabilities are altered. For instance, the random simulations for a move may look bad at first, but if it turns out that this move can be followed up by a killer move, its evaluation may increase when it is searched more deeply.

In order to avoid the dangers of completely pruning a move, it is possible to design schemes for the allocation of simulations that reduce the probability of exploring a bad move, without ever letting this probability go to zero. Ideas for this kind of algorithm can be found in two fields of research: n-armed bandit problems, and discrete stochastic optimization. The n-armed bandit techniques (Sutton and Barto's book [25] provides a suitable introduction) are the basis for the Monte-Carlo tree-search algorithm by Chang, Fu, and Marcus [12]. Optimal schemes for the allocation of simulations in discrete stochastic optimization [3,13,16], could also be applied to Monte-Carlo tree search.

Although they provide interesting sources of inspiration, the theoretical frameworks of n-armed bandit problems and discrete stochastic optimization do not fit Monte-Carlo tree search perfectly. We provide two reasons: First, and most importantly, n-armed bandit algorithms and stochastic optimization assume stationary distributions of evaluations, which is not the case when searching recursively. Second, in n-armed bandit problems, the objective is to allocate simulations in order to minimize the number of selections of non-optimal moves during simulations. This is not the objective of Monte-Carlo search, since it does not matter when bad moves are searched, as long a good move is finally selected.

The field of discrete stochastic optimization is more interesting in this respect, since its objective is to optimize the final decision, either by maximizing the probability of selecting the best move [13], or by maximizing the expected value of the final choice [16]. This maximizing principle should be the objective at the root of the tree, but not in internal nodes, where the true objective in Monte-Carlo search is to estimate the value of the node as accurately as possible. For instance, let us take Chen's formula [13], with the choice between two moves, and let the simulations of these two moves have the same variance, then the optimal allocation consists in exploring both moves equally more deeply, regardless of their estimated values. This does indeed optimize the probability of selecting the best move, but is not at all what we wish to do inside a search tree: the best move should be searched more than the other move, since it will influence the backed-up value more.

3.2 Crazy Stone's Algorithm

The basic principle of CRAZY STONE's selectivity algorithm is to allocate simulations to each move according to its probability of being better than the current best move. This scheme seems to be sound when the objective is to obtain an accurate backed-up value, since the probability of being best corresponds to the probability that this simulation would have an influence on the final backed up value if the algorithm had enough time to converge.

Assuming each move has an estimated value of μ_i with a variance of σ_i^2, and moves are ordered so that $\mu_0 > \mu_1 > \ldots > \mu_N$, each move is selected with a probability proportional to

$$u_i = \exp\left(-2.4\frac{\mu_0 - \mu_i}{\sqrt{2(\sigma_0^2 + \sigma_i^2)}}\right) + \epsilon_i. \tag{1}$$

This formula is an approximation of what would be obtained assuming Gaussian distributions (the 2.4 constant was chosen to match the normal distribution function). The formula is very similar to the Boltzmann distributions that are often used in n-armed bandits problems. The value ϵ_i is a constant. It ensures that the urgency of a move never goes to zero, and is defined by

$$\epsilon_i = \frac{0.1 + 2^{-i} + a_i}{N}, \tag{2}$$

where a_i is 1 when move i is an atari, and 0 otherwise. This formula for ϵ_i was determined empirically by trial and error from the analysis of tactical mistakes by CRAZY STONE. It is important to increase the urgency of atari moves, because they are likely to force an answer by the opponent, and may be underestimated because their true value requires another follow-up move.

For each move i, μ_i is the opposite of the value μ of the successor node, and σ_i^2 is its variance σ^2. For internal nodes of the search tree, μ and σ^2 are computed with the backup method described in Section 4. For external nodes, i.e., nodes that have been visited fewer times than the threshold defined in Section 2, μ and σ^2 are computed as $\mu = \Sigma/S$, and

$$\sigma^2 = \frac{\Sigma_2 - S\mu^2 + 4P^2}{S + 1}, \tag{3}$$

where P is the number of points of the board, Σ_2 is the sum of squared values of this node, Σ is the sum of values, and S is the number of simulations. The formula for σ^2 does as if a virtual game with a very high variance had been played. This high prior variance is necessary to make sure that nodes that have been rarely explored are considered very uncertain.

4   Backup Method

The most straightforward method to backup node values and uncertainties consists in applying the formula of external nodes to internal nodes as well. As the number of simulations grows, the frequency of the best move will dominate the others, so the mean value of this node will converge to the maximum value of all its moves, and the whole tree will become a negamax tree. This is the principle of the algorithm of Chang, Fu, and Marcus [12].

This approach is simple but very inefficient. If we consider N independent random variables, then the expected maximum of these variables is not equal, in general, to the sum of the expected values weighted by the probabilities of each variable to be the best. This weighted sum underestimates the best move.

Backing up the maximum ($\max_i \mu_i$) is not a good method either. When the number of moves is high, and the number of simulations is low, move estimates are noisy. So, instead of being really the best move, it is likely that the move with the best value is simply the most lucky move. Backing up the maximum evaluation overestimates the best move, and generates a great amount of instability in the search.

Other candidates for a backup method would be algorithms that operate on probability distributions [4,21]. The weakness of these methods is that they have to assume some degree of independence between probability distributions. This assumption of independence is wrong in the case of Monte-Carlo evaluation because, as explained in the previous paragraph, the move with the highest value is more likely to be overestimated than other moves. Also, a refutation of a move is likely to refute other moves of a node, too.

Since formal methods seem difficult to apply, the backup operator of CRAZY STONE was determined empirically, by an algorithm similar to the temporal difference method [24]. In the beginning, the backup method for internal nodes was the external-node method. An amount of 1,500 positions were sampled at random from self-play games. For each of these 1,500 positions, the tree search was run for 2^{19} simulations. The estimated value of the position was recorded every 2^n simulations, along with useful features to compute the backed-up value. Backup formulas were tuned so that the estimated value after 2^n simulations matches the estimated value after 2^{n+1} simulations. This process was iterated a few times during the development of CRAZY STONE.

4.1 Value Backup

Numerical values of the last iteration are provided in Table 1. This table contains the error measures for different value-backup methods. $\langle \delta^2 \rangle$ is the mean squared error and $\langle \delta \rangle$ is the mean error. The error δ is measured as the difference between the value obtained by the value-backup operator on the data available after S simulations, with the "true" value obtained after $2S$ simulations. The "true" value is the value obtained by searching with the "Mix" operator, described in Figure 1.

Table 1. Backup experiments

Simulations	Mean		Max		Robust Max		Mix	
	$\sqrt{\langle \delta^2 \rangle}$	$\langle \delta \rangle$	$\sqrt{\langle \delta^2 \rangle}$	$\langle \delta \rangle$	$\sqrt{\langle \delta^2 \rangle}$	$\langle \delta \rangle$	$\sqrt{\langle \delta^2 \rangle}$	$\langle \delta \rangle$
128	6.44	−3.32	41.70	37.00	39.60	35.30	5.29	−1.43
256	7.17	−4.78	25.00	22.00	23.60	20.90	4.72	−1.89
512	7.56	−5.84	14.90	12.70	13.90	11.90	4.08	−1.70
1,024	6.26	−4.86	9.48	7.91	8.82	7.41	3.06	0.13
2,048	4.38	−3.15	6.72	5.37	6.11	4.91	2.63	0.77
4,096	2.84	−1.55	4.48	3.33	3.94	2.91	2.05	0.69
8,192	2.23	−0.62	2.78	1.47	2.42	1.07	1.85	0.32
16,384	2.34	−0.57	2.45	0.01	2.40	−0.30	2.10	−0.19
32,768	2.15	−0.52	2.19	0.10	2.26	−0.12	1.93	−0.02
65,536	2.03	−0.50	1.83	0.23	1.88	0.01	1.70	0.01
131,072	2.07	−0.54	1.80	0.25	1.94	0.02	1.80	−0.02
262,144	1.85	−0.58	1.49	0.25	1.51	0.07	1.39	−0.02

These data clearly demonstrate what was suggested intuitively in the beginning of this section: the mean operator (Σ/S) under-estimates the node value, whereas the max operator over-estimates it. Also, the mean operator is more accurate when the number of simulations is low, and the max operator is more accurate when the number of simulations is high.

The robust max operator consists in returning the value of the move that has the maximum number of games. Most of the time, it will be the move with the best value. In case it is not the move with the best value, it is wiser not to back up the value of a move that has been searched less frequently. A similar idea had been used by Alrefaei and Andradóttir [3] in their stochastic simulated annealing algorithm.

```
float MeanWeight = 2 * WIDTH * HEIGHT;
if (Simulations > 16 * WIDTH * HEIGHT)
 MeanWeight *= float(Simulations) / (16 * WIDTH * HEIGHT);

float Value = MeanValue;
if (tGames[1] && tGames[0])
{
 float tAveragedValue[2];
 for (int i = 2; --i >= 0;)
  tAveragedValue[i] =
  (tGames[i] * tValue[i] + MeanWeight * Value) / (tGames[i] + MeanWeight);

 if (tGames[0] < tGames[1])
 {
  if (tValue[1] > Value)
  Value = tAveragedValue[1];
  else if (tValue[0] < Value)
  Value = tAveragedValue[0];
 }
 else
  Value = tAveragedValue[0];
}
else
 Value = tValue[0].
return Value;
```

Fig. 1. Value-backup algorithm. The size of the goban is given by "WIDTH" and "HEIGHT". "Simulations" is the number of random games that were run from this node, and "MeanValue" the mean value of these simulations. Move number 0 is the best move, move number 1 is the second best move or the move with the highest number of games, if it is different from the two best moves. tValue[i] are the backed-up values of the moves and tGames[i] their numbers of simulations.

Figure 1 describes the "Mix" operator, that was found to provide the best value back up. It is a linear combination between the robust max operator and the mean operator, with some refinements to handle situations where the mean is superior to the max (this may actually happen, because of the non-stationarity of evaluations).

4.2 Uncertainty Backup

Uncertainty back-up in CRAZY STONE is also based on the data presented in the previous section. These data were used to compute the mean squared difference between the backed-up value after S simulations and the backed-up value after $2S$ simulations. To approximate the shape of this squared difference, the backed-up variance was chosen to be $\sigma^2 / \min(500, S)$ instead of σ^2/S. This is an extremely primitive and inaccurate way to back up uncertainty. It seems possible to find better methods.

5   Game Results

As indicated in the abstract, CRAZY STONE won the 10th KGS computer-Go tournament, ahead of 8 participants, including GNU GO, NEURO GO, VIKING

5, and AYA [28]. This is a spectacular result, but this was only a 6-round tournament, and luck was probably one of the main factors in this victory.

In order to test the strength of CRAZY STONE more accurately, 100-game matches were run against GNU GO, and the latest version of INDIGO (I thank Bruno Bouzy for providing it to me), performing a search at depth 3, with a width of 7. Games were run on an AMD Athlon 3400+ PC running Linux. Results are summarized in Table 2.

Table 2. Match results with 95% confidence intervals

Player	Opponent	Winning Rate	Komi
CRAZYSTONE (5 min/game)	INDIGO 2005 (8 min/game)	61% (±4.9)	6.5
INDIGO 2005 (8 min/game)	GNU GO 3.6 (level 10)	28% (±4.4)	6.5
CRAZYSTONE (4 min/game)	GNU GO 3.6 (level 10)	25% (±4.3)	7.5
CRAZYSTONE (8 min/game)	GNU GO 3.6 (level 10)	32% (±4.7)	7.5
CRAZYSTONE (16 min/game)	GNU GO 3.6 (level 10)	36% (±4.8)	7.5

These results show that CRAZY STONE clearly outperforms INDIGO. This is a good indication that the tree search algorithm presented in this paper is more efficient than Bouzy's algorithm. Nevertheless, it is difficult to draw definitive conclusions from this match, since INDIGO's algorithm differs from CRAZY STONE's in many points. First, it relies on a knowledge-based move pre-selector, that CRAZY STONE does not have. Also, the random simulations are different. CRAZY STONE's simulations probably have better handling of the urgency of captures. INDIGO's simulations use patterns, while CRAZY STONE's simulations are based on an uniform distribution. All in all, this victory is still a rather convincing indication of the power of the algorithm presented in this paper.

The results against GNU GO indicate that CRAZY STONE is still weaker, especially at equal time control (GNU GO used about 22 seconds per game, on average). The progression of results with a longer time control indicates that the strength of CRAZY STONE scales well with the amount of CPU time it is given.

Beyond the raw numbers, it is interesting to take a look at the games, and the playing styles of the different players[1]. Most of the losses of CRAZY STONE against GNU Go are due to tactics that are too deep, such as ladders, long semeais, and monkey jumps, that GNU Go has no difficulty to see. The wins of CRAZY STONE over GNU Go are based on a better global understanding of the position. Because they are based on the same principles, the styles of CRAZY STONE and INDIGO are quite similar. It seems that the excessive pruning of INDIGO cause it to play tactical errors that CRAZY STONE knows how to exploit.

6   Conclusion

In this paper we have presented a new algorithm for Monte-Carlo tree search. It is an improvement over previous algorithms, mainly thanks to a new efficient

---

[1] Games of the matches are available at http://remi.coulom.free.fr/CG2006/

backup method. It was implemented in the computer-Go program CRAZYSTONE that performed very well in tournaments, and won a 100-game match convincingly against a state-of-the-art Monte-Carlo Go-playing program. Directions for future research include these items.

1. Improving the selectivity algorithm and uncertainty-backup operator. In particular, it might be a good idea to use stochastic optimization algorithms at the root of the search tree.
2. Trying to overcome tactical weaknesses by incorporating game-specific knowledge into random simulations.
3. Scaling the approach to larger boards. For 19x19, an approach based on a global tree search does not seem reasonable. Generalizing the tree search with high-level tactical objectives such as Cazenave and Helmstetter's algorithm [11] might be an interesting solution.

Acknowledgments

I thank Bruno Bouzy and Guillaume Chaslot, for introducing me to Monte-Carlo Go. Much of the inspiration for the research presented in this paper came from our discussions. I also thank the referees of this paper for their feedback that helped to improve this paper.

References

1. Abramson, B.: Expected-Outcome: A General Model of Static Evaluation. IEEE Transactions on Pattern Analysis and Machine Intelligence 12(2), 182–193 (1990)
2. Allis, L.V.: Searching for Solutions in Games and Artificial Intelligence. PhD thesis, Universiteit Maastricht, Maastricht, The Netherlands (1994)
3. Alrefaei, M.H., Andradóttir, S.: A Simulated Annealing Algorithm with Constant Temperature for Discrete Stochastic Optimization. Management Science 45(5), 748–764 (1999)
4. Baum, E.B., Smith, W.D.: A Bayesian Approach to Relevance in Game Playing. Artificial Intelligence 97(1–2), 195–242 (1997)
5. Billings, D., Papp, D., Peña, L., Schaeffer, J., Szafron, D.: Using Selective-Sampling Simulations in Poker. In: Proceedings of the AAAI Spring Symposium on Search Techniques for Problem Solving under Uncertainty and Incomplete Information (1999)
6. Bouzy, B.: Associating Shallow and Selective Global Tree Search with Monte Carlo for 9 × 9 Go. In: van den Herik, H.J., Björnsson, Y., Netanyahu, N.S. (eds.) CG 2004. LNCS, vol. 3846, pp. 67–80. Springer, Heidelberg (2006)
7. Bouzy, B.: Move Pruning Techniques for Monte-Carlo Go. In: van den Herik, H.J., Hsu, S.-C., Hsu, T.-s., Donkers, H.H.L.M. (eds.) CG 2005. LNCS, vol. 4250, pp. 104–119. Springer, Heidelberg (2006)
8. Bouzy, B., Cazenave, T.: Computer Go: an AI-oriented Survey. Artificial Intelligence 132, 39–103 (2001)

9. Bouzy, B., Helmstetter, B.: Monte Carlo Go Developments. In: van den Herik, H.J., Iida, H., Heinz, E.A. (eds.) 10th Advances in Computer Games (ACG10), Many Games, Many Challenges, pp. 159–174. Kluwer Academic Publishers, Boston (2004)

10. Brügmann, B.: Monte Carlo Go, Unpublished technical report (1993)

11. Cazenave, T., Helmstetter, B.: Combining Tactical Search and Monte-Carlo in the Game of Go. In: Kendall, G., Lucas, S. (eds.) Proceedings of the IEEE Symposium on Computational Intelligence and Games, pp. 117–124. IEEE Computer Society Press, Los Alamitos (2005)

12. Chang, H.S., Fu, M.C., Hu, J., Marcus, S.I.: An Adaptive Sampling Algorithm for Solving Markov Decision Processes. Operations Research 53(1), 126–139 (2005)

13. Chen, C.-H., Lin, J., Yücesan, E., Chick, S.E.: Simulation Budget Allocation for Further Enhancing the Efficiency of Ordinal Optimization. Journal of Discrete Event Dynamic Systems: Theory and Applications 10(3), 251–270 (2000)

14. Chung, M., Buro, M., Schaeffer, J.: Monte-Carlo Planning in RTS Games. In: Kendall, G., Lucas, S. (eds.) Proceedings of the IEEE Symposium on Computational Intelligence and Games, pp. 117–124. IEEE Computer Society Press, Los Alamitos (2005)

15. Enzenberger, M.: Evaluation in Go by a Neural Network Using Soft Segmentation. In: van den Herik, H.J., Iida, H., Heinz, E.A. (eds.) 10th Advances in Computer Games (ACG10), Many Games, Many Challenges, pp. 97–108. Kluwer Academic Publishers, Boston (2004)

16. Futschik, A., Pflug, G.Ch.: Optimal Allocation of Simulation Experiments in Discrete Stochastic Optimization and Approximative Algorithms. European Journal of Operational Research 101, 245–260 (1997)

17. Ginsberg, M.L.: GIB: Steps Toward an Expert-Level Bridge-Playing Program. In: Dean, Th. (ed.) Proceedings of the Sixteenth International Joint Conference on Artificial Intelligence, pp. 584–593. Morgan Kaufmann, Los Altos, CA (1999)

18. Juillé, H.: Methods for Statistical Inference: Extending the Evolutionary Computation Paradigm. PhD thesis, Brandeis University, Department of Computer Science (May 1999)

19. Kearns, M., Mansour, Y., Ng, A.Y.: A Sparse Sampling Algorithm for Near-Optimal Planning in Large Markov Decision Processes. In: Dean, Th. (ed.) Proceedings of the Sixteenth Internation Joint Conference on Artificial Intelligence, pp. 1324–1331. Morgan Kaufmann, Los Altos, CA (1999)

20. Knuth, D.E., Moore, R.W.: An Analysis of Alpha-Beta Pruning. Artificial Intelligence 6, 293–326 (1975)

21. Palay, A.J.: Searching with Probabilities. Pitman, Marshfield, MA (1984)

22. Péret, L., Garcia, F.: On-line Search for Solving Large Markov Decision Processes. In: De Mantaras, R.L., Saitta, L. (eds.) Proceedings of the 16th European Conference on Artificial Intelligence (2004)

23. Sheppard, B.: Efficient Control of Selective Simulations. ICGA Journal 27(2), 67–79 (2004)

24. Sutton, R.S.: Learning to Predict by the Methods of Temporal Differences. Machine Learning 3, 9–44 (1988)

25. Sutton, R.S., Barto, A.G.: Reinforcement Learning: An Introduction. MIT Press, Cambridge, MA (1998)

26. Tesauro, G.: Programming Backgammon Using Self-Teaching Neural Nets. Artificial Intelligence 134, 181–199 (2002)

27. Tromp, J., Farnebäck, G.: Combinatorics of Go. In: van den Herik, H.J., Ciancarini, P., Donkers, H.L.L.M. (eds.) 5th Computers and Games Conference (CG 2006). LNCS, vol. 4630, pp. 85–100. Springer, Heidelberg (2007)
28. Wedd, N.: Computer Go Tournaments on KGS (2005),
    http://www.weddslist.com/kgs/

A   Random Simulations in Crazy Stone

The most basic method to perform random simulations in computer Go consists in selecting legal moves uniformly at random, with the exception of eye-filling moves that are forbidden. The choice of a more clever probability distribution can improve the quality of the Monte-Carlo estimation. This section describes domain-specific heuristics used in CRAZY STONE.

A.1   Urgencies

At each point of the goban, an urgency is computed for each player. The urgency of the black player on a particular point is computed as follows.

- If playing at this point is illegal, or this point is completely surrounded by black stones that are not in atari, then the urgency is set to zero, and processing of this urgency is stopped. This rule will prevent some needed connection moves, but distinguishing false eyes from true eyes was found to be too difficult to be done fast enough during simulations.
- Otherwise, the urgency is set to 1.
- If this point is the only liberty of a black string[2] of size S, then $1,000 \times S$ is added to the urgency, unless it is possible to determine that this point is a hopeless extension. A point is considered a hopeless extension when
  - there is at most one contiguous empty intersection, and
  - there is no contiguous white string in atari, and
  - there is no contiguous black string not in atari.
- If this point is the only liberty of a white string of size S, and it is not considered a hopeless extension for White, then $10,000 \times S$ is added to the urgency. Also, if the white string in question is contiguous to a black string in atari, then $100,000 \times S$ is added to the urgency (regardless of whether this point is considered a hopeless extension for White).

The numerical values for urgencies are arbitrary. No effort was made to try other values and measure their effects. They could probably be improved.

A.2   Useless Moves

Once urgencies have been computed, a move is selected at random with a probability proportional to its urgency. This move may be undone and another may be selected instead, in the following situations.

---

[2] A string is a maximal set of orthogonally-connected stones of the same color.

- If the move is surrounded by stones of the same color except for one empty contiguous point, and these stones are part of the same string, and the empty contiguous point is also contiguous to this string, then play in the contiguous point instead. Playing in the contiguous point is better since it creates an eye. With this heuristic, a player will always play in the middle of a 3-point eye (I thank Peter McKenzie for suggesting this idea to me).
- If a move is surrounded by stones of the opponent except for one empty contiguous point, and this move does not change the atari status of any opponent string, then play in the empty contiguous point instead.
- If a move creates a string in atari of more than one stone then
  - if this move had an urgency that is more than or equal to 1,000, then this move is undone, its urgency is reset to 1, and a new move is selected at random (it may be the same move);
  - if this string had a contiguous string in atari before the move, then capture the contiguous string in atari instead (doing this is very important, since capturing the contiguous string may not have a high urgency);
  - otherwise, if the string had two liberties before the move, play in the other liberty instead.

A.3   Performance

On an AMD Athlon 3400+, compiled with 64-bit gcc 4.0.3, CRAZY STONE simulates about 17,000 random games per second from the empty 9×9 goban.

Combinatorics of Go

John Tromp[1] and Gunnar Farnebäck[2]

[1] CWI, Amsterdam
john.tromp@gmail.com
[2] Laboratory of Mathematics in Imaging,
Harvard Medical School, Boston, Massachusetts
gunnar@lysator.liu.se

Abstract. We present several results concerning the number of positions and games of Go. We derive recurrences for $L(m,n)$, the number of legal positions on an $m \times n$ board, and develop a dynamic programming algorithm which computes $L(m,n)$ in time $O(m^3 n^2 \lambda^m)$ and space $O(m\lambda^m)$, for some constant $\lambda < 5.4$. An implementation of this algorithm enables us to list $L(n,n)$ for $n \le 17$. For larger boards, we prove the existence of a *base of liberties* $\lim \sqrt[mn]{L(m,n)}$ of $\sim 2.9757341920433572493$. Based on a conjecture about vanishing error terms, we derive an asymptotic formula for $L(m,n)$, which is shown to be highly accurate.

We also study the Game Tree complexity of Go, proving an upper bound on the number of possible games of $(mn)^{L(m,n)}$ and a lower bound of $2^{2^{n^2/2 - O(n)}}$ on $n \times n$ boards and $2^{2^{n-1}}$ on $1 \times n$ boards, in addition to exact counts for $mn \le 4$ and estimates up to $mn = 9$. Most proofs and some additional results had to be left out to observe the page limit. They may be found in the full version available at [8].

1 Introduction

Go, originating over 3,000 years ago in China, is perhaps the oldest board game in the world. It is enjoyed by millions of players worldwide. Its deceptively simple rules [7] give rise to an amazing strategic depth. The first results on the computational complexity of Go date back some 25 years. In 1980, Lichtenstein and Sipser [5] proved Go to be PSPACE-hard, while 3 years later, Robson [6] showed Go with the basic ko rule to be EXPTIME-complete. More recently, certain subproblems of the game have been shown PSPACE-complete, like endgames [10] and ladders [3]. This paper focuses on the *state complexity* of Go. We are motivated by the fact that the number of legal positions is a fundamental property of a game. The notion of a legal position is unambiguously defined for Go, despite many variances in rule sets. However, the computation of the number of legal positions turns out to be almost, but not quite, impossible. In this contribution, we demonstrate in particular how computing the number of legal 19×19 positions, a number of 171 digits, will be feasible within a decade.

H.J. van den Herik et al. (Eds.): CG 2006, LNCS 4630, pp. 84–99, 2007.

2   Previous Work

Results on the state complexity of Go have been mostly published by the online newsgroup `rec.games.go` and on the `computer-go` mailing list. In September 1992, a `rec.games.go` thread "complexity of Go" raised the question of how many positions are legal. It was noted that a trivial upper bound is 3^{mn}, since every point on the board may be empty, black, or white. A position is legal if and only if every string of connected stones of the same color has an empty point adjacent to it. Achim Flammenkamp was the first to post simulation results, showing that $L(19,19) \sim 0.012 \times 3^{361} \sim 2.089 \times 10^{170}$. In August 1994, a thread "Complexity of Chess and Go" revisited the problem. Jack Hahn, Jonathan Cano, and John Tromp all posted programs to compute the number of legal positions by brute-force enumeration. The largest count published at the time was $L(4,5) = 1,840,058,693$. A week's worth of computation would have found $L(5,5)$ as well, but enumerating $L(6,6)$ would take over 10,000 times longer, severely limiting this approach.

In a January 2000 thread "Number of Legal Positions on Almost Rectangular Boards", (inspired by earlier remarks by John Tromp and Hans Zschintzsch), Les Fables first explained in detail how to count using dynamic programming. His remark "*Calculation for* 9×9 *should be possible on any PC, and a supercomputer should easily be able to handle* 13×13." proves to be exactly right. Much later, on January 23, 2005, Eric Boesch independently discovered this method on the `computer-go` mailing list. His method was implemented the next day by Tapani Raiko, but a bug led him to post a wrong count for $L(5,5)$. Later that day Jeffrey Rainy, based on his own implementation, gave the correct values for $L(5,5)$ and $L(6,6)$, but wrong values for $L(7,7)$ and $L(8,8)$. Finally, the next day, Gunnar Farnebäck posted the first bug free program, providing counts up to $L(10,10)$.

In June 1999, a thread "Math and Go" discussed the number of games of Go. Robert Jasiek claimed an upper bound of n^{3^n}, which still needs to be corrected for intermediate passes. John Tromp showed how to obtain a double exponential lower bound, which we formalize in this paper, while fixing a slight flaw. In the same month, John Tromp started a thread "number of 2×2 games", noting that the number is 386,356,909,593, as was recently independently verified.

3   Preliminaries

A position on an $m \times n$ Go board is a mapping from the set of *points* $\{0, \ldots, m-1\} \times \{0, \ldots, n-1\}$ to the set of colors {empty, black, white}. Points are *adjacent* in the usual grid sense – equal in one coordinate and differing by one in the other. A point colored black or white is called a *stone*. Adjacent stones of the same color form connected components called *strings*. An empty point adjacent to a string is called a *liberty* of that string. A game of Go starts with an empty board. The players, Black and White, alternate turns, starting with Black. On his turn, a player can either *pass*, or make a move which does not repeat an earlier position. This is the so-called Positional SuperKo (PSK) rule. Some rule sets, notably the

American Go Association's AGA rules, use the Situational SuperKo (SSK) rule, which only forbids repeating a position with the same player to move. A move consists of coloring an empty point by your color, followed by emptying all opponent strings without liberties (capture), followed by emptying all your own strings which then have no liberties (suicide). Thus, in positions arising in a Go game, strings always have liberties. Such positions are called *legal*. The number of legal $m \times n$ positions is denoted $L(m, n)$. Finally, two consecutive passes end the game.

Definition 1 (Game Graph). *Let $G(m, n)$ be the directed graph whose vertices are the legal $m \times n$ positions, and which has a directed edge from position p to a different position q, whenever q is the result of a white or black move from p.*

Note that we exclude self-loops, corresponding to single-stone suicides. This will prove useful in Lemma 1 below, and the PSK rule forbids them anyway.

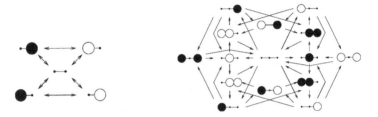

Fig. 1. game graph $G(1, 2)$ **Fig. 2.** game graph $G(1, 3)$

Figure 1 shows $G(1, 2)$, consisting of 5 nodes and 12 edges; Figure 2 shows $G(1, 3)$, consisting of 15 nodes and 42 edges. Below we establish some basic properties of Go game graphs.

Lemma 1. *Outgoing edges from a position are in 1-1 correspondence with the moves that are not single-stone suicides.*

Corollary 1. *A node with k empty points has outdegree at most $2k$.*

Corollary 2. *Each edge has an implied black or white color.*

We recall that a *simple* path is one that has no repeated vertices.

Lemma 2. *Go games are in 1-1 correspondence with simple paths starting at the all-empty node in the game graph.*

Lemma 2 applies only to rules with Positional SuperKo. With Situational SuperKo, the corresponding paths are not necessarily simple, and a position can be visited twice (once by each player) if the first visit is not followed by a pass.

Lemma 3. *The game graph is strongly connected.*

We note that this result depends on the possibility of suicide, and fails to hold for alternative rule sets, such as the Japanese rules of Go, which forbid suicide. Under such rules, a slightly weaker property can be shown.

Lemma 4. *On all boards except* 1×1, 1×2 *and* 2×1, *the game subgraph obtained by removing the all-empty node and all suicide edges, is strongly connected.*

4   Counting Legal Positions

The easiest way to count $L(m, n)$, the number of legal $m \times n$ positions, is by brute force, just trying all 3^{mn} positions and testing each one for legality. However, a 5×5 board already has over 400 billion possible positions, and 9×9 has over 10^{38}. Instead, we establish a correspondence between legal positions and paths in the so-called *border state graph*, whose size is much more manageable. The problem thus reduces to that of counting paths of a certain length in a graph, which can be done efficiently by a method known as Dynamic Programming. To achieve this goal we introduce the notion of partial boards, from which the border states naturally arise.

4.1   Partial Boards

We recall from the preliminaries that we number the points $(x, y) \in \{0, \ldots, n - 1\} \times \{0, \ldots, m - 1\}$. We picture a Go board with the point $(0, 0)$ in the top-left, x-coordinates increasing to the right, and y-coordinates increasing downward. For $0 \leq x < n$ and $0 \leq y < m$, let a partial Go board up to column x and row y consist of all the points to the left of and above (x, y). It has x full columns and, if $y > 0$, one partial column of y points. Figure 3 shows two example partial $7 \times n$ positions up to $(3, 3)$.

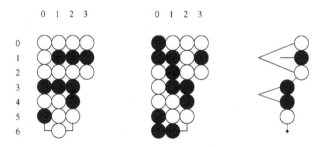

Fig. 3. Two partial positions up to $(3, 3)$ and their common border state

What these positions have in common is that the set of possible completions into legal full-board positions is identical. In either case, the remainder of the position has to provide a liberty to the top white group, to the black group it surrounds, and to the middle black group. We say that the positions share the same *border state*.

4.2   Border States

Definition 2 (Border State). *A border state, or state for short, consists of the following information (x acting only as a symbol whose value is immaterial to the state):*

(1) the board height m,
(2) the size $0 \leq y < m$ of the partial column,
(3) the color of border points $(x, 0), \ldots, (x, y - 1), (x - 1, y), \ldots, (x - 1, m - 1)$,
(4) for each stone on the border, whether it has liberties,
(5) connections among libertyless stones.

A state with height m and partial column size y is called an $\binom{m}{y}$-state, or simply y-state if m is clear from context. A partial position is pseudolegal if all libertyless stones are on, or connected to, the border. A state is called constructible *if it is the border/edge state of some pseudolegal partial position of arbitrary width.*

Information of liberties and connections is assumed to be consistent within the border.

In addition to these border states, we also have *edge states* for $x = 0$. These are like border states, except that points $(x - 1, y), \ldots, (x - 1, m - 1)$ have the special value 'edge'.

In figures, libertyless stones and their connections are indicated with lines emanating to the left.

4.3   The Border State Graph

The border state of a partial board up to $(x, y + 1)$ is uniquely determined by the border state up to (x, y) and the color of (x, y). If (x, y) is empty, then any libertyless stones at $(x, y - 1)$ and $(x - 1, y)$ as well as all stones connected to them, become stones with liberties. Now suppose (x, y) is black. If $(x - 1, y)$ is a libertyless white stone without connections, then it can no longer get any liberties. This case cannot arise for a pseudolegal partial position. If either $(x, y - 1)$ or $(x - 1, y)$ is black with liberties then so is (x, y), and it may provide liberties to the other neighbor. Finally, if neither $(x, y - 1)$ nor $(x - 1, y)$ is black with liberties, then (x, y) is without liberties and connects any black neighbors. The cases for (x, y) being white are symmetric.

Definition 3 ((Augmented) Border State Graph). *Let $B(m)$ be the directed graph whose vertices are the constructible border states of height m, and which has edges from each y-state to its 2 or 3 successor $((y + 1) \bmod m)$ states. The augmented border state graph $AB(m)$ has additional vertices and outgoing edges for all edge states.*

Figure 4 shows the 3 successors of a border state, while Figure 5 shows the augmented border state graph $AB(1)$.

Lemma 5. *There is a 1-1 correspondence between pseudolegal partial positions up to (n, y) and paths of length $mn + y$ through the augmented border state graph that start at the all-edge state.*

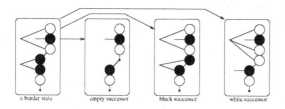

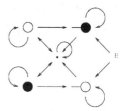

Fig. 4. A border state and its 3 successors

Fig. 5. Border state graph $AB(1)$

4.4 Recurrences

Definition 4 (State Counts). *For an $\binom{m}{y}$-state s, denote by $L(m,n,y,s)$ the number of pseudolegal partial positions up to (n,y) that have border/edge state s. Call a y-state s legal if $y = 0$ and s has no libertyless stones.*

This definition obviously implies Lemma 6.

Lemma 6 (color symmetry). *Let state s' be derived from state s by reversing the colors of all stones. Then $L(m,n,y,s) = L(m,n,y,s')$.*

Definition 5 (state classes). *We define a state class, denoted $[s]$, as the equivalence class of state s under color reversal. Call a state class legal when its members are, and define $L(m,n,y,[s]) = \sum_{s' \in [s]} L(m,n,y,s')$.*

Note that all equivalence classes, except for stoneless states, consist of exactly 2 states. These definitions immediately imply Lemma 7.

Lemma 7. $L(m,n) = \sum_{\text{legal } s} L(m,n,0,s) = \sum_{\text{legal } [s]} L(m,n,0,[s])$.

Definition 6 (state count vector). *Denote by $\mathbf{L}(m,n,y)$ the state-indexed vector with elements $L(m,n,y,s)$ for all constructible y-states s, and by 1_m the characteristic vector of legal states of height m.*

Now Lemma 7 can be expressed as

$$L(m,n) = 1_m^T \mathbf{L}(m,n,0) \ . \tag{1}$$

The following crucial observation forms the basis for the recurrences we derive. Since $L(m,n,y+1,s')$ equals the sum of $L(m,n,y,s')$ over all predecessor states s' of s in the augmented border state graph, it follows that the border state vectors $\mathbf{L}(m,n,y), n > 0$ are related by linear transformations $\mathbf{T}_{m,y}$, such that $\mathbf{L}(m,n,y+1) = \mathbf{T}_{m,y}\mathbf{L}(m,n,y)$ (we abuse notation by taking $\mathbf{L}(m,n,m)$ as a synonym for $\mathbf{L}(m,n+1,0)$). Indeed, the $\mathbf{T}_{m,y}$ appear as submatrices of the transposed adjacency matrix of the border state graph. As a consequence, successive 0-state vectors are related as

$$\mathbf{L}(m,n+1,0) = \mathbf{T}_{m,m-1}\mathbf{T}_{m,m-2}\cdots\mathbf{T}_{m,1}\mathbf{T}_{m,0}\mathbf{L}(m,n,0) \ . \tag{2}$$

Thus we are led to define the recurrence matrix.

Definition 7 (Recurrence matrix). *Let* $\mathbf{T}_m = \mathbf{T}_{m,m-1} \cdots \mathbf{T}_{m,0}$.

This leads to a matrix power expression for $L(m,n)$:

$$L(m,n) = \mathbf{l}_m^T \mathbf{T}_m^{n-1} \mathbf{L}(m,1,0) \ . \tag{3}$$

Furthermore, $L(m,n)$ can be shown to satisfy a recurrence not involving the border state counts. To simplify the following derivations, m is understood to be fixed and is dropped from the notation, so that $L(m,n) = \mathbf{l}^T \mathbf{T}^{n-1} \mathbf{L}(1,0)$.

Theorem 1. *For fixed* m, $L(m,n)$ *satisfies a linear recurrence whose order is at most the number of constructible 0-state classes.*

The structure of solutions to linear recurrences is well known [4]. Theorem 1 implies Corollary 3.

Corollary 3. *For fixed* m, $L(m,n)$ *can be written in the form*

$$L(m,n) = \sum_k q_k(n) l_k^n \ , \tag{4}$$

where l_k *are the distinct eigenvalues of* $\mathbf{T}$, *and* q_k *are polynomials of degree at most multiplicity*$(l_k) - 1$.

Notice that some of the terms may vanish but not the largest eigenvalue, about which we have additional information.

Theorem 2. *There exist* $a_m > 0$, $0 < \lambda_m \le 3^m$, *and* $0 < \phi_m < 1$ *such that*

$$L(m,n) = a_m \lambda_m^n (1 + r(m,n)) \, with \, r(m,n) = O(\phi_m^n) \ . \tag{5}$$

4.5 $1 \times n$ Boards

For one-dimensional boards, with $m = 1$, Figure 5 shows the five possible border states "empty", "black with liberty", "white with liberty", "black without liberty", and "white without liberty". The first three are legal, so $\mathbf{l} = (1,1,1,0,0)^T$. The state count transformation is given by the transposed adjacency matrix $\mathbf{T} = ((1,1,1,1,1),(1,1,0,0,0),(1,0,1,0,0),(0,0,1,1,0),(0,1,0,0,1))$ and the initial state count with one column gives $\mathbf{L}(1,0) = (1,0,0,1,1)^T$. It follows that $L(1,n) = \mathbf{l}^T \mathbf{T}^{n-1} \mathbf{L}(1,0)$, which gives the sequence 1,5,15,41,113,313,867,2401, 6649,18413,...

The characteristic polynomial of $\mathbf{T}$ is $p(\lambda) = \det(\lambda \mathbf{I} - \mathbf{T}) = \lambda^5 - 5\lambda^4 + 8\lambda^3 - 6\lambda^2 + 3\lambda - 1$. It follows that $L(1,n)$ satisfies the recurrence

$$L(1,k+5) = 5L(1,k+4) - 8L(1,k+3) + 6L(1,k+2) - 3L(1,k+1) + L(1,k) \ . \tag{6}$$

This is not a minimal order recurrence, however. Using state classes instead yields

$$\mathbf{l} = \begin{pmatrix} 1 \\ 1 \\ 0 \end{pmatrix}, \mathbf{T} = \begin{pmatrix} 1 & 1 & 1 \\ 2 & 1 & 0 \\ 0 & 1 & 1 \end{pmatrix}, \mathbf{L}(1,0) = \begin{pmatrix} 1 \\ 0 \\ 2 \end{pmatrix} \tag{7}$$

and a characteristic polynomial $p(\lambda) = \lambda^3 - 3\lambda^2 + \lambda - 1$, leading to the minimal recurrence

$$L(1, k+3) = 3L(1, k+2) - L(1, k+1) + L(1, k) . \tag{8}$$

The largest eigenvalue can be written in closed form as

$$\lambda_1 = 1 + \frac{1}{3}\left((27 + 3\sqrt{57})^{\frac{1}{3}} + (27 - 3\sqrt{57})^{\frac{1}{3}}\right) \sim 2.769292354 . \tag{9}$$

4.6 $2 \times n$ Up to $9 \times n$ Boards

Minimal orders and the λ_m values from theorem 2 are listed in Table 1 for $1 \le m \le 9$.

Table 1. Small board recurrences

size	order	a_m	λ_m
$1 \times n$	3	0.69412340909080771809	2.76929235423863141524
$2 \times n$	7	0.77605920648443217564	8.53365251207176310397
$3 \times n$	19	0.76692462372625158688	25.44501470555814081494
$4 \times n$	57	0.73972591465609392167	75.70934113501819973789
$5 \times n$	217	0.71384057986002504205	225.28834590398701930674
$6 \times n$	791	0.68921150040083474629	670.39821492744590475404
$7 \times n$	3,107	0.66545979340188479816	1,994.92693537832618289977
$8 \times n$	12,110	0.64252516474515096185	5,936.37229306818075324832
$9 \times n$	49,361	0.62038058380200867949	17,665.06600837227629766227

4.7   The Dynamic Programming Algorithm

The algorithm first computes the state vector $\mathbf{L}(m, 1, 0)$, and then performs $m(n-1)$ linear transformations $\mathbf{T}_{m,y}$ to obtain $\mathbf{L}(m, n, 0)$. Instead of keeping exact counts $L(m, n, y, s)$ as vector elements, we use state class counts modulo some number M close to 2^{64}. Running the algorithm repeatedly ($\lceil mn \log_2(3)/64 \rceil$ times suffices) with different, relatively prime, moduli gives us a set of equations

$$L(m, n) = a_i \bmod M_i , \tag{10}$$

which is readily solved using the Chinese Remainder Theorem. This technique trades off memory and disk space (which are more constrained) for time.

The heart of the algorithm is an efficient representation of border state classes, using just 3 bits per point, or $3m$ bits for a state class. This makes the standard height of $m = 19$ fit comfortably in 64-bit integers. The non-crossing connections can be represented with just 2 booleans per libertyless stone: whether it has a connection above it, and whether it has a connection below it. The representation further exploits the fact that neighboring points in the border highly constrain each other. Figure 6 shows possible transitions from one point to the next in

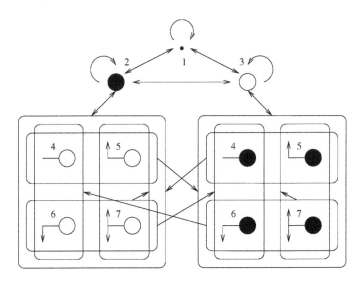

Fig. 6. Intra-border transitions

a "bump-free" 0-state. Upward and downward pointing arrows from the line indicating lack of liberties represent the two boolean flags.

Edges between boxed sets of points indicate the presence of edges from any point in one set to any point in the other. Next to each point is shown its 3-bit code. Note that no point has two different transitions to same-numbered points. This reflects the fact that two libertyless adjacent stones have the same color if and only if they are connected, and a stone with liberties cannot be adjacent to a libertyless stone of the same color. The algorithm uses code 0 for 'edge' points in edge states. Two pieces of information are still lacking; the color of a libertyless stone at $(x, 0)$, and the color of a libertyless stone on $(x - 1, y)$, which is not adjacent to the previous border point at $(x, y - 1)$. However, since we represent state classes rather than states, we may assume that $(x - 1, y)$, if non-empty, is always white. In this case, the color of a libertyless stone at $(x, 0)$ can be stored in the boolean indicating connections above, since the latter is always false. If $(x - 1, y)$ is empty, then we can assume any stone on $(x, 0)$ is white. If both $(x - 1, y)$ and $(x, 0)$ are empty then we can normalize the color of, e.g., the bottom-most stone on the border.

In order to compute $\mathbf{L}(m, n, y+1) = \mathbf{T}_{m,y} \mathbf{L}(m, n, y)$, the algorithm loops over all state-count pairs (s, i) in $\mathbf{L}(m, n, y)$, computes the 2 or 3 successors s' of s, and stores the new pairs (s', i) in some datastructure. Matching pairs (s', i) and (s', j) need to be combined into a single pair $(s', i + j \bmod M)$, which is easy if all states can be kept in memory. For large computations, like $m = 19$, this is not possible and the state-count pairs need to be stored on disk. To allow for efficient combining of states, we collect as many pairs as possible in memory, before flushing them all to disk in sorted order. To save space, we store only the differences between successive states (using 7 bits per byte, with the 8th bit indicating the last byte). Note that a state can appear in as many files as it has predecessors in $\mathbf{L}(m, n, y)$. The total number of state-count-pairs in all output files is thus larger than the number of $(y+1)$-states by a factor in the range $[1, 3)$. This *redundancy* is the average number of files in which a $(y + 1)$-state occurs. Merging all output files while combining like pairs removes this redundancy and produces the required $\mathbf{L}(m, n, y + 1)$. To keep redundancy small, we want the different predecessors of a state to be close together in the input ordering, so that combination can take place before a memory flush. This ordering depends on how the m 3-bit fields are joined into a $3m$ bit integer. We do this in different ways, such that the most variable fields, namely, those close to y, end up in the less significant bits of the $3m$ bit integer.

Finally, if $\mathbf{L}(m, n, 0)$ is computed, then we just sum all counters of legal states to obtain the desired $L(m, n) \bmod M$ result.

4.8   Complexity

The main factor in both time and space complexity is the number of state classes s. This may be upperbounded by ignoring the connection constraints (equivalent to balancing parentheses) and computing the largest eigenvalue of the intra-border transition matrix, which turns out to be $\lambda \sim 5.372$ (the largest root of $(\lambda - 2)(\lambda^2 - 5\lambda - 2)$). The number of paths of length m through the intra-border transition graph, and hence s, is therefore bounded by $O(\lambda^m)$. We thus need $\lambda^m(3m + 64)$ bits plus the overhead for the datastructure which is at most linear, for a space complexity of $O(m\lambda^m)$.

For time complexity, we have the product of the number of moduli, which is $\lceil mn \log_2(3)/64 \rceil$, the number mn of passes, the number $O(\lambda^m)$ of states, and the amount of work $O(m)$ per state, which results in time $O(m^3 n^2 \lambda^m)$.

4.9   Results

Table 2 shows the number of legal positions. The computation for 17×17 took about 8,000 CPU-hours and 400 GB of disk-space.

4.10   Asymptotic Bounds

Let K denote the set of points on an $m \times n$ board, reachable from one (3,3) point with 'orthogonal' knight moves, as shown in Figure 7. For simplicity, assume that m and n are divisible by 5, so that set K has size $mn/5$ (for other m, n, $|K|$ can be shown equal to $= mn/5 + \delta, |\delta| \leq 3/5$). We use K to derive both lower and upper bounds on $L(m, n)$.

Theorem 3. *For m, n divisible by 5,*

$$3^{\frac{4mn}{5}}(1 - \frac{2}{81})^{\frac{2(m+n)}{5}} \leq L(m, n) \leq 3^{mn}(1 - \frac{2}{81})^{\frac{2(m+n)}{5}}(1 - \frac{2}{243})^{\frac{mn - 2(m+n)}{5}} . \quad (11)$$

Proof. For the lower bound, color the points in K empty, and all other points randomly. Then illegality can only arise at the $2(m + n)/5$ points on the edge that neither belong to K nor neighbor K. For each such point, the probability

Table 2. Number of legal positions

n	digits	$L(n,n)$
1	1	1
2	2	57
3	5	12,675
4	8	24,318,165
5	12	414,295,148,741
6	17	62,567,386,502,084,877
7	23	83,677,847,847,984,287,628,595
8	30	990,966,953,618,170,260,281,935,463,385
9	39	103,919,148,791,293,834,318,983,090,438,798,793,469
10	47	96,498,428,501,909,654,589,630,887,978,835,098,088,148,177,857
11	57	793,474,866,816,582,266,820,936,671,790,189,132,321,673,383,112,185,151,899
12	68	57,774,258,489,513,238,998,237,970,307,483,999,327,287,210,756,991,189,655, 942,651,331,169
13	80	37,249,792,307,686,396,442,294,904,767,024,517,674,249,157,948,208,717,533, 254,799,550,970,595,875,237,705
14	93	212,667,732,900,366,224,249,789,357,650,440,598,098,805,861,083,269,127,196, 623,872,213,228,196,352,455,447,575,029,701,325
15	107	10,751,464,308,361,383,118,768,413,754,866,123,809,733,788,820,327,844,402, 764,601,662,870,883,601,711,298,309,339,239,868,998,337,801,509,491
16	121	4,813,066,963,822,755,416,429,056,022,484,299,646,486,874,100,967,249,263, 944,719,599,975,607,459,850,502,222,039,591,149,331,431,805,524,655,467,453, 067,042,377
17	137	19,079,388,919,628,199,204,605,726,181,850,465,220,151,058,338,147,922,243, 967,269,231,944,059,187,214,767,997,105,992,341,735,209,230,667,288,462,179, 090,073,659,712,583,262,087,437

of being a libertyless stone is $2 \cdot 3^{-4} = 2/81$, and these events are independent, so the whole position is legal with probability $(1 - 2/81)^{2(m+n)/5}$.

For the upper bound, we color all points randomly and only check if any point in K is a one-stone string without liberties. For each of the $2(m+n)/5$ edge points of K this happens with probability $2/81$ and for each of the $mn/5 - 2(m+n)/5$ interior points of K this happens with probability $2 \cdot 3^{-5} = 2/243$. Again, these events are all independent.

4.11   The Base of Liberties

The previous section shows that $L(m,n)^{1/mn}$ is roughly between $3^{4/5} \sim 2.4$ and $3(1 - \frac{2}{243})^{1/5} \sim 2.995$. In this section we prove that $L(m,n)^{1/mn}$ in fact converges to a specific value L, which we call the *base of liberties*. This is the 2-dimensional analogue of the 1-dimensional growth rate $\lambda_1 \sim 2.7693$ derived in Subsection 4.5.

Fix m and n. Consider any $M = q_m m + r_m$ and $N = q_n n + r_n$ with $0 \leq r_m < m, 0 \leq r_n < n$. Since tiling legal positions together preserves legality, we have

$$L(M,N) \geq L(m,n)^{q_m q_n} L(m,r_n)^{q_m} L(r_m,n)^{q_n} L(r_m,r_n) \ . \tag{12}$$

This proves Theorem 4 and 5.

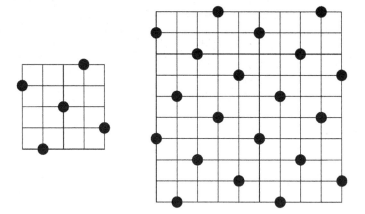

Fig. 7. Knight subset of points on 5×5 and 10×10 boards

Theorem 4. $\ln L(m,n)$ *is superadditive in both arguments.*

Theorem 5. $\lim_{\min(m,n)\to\infty} L(m,n)^{1/mn}$ *converges to some value* L.

4.12 An Asymptotic Formula

Theorems 5 and 2 together imply that $a_m^{1/mn}\lambda_m^{1/m}$, and hence $\lambda_m^{1/m}$ (since n can go to infinity arbitrarily faster than m), converge to L. Table 1 confirms that the λ_m values behave roughly as L^m for some L. In this Subsection we extend Theorem 2 to derive a much stronger result, albeit contingent on a conjecture about how fast the subdominant terms disappear. The central idea is that since $\ln(L(m,n))$ is asymptotically linear in n for each m, and symmetric, it can be expected to be asymptotically bilinear in m and n.

Conjecture 1

$$r(m,n) = O(\phi^m) \ \text{for some} \ \phi < 1 \ \text{and} \ n = \Theta(m) \ . \tag{13}$$

(This should be read as: for any constant $c \geq 1$, there exists a $\phi < 1$ such that $r(m,n) = O(\phi^m)$ for all $m/c \leq n \leq cm$.) This assumption suffices to prove Theorem 6.

Theorem 6. *Conjecture 1 implies*

$$L(m,n) = A \ B^{m+n} L^{mn}(1 + O(m\phi^m)) \tag{14}$$

for some constants $A, B, \phi < 1$, *and* $n = \Theta(m)$.

The constants A, B, and L can all be computed as limits of expressions involving legal counts of square and almost-square boards.

Corollary 4 (contingent on Conjecture 1)

$$L = \lim_{n \to \infty} \frac{L(n,n)L(n+1,n+1)}{L(n,n+1)^2} \,,$$

$$B = \lim_{n \to \infty} \frac{L(n,n+1)}{L(n,n)L^n} = \lim_{n \to \infty} \frac{L(n,n)}{L(n,n-1)L^n} \,, \tag{15}$$

$$A = \lim_{n \to \infty} \frac{L(n,n)}{B^{2n}L^{n^2}} \,.$$

Of course L could also be approximated according to its definition as $L(n,n)^{n^{-2}}$ but the above formula offers much better convergence. Using almost-square legal counts, as computed by our algorithm, our best estimates using $L(17,17)$, $L(17,16)$, and $L(16,16)$ are

$$L \approx 2.9757341920433572493 \,,$$
$$B \approx 0.965535059338374 \,, \tag{16}$$
$$A \approx 0.8506399258457 \,.$$

Although the formula for $L(m,n)$ is only asymptotic, the convergence turns out to be quite fast. Compared to the exact results in Table 2, it achieves relative accuracy 0.99993 at $n = 5$, 0.99999999 at $n = 9$, and 1.00000000000025 at $n = 13$. It is consistent with all the simulated results. For $n = 99$ it gives the same result of $4 \cdot 10^{4638}$. Accuracy is also excellent far away from the diagonal. For instance, at $L(7,268)$, the relative accuracy is still 1.0000007, witnessing the wide range of application of Theorem 6.

For 19×19, the formula gives $2.08168199382 \cdot 10^{170}$, of which we can expect all digits to be correct.

5   Counting Games

5.1   Exact Values

By Lemma 2, the number of games equals the number of simple paths in the game graph. For very small boards, we can find these numbers by brute-force enumeration, as shown in Table 3.

5.2   Upper Bounds

We can relate the number of simple paths to the product of outdegrees. First, we need a technical lemma.

Lemma 8. *On boards larger than 1×1, every node in the game graph has outdegree at least 2.*

Now consider the game tree, consisting of all simple paths. We want to avoid internal nodes with only one child, so we make them binary by duplicating

Table 3. Exact and estimated number of games on small boards

$m \setminus n$	1	2	3	4	5	6
1	1	9	907	$\sim 2.1 \cdot 10^{9*}$	$\sim 10^{31}$	$\sim 10^{100}$
2	n.a.	$\sim 3.9 \cdot 10^{11**}$	$\sim 10^{86}$	$10^{\sim 5.3 \cdot 10^{2}}$	n.a.	n.a.
3	n.a.	n.a.	$10^{\sim 1.1 \cdot 10^{3}}$	n.a.	n.a.	n.a.

* The exact number is 2,098,407,841.
** The exact number is 386,356,909,593.

their child subtree. By Lemma 8, the resulting tree still has no more than $(\prod_v \text{outdeg}(v))/\text{mindeg}$ leaves, where mindeg ≥ 2 is the minimum outdegree. Furthermore, since the tree is at least binary, it must have fewer internal nodes than leaves. This proves Lemma 9.

Lemma 9. *The number of games on an $m \times n$ board with, $mn > 1$, is at most $\prod_v \text{outdeg}(v)$.*

By Corollary 1, this is in turn bounded by $(2mn)^{L(m,n)}$. Most positions have about $mn/3$ empty points though, and some of the moves are illegal self-loops, so the average outdegree is much less than $2mn$.

Theorem 7. *The number of games on an $m \times n$ board is at most $(mn)^{L(m,n)}$.*

This bound is quite crude for small boards. For example, the 1×3 board has an average outdegree of $\frac{42}{15} = 2.8$, an outdegree product of $3 \cdot 2^{19} = 1,572,864$ which is bounded by $3^{15} = 14,348,907$, while the actual number of games is only 907.

We conjecture that for any $mn \geq 3$, the number of games is less than $(2mn/3)^{L(m,n)}$. The fact that legal positions have on average more empty points than arbitrary positions should be amply offset by the removal of self-loops and, more importantly, the widening gap between geometric and arithmetic averages.

5.3 Lower Bounds

Note that the game graph need not be Hamiltonian, and constructing a game visiting even 2^{mn} legal positions is a major challenge (achievable for one-dimensional boards as we will see later). We can still obtain a nontrivial lower bound by visiting only a highly structured subset of legal positions.

Theorem 8. *Suppose the mn points on the board can be partitioned into 3 sets B, W, E such that*

(1) $|B| = |W| = k, |E| = l = mn - 2k$,
(2) B and W are connected,
(3) each point in E is adjacent to both B and W.

Then there are at least $(k!)^{2^{l-1}}$ possible games, all lasting over $k2^{l-1}$ moves.

Corollary 5. *There are between* $2^{2^{n^2/2 - O(n)}}$ *and* $2^{2^{n^2 \log 3 + \log \log n + O(1)}}$ *Go games on an* $n \times n$ *board,*

Corollary 6. *The number* N *of* 19×19 *Go games is*

$$(103!)^{2^{154}} \leq N \leq 361^{0.012 \cdot 3^{361}} \ , \tag{17}$$

in binary

$$2^{2^{163}} < N < 2^{2^{569}} \ , \tag{18}$$

and in decimal

$$10^{10^{48}} < N < 10^{10^{171}} \ . \tag{19}$$

In one dimension, the conditions of Theorem 8 can only be met by taking E a singleton set, giving a useless bound. Fortunately, the highly structured nature of one-dimensional boards allows us to prove much better bounds.

Theorem 9. *There are at least* $2^{2^{n-1}}$ *games on an* $1 \times n$ *board with* $n \geq 2$, *which last from* $3 \cdot 2^{n-1} - 5$ *up to* $2^{n+1} - 4$ *moves.*

6   Open Problems

Computing $L(19, 19)$, the number of legal positions on a standard size Go board, remains the main open problem. The algorithm presented should suffice to compute it within the next decade. Still, a more space efficient algorithm would be welcome.

Theorem 6 and its corollaries are contingent on Conjecture 1. Proving this would be important but might require a deep understanding of the structure of the border state graphs and their spectral properties.

Game graphs are an interesting object of study for graph theorists. We conjecture that all $G(1, n)$ with $n > 2$ have color symmetric cycles.

Finally, a significant gap remains in the double exponent between the upper and lower bound on the number of games.

Acknowledgments

We are indebted to Martin Müller for suggesting publication of these results, to Piet Hut for extensive commentary on preliminary versions, and especially to Michal Koucký for the elegant idea of using Chinese Remaindering and for his extensive help with developing and running the file-based implementations that provided the counts for $n = 14, \ldots, 17$.

References

1. Blahut, R.E.: Theory and Practice of Error Control Codes. Addison-Wesley, London (1983)
2. Chen, P.: Heuristic Sampling: A Method for Predicting the Performance of Tree Searching Programs. SIAM J. Comput. 21(2), 295–315 (1992)

3. Crâşmaru, M., Tromp, J.: Ladders are PSPACE-complete. In: Marsland, T., Frank, I. (eds.) CG 2001. LNCS, vol. 2063, pp. 241–249. Springer, Heidelberg (2002)
4. Graham, R.L., Knuth, D.E., Patashnik, O.: Concrete Mathematics, 2nd edn. Addison-Wesley, London (1994)
5. Lichtenstein, D., Sipser, M.: GO is Polynomial-Space Hard. Journal of the ACM 27(2), 393–401 (1980)
6. Robson, J.M.: The Complexity of Go. In: Proc. IFIP (International Federation of Information Processing), pp. 413–417 (1983)
7. Tromp, J.: The Game of Go (2006), http://www.cwi.nl/~tromp/go.html
8. Tromp, J.: Number of Legal Go Positions (2006), http://www.cwi.nl/~tromp/go/legal.html
9. Woeginger, G.: Personal communication
10. Wolfe, D.: Go endgames are PSPACE-hard. In: Nowakowski, R.J. (ed.) More Games of No Chance, vol. 42, pp. 125–136. MSRI Publications (2002)

Abstracting Knowledge from Annotated Chinese-Chess Game Records

Bo-Nian Chen[1,*], Pangfang Liu[1], Shun-Chin Hsu[2], and Tsan-sheng Hsu[3,**]

[1] Department of Computer Science and Information Engineering,
National Taiwan University, Taipei, Taiwan
{r92025,pangfeng}@csie.ntu.edu.tw
[2] Department of Information Management,
Chang Jung Christian University, Tainan, Taiwan
schsu@mail.cju.edu.tw
[3] Institute of Information Science, Academia Sinica, Taipei, Taiwan
tshsu@iis.sinica.edu.tw

Abstract. Expert knowledge is crucial for improving the strength of computer Chinese-Chess programs. Although a great deal of expert knowledge is available in text format that using natural languages, manually transforming it into computer readable forms is time consuming and difficult. Written expert annotations of Chinese-Chess games show different styles. By analyzing and collecting commonly used phrases and patterns from experts' annotations, we introduce a novel pattern matching strategy. It automatically epitomises knowledge from a large number of annotated game records. The results of the experiments on the analysis of the middle phase of games indicate that our strategy achieves a low error rate. We hope to exploit this approach to collect automatically a great diversity of Chinese-Chess knowledge that is currently in text format.[1]

1   Introduction

Computer Chinese-Chess has developed well over the last 20 years. There are now several computer Chinese-Chess programs that show a human master level of playing expertise [19]. A popular strategy for program designers is to use variations of the α-β pruning search algorithm with rule-based evaluation functions. The algorithm is the core component that identifies the best move, while the evaluation function provides a standard measurement of the given position. In this game-playing model, knowledge of Chinese Chess is embedded in the evaluation functions.

---

* Supported in part by National Science Council (Taiwan) Grants 94-2213-E-001-014.
** Corresponding author.
[1] Early concepts of this paper appeared in *"Automatic Expert Knowledge Retrieval Strategy"*, the 10th Conference on Artificial Intelligence and Applications, Taiwan, 2005, p. 11 (one page abstract only).

H.J. van den Herik et al. (Eds.): CG 2006, LNCS 4630, pp. 100–111, 2007.

For the program to be strong enough to play against human masters, extensive expert knowledge must be embedded in the evaluation function [17]. A simple strategy is to add the expert knowledge as rules to the evaluation function manually; however, this strategy has a number of drawbacks. First, there is a bound on the number of rules that can be accommodated by the evaluation function. The more rules there are, the greater the likelihood of conflict between the rules. Second, for the sake of efficiency, we cannot afford to use high-level knowledge in the evaluation function. Third, adding rules manually is time consuming and subject to human error.

In this paper, we propose a strategy that transforms expert knowledge into information in a form that can be easily exploited by computer programs. In Section 2, we introduce our approach. In Section 3, we describe our strategy for retrieving expert knowledge automatically. In Section 4, we present the experimental results. Finally, in Section 5, we present our conclusions.

2   Preliminaries

In this section, we first define expert knowledge in computer Chinese Chess [19]. Then, we introduce traditional knowledge retrieval schemes and compare them with the automatic expert knowledge retrieval scheme proposed in this paper.

2.1   Expert Knowledge in Computer Chinese Chess

Chinese Chess, like Western Chess, is a two-player game in which each player alternately moves his[2] respective pieces toward a destination. Many theories about playing Chinese Chess have been developed and a large amount of knowledge has been accumulated. Some of this knowledge has been recorded in books or on web sites, but expert knowledge available to researchers is mostly in text format.

There are three phases in a Chinese Chess game. The first phase is the *opening phase*. It consists of the first 20 or so moves in the game. Opening strategies can be divided into two types: popular opening strategies [3] and rarely used opening strategies [2]. Many of the opening strategies developed by human masters have been published in opening game books [4]. There are also books that discuss the theories of opening games [8]. In the second phase of a game, called the *middle phase*, each player decides the tactics he will use to strengthen his position and thereby threaten the other player. Evaluating the advantage or disadvantage of a middle-phase position is difficult, even for a master [1]. There are also many books that teach the middle game knowledge by examples [7]. Third, we have the *end phase* when the players have very few pieces left. In this phase, heuristics and rules are used to play the game adequately as possible using the remaining pieces. There are many books about endgame heuristics and rules [16]. There are many hard endgame problems that cannot be solved by current computer programs.

---

[2] For brevity and readability we use 'he' and 'his' wherever we mean 'he or she' and 'his or her'.

The positions of the opening games and the endgames have been stored in opening game databases [14] and endgame databases [6,10], respectively. However, expert knowledge about the middle game is not clearly defined. Instead, there are only annotated game records, which are just examples of human knowledge about the middle game. Thus, in this paper, we focus on expert knowledge about the middle game [13].

2.2 The Importance of Automatic Expert Knowledge Retrieval

Existing expert knowledge retrieval schemes have two potential drawbacks: (1) designers may need to spend a large amount of time implementing the rules one by one; and (2) the position sets of different expert knowledge schemes may be different. Consequently, if two schemes are applied simultaneously, their knowledge may overlap or conflict. The advantages of automatic expert knowledge retrieval are; (1) reducing the implementation time and (2) partitioning the position set into several classes so that each class can be marked as a rule.

3 Automatic Expert Knowledge Retrieval

In this section, we define our strategy for automatically retrieving expert knowledge from a large number of annotated game records in text format.

3.1 Expert Knowledge and Position Value

We focus on expert knowledge in the middle phase of games. We use a program that automatically transforms a text-formatted annotated move in a game record into a *position value*, which is the score of that position. Position value assignment has been used in many applications, such as [15,11]. The position value is an integer in the range of 0 to 9, where each number represents a class used to measure the advantage or disadvantage of a given position to the red side, as shown in Table 1.

Thus, a position that is advantageous to the red or black player is classified into the red-advantage or red-disadvantage class respectively.

3.2 Novel Pattern Matching Algorithm

Text data used to annotate game records is written in a special form of natural language. Instead of using classical natural language processing techniques, with emphasis on efficiency and accuracy, we use a novel pattern matching algorithm (see Figure 1) to parse the annotations of game records. The process uses a processing algorithm and a keyword set. The latter is separated into several classes, each of which is mapped to a position value. When the sentences of annotations are recognized by one of the grammar rules or several consistent rules, the score-related element of the current state is mapped to a class of position values.

The algorithm recognizes the following classes of keywords: *subject, score-related, passive, passive-exception, condition, condition-exception, negation,* and

Table 1. The position values with respect to the red side and their meanings

Value	Class	Description
0	excellent	Red is sure to win.
1	very good	Red has very little chance of losing, and a better chance of winning than drawing.
2	good	Red has more chances of winning than losing or drawing.
3	advantage	The position of Red is a little better than that of the Black
4	even	Both sides have an equal chance of winning.
5	draw	Neither player can win.
6	disadvantage	The red position is not as good as that of Black.
7	bad	Red has more chances of losing than winning or drawing.
8	very bad	Red has very little chance of winning, and many more chances of losing than drawing.
9	worse	Red is sure to lose.

negation-exception. In the following, we present a brief definition of each class of keywords. We note that the sentences are originally written in Chinese. In some of the examples, we used a word-by-word translation, which may not be grammatically correct in English.

Subject element: a phrase that indicates which player is the main subject of the discussion. For example:

"*For the current position, the black side has material advantage.*" (局面到 此黑方多子)

"*The black side*" is a subject element.

Score-related element: a phrase that describes the property of a given position, which can be mapped to the position's value. For example:

"*For the current position, the black side has an advantage, and the red side has disadvantage.*" (局面到此，黑方優勢，紅方劣勢)

"*Advantage*" is a score-related element.

Passive element: a phrase that indicates a sentence is in the passive form, i.e., the subject and object are exchanged. If there is a passive element, the subject should be found after the score-related element. For example:

"*The rook of the red player is threatened.*" (紅方的六路車就受到威脅啦)

"*Threatened*" is a passive element. The difference between English and Chinese sentences is that the words indicating a passive meaning in English are always verbs, but in Chinese there are special words that indicate a passive meaning.

Condition element: a phrase that indicates the subsequent sentence is true, subject to some conditions, i.e., it is not a true statement now, but it will be true after the condition occurs. For example:

"*If you move the advisor, the rook on the right will not have a good place to move to and the situation would be more severe.*" (如果改走補仕則右車難 以出頭，反而形勢更加不利)

"*If*" is a condition element; thus, we may not conclude that the red side will win. For simplicity, we ignore sentences with condition elements.

Negation element: a phrase that invalidates a score-related element. For example:

"*The red side has no advantage.*" (紅方並沒有佔到便宜)

"*No*" is a negation element, which means the score-related element "*advantage*" has no effect in this sentence.

Exception element: passive-exception elements, condition-exception elements, and negation-exception elements are all called exception elements. They are used to filter phrases that look similar to an element, but they have different meanings. These classes of phrases usually only occur in Chinese. For example:

"*The position of the black side is not bad.*" (這樣黑方局面不錯)

Here "*not bad*", which contains a negative word even if it is translated into Chinese, does not really have a negative meaning; thus, it is a negative-exception element.

3.3   Chinese-Chess Annotation Abstraction Algorithm

This processing algorithm (Fig. 1) is a novel pattern matching strategy, designed especially for Chinese-Chess annotation abstraction. When the algorithm is applied to a sentence, it finds the elements and uses the grammar rules below to extract the meaning of the input sentence. We note that these rules are designed specifically for the annotations written in Chinese.

1. $Stmt \rightarrow Subject \quad Score \quad [Object]$
   This rule means that the player defined in the subject element receives the score defined in the score-related element. In Chinese Chess, there may be an *object* after the score-related element, which must be the opponent of the subject element. For example: "*In the current position, black has advantage, red has a disadvantage.*" (局面到此，黑方優勢，紅方劣勢。)

2. $Stmt \rightarrow Subject \quad Negation \quad Score \quad [Object]$
   This rule has a negation element, so the player defined in the subject element does not receive any points. For example: "*Black uses cannon to threaten red's knight, thus red loses the advantage.*" (但被黑方造包打偶，紅方先走之利的優勢沒有了。)

3. $Stmt \rightarrow Object \quad Passive \quad Score \quad [Subject]$
   This rule has a passive element, so the object element occurs first. Thus, the opponent of the player defined in the object element gets the score defined in the score-related element. For example: "*Black moves cannon, red's advantage is lost. Because black threatens red's knight, it also threatens red's cannon.*" (後黑方進過河包，紅方還是被反先了，因為黑方有平包射偶的棋，或者出貼身車抓士角炮的棋。)

4. $Stmt \rightarrow Cond \quad Stmt$
   This rule has a condition element so that the subsequent sentence is ignored. For example: "*If the red side moves his rook to exchange the black's rook, the formation of the black side will be flexible, because the red pawn in the 1st column is threatened by the black cannon in the 9th column.*" (如果紅方進車兌車反而黑方陣容具有彈性，因為紅方的一路兵要被黑方的九路包射。)

5. *Stmt → Score*

This rule does not have a subject element. There are two cases for this rule, depending on the score-related element.

(a) The score-related element indicates a draw position value. We therefore assign a draw score. For example: *"After exchanging rooks, a draw would be very likely."* (兌車之後和棋的成份很大。)

(b) In Chinese, a score-related element is a combination of a subject element and a score-related element. We therefore assign the score to the subject. For example: *"Red wins."* (紅勝。)

The algorithm returns a position value when the sentences match one of the grammar rules, or several consistent rules. Conversely, it reports an error when a sentence does not match any grammar rule, or it matches several rules with contradictory results. Note, if an annotation consists of more than one sentence, we parse each sentence individually. A conflict between two sentences means that their scores are not consistent. In this situation, the algorithm chooses the last score of the sentences, since the conclusion is more likely to be in the last few sentences.

```
procedure CCAA(text, Score, Error)      // output score and error
      player = 0, subject = 0, score = 0, inconsistent_error = false
      active = 1, positive = 1, condition = not found
      keyword_found = not found
      for i < Len(text) do
          if(a new sentence is starting)
                subject = 0, active = 1, positive = 1
                condition = not found
          if(condition == not found)
                key_score = FindScoreKeyword (i, text)
                if(keyword is found)
                    if(keyword_found == found)
                         if( ! IsSlightError(key_score, score) )
                               inconsistent_error = true
                    if(keyword doesn't need subject)
                         score = active * positive * key_score
                    else
                         score = player * active * positive * key_score
                    keyword_found = found
          if(FindPassiveKeyword(i, text) = found)
                active = -1
          if(subject not found or active == -1)
                subject = FindPlayerKeyword(i, text)
          if(FindConditionException(i, text) == not found)
                condition = FindConditionKeyword(i, text)
          if(FindNegativeException(i, text) == not found)
                positive = FindNegativeKeyword(i, text)
      if(keyword_found == not found)
          Score = 0, Error = not found
      else
          Score = score, Error = inconsistent_error
end procedure
```

Fig. 1. The Chinese-Chess Annotation Abstraction Algorithm

3.4 Automatic Procedure

The procedure for automatically retrieving knowledge from a large number of annotated game records uses an auto-feeder to read an annotated list of game records and then feeds them into the Chinese-Chess annotation abstraction algorithm one at a time. The keywords provided by the algorithm are stored in a file, called the *keyword pattern base*. Each output from the language processor is the position value for the corresponding record. The value is classified by the distributor, which automatically puts the position into the corresponding class. The collected data is then saved in a Position Evaluation Database. Note that the data in the database consists of scored positions. An advantage of the database is that it can be used to train a reliable evaluation function for search algorithms, or to measure the performance of an evaluation function.

4 Experimental Results

4.1 Experimental Design

The work in [17] describes many popular opening games and discusses games that even may be extended to the endgame. The author, G. L. Wu., is considered the best player in Taiwan. He uses several examples to formulate as precisely as possible the knowledge about the opening, the middle game, and the end game. An important issue is how to determine which position to select. To achieve this, Grandmaster Wu has listed explicitly the advantages and the disadvantages of a critical position in text format. With these lists, we first conducted an experiment on expert knowledge data taken from all of the annotated game records in all volumes of [17]. Since the leaf nodes of Wu's game tree are the most critical positions, we collected the annotations on the leaf nodes for our first experiment. There are 26,831 nodes and 2,043 leaf nodes on the game tree in [17]; each node represents a single position. We deleted 178 positions, which were not relevant to the measurement of the position.

The remaining 1,865 positions cover the middle phase, and in some cases the opening phase and the end phase. We first identify the phase of a game that is assigned to a position. There are many different definitions for the phases of a game. In [8], the opening game consists of about 10 plies; of course, there may be more or fewer plies, depending on the state of the position. In [9], the middle game starts before the actual battle proceeds. None of the phases is clearly defined. Hence, for ease of implementation, we adopted the following definitions.

Opening game. None of the rooks, cannons, and horses, (which are the *strong pieces*), are captured, but some pawns may be taken.

Middle game. For clarity, the middle game is divided into two cases: (1) at least one player has more than four strong pieces, and at least one strong piece of either player is captured; and (2) at least one player has exactly four strong pieces, one of which is a rook. Note that the rook is considered as two strong pieces, since its value is equal to two cannons or two horses, or one of each.

Endgame. There are two cases in the definition of the endgame: (1) both players have less than or exactly three strong pieces, and (2) neither player has a rook.

Based on the above definitions, the 1,865 positions consist of 424 opening game positions, 1,377 middle game positions, and 64 endgame positions. All the positions were annotated and the sentences were verified manually to ensure they were free from grammatical errors and scoring conflicts. This set of test data, called WU1865, is used to fine tune our algorithm and find various elements. In the current algorithm, there are 306 score-related elements, 6 subject elements, 5 passive elements, 13 condition elements, 7 condition-exception elements, 5 negation elements, and 21 negation-exception elements.

The position values of the 1,865 annotated positions were manually assigned in one week by the first author, who is a certified Chinese-Chess 2-Dan[3] player.

By comparing the answers annotated manually and the results from our algorithm, we were able to analyze the effectiveness of our approach statistically.

The statistical analysis of WU1865 is shown in Table 2. The numbers in the first row are position values of the manually assigned values, while the numbers in the first column are position values of the program results. Each grid in the i-th row and the j-th column represents the number of cases for which the answer is position value j, but our algorithm gives position value i, denoted by $val(i, j)$. The diagonal grids are correct cases, denoted by $val(i, i)$. If the numbers on the diagonal grids are much larger than on the vertical grids and horizontal grids, it means the algorithm is very reliable. The numbers in the last row and the last column are the summations of the rows or columns , denoted by $row\_sum(j)$ for summation of the i-th row and $col\_sum(i)$ for summation of the j-th column, respectively. We use $val(i, j)/row\_sum(i)$ and $val(i, j)/col\_sum(j)$ to measure the error rate; the lower the error rate, the better the performance of the algorithm.

As the experimental results show, the error rate is less than 1/10 for all rows and columns. Our algorithm can identify the following three errors.

1. No keyword error, named as *E1* error. There are two possible cases of this error: (1) the input sentences are not relevant to the measurement of a position, and (2) the keyword cannot be found in the keyword pattern base.
   For example: *"Moving the red rook ahead is considered a traditional and progressive move."* (紅方進橫車是傳統的穩健走法)
   There are no keywords in this example, i.e., it does not contain any words relevant to the measurement of the position.
   For case 1, the input sentence should be discarded and is not discussed here.
   For case 2, we can increase our keyword pattern base to solve the problem.
   In Table 2, positions that cause no keyword error are classified as class E1.

---

[3] Dan is a measurement of the playing expertise of a Chinese-Chess player in Taiwan. The range of Dan is from 1-Dan to 9-Dan. 1-Dan to 3-Dan are roughly considered experts, 4-Dan to 6-Dan are roughly masters, and 7-Dan to 9-Dan are roughly grandmasters.

2. Inconsistent element error, named as *E2* error. This error occurs when two or more scores can apply to a position. As noted earlier, the algorithm chooses the last score as the score of the position. In Table 2, positions that cause inconsistent element errors are classified as class E2.

For example: "*The black side was careless to check the red king and let a piece be captured. He should have moved k6+1, which would have given him an advantage.*" (黑方隨手下將造成白去子，應走將6進1，黑方而來較為有 好)

In this example, there are two elements: let a piece be captured and advantage, which have different meanings for the position.

3. Scoring error: This occurs when the score generated by the algorithm is different from that of manual annotation and, of course, the scoring error is free from the first two errors. This type of error arises when the text uses complicated grammar that our algorithm cannot recognize. It is also difficult to formulate the underlying grammatical structure with our approach. This error is relevant to the correctness of the algorithm.

For example: "*The black side moves his rook in front of the river, using the rightmost line containment to prevent the red side moving H4+6.*" (黑方進車過 河，利用邊線牽制，防止紅方傌四進六快攻)

In this example, the keywords "containment" and "prevent", which refer to the black side are considered advantageous keywords in our algorithm, but the actual annotation of the position should be even.

Furthermore, it is noteworthy that $val(i, i + 1)$, $val(i, i - 1)$, $val(i + 1, i)$, or $val(i - 1, i)$ and the advantages or disadvantages are is not changed. Also in the *drawing state*, which comprises even and draw, changes between the two are called *slight errors*. This kind of error often occurs because of subjective judgments of the annotators. Even so, such errors are tolerable in computer Chinese-Chess applications.

We define the number of *significant errors* as the measurement of the total errors due to the limitations of the algorithm: $number\_of\_significant\_errors = scoring\_errors - slight\_errors$ According to the statistical results, there are 11 no-keyword errors, 110 keyword-inconsistent errors, and 129 scoring errors. There are also 75 slight errors in the training set. The number of significant errors is 54.

It took less than 1 minute to parse the annotation of 1,865 positions and generate a statistical analysis on an Intel Pentium IV 1.8GHz CPU with a 512MB RAM. The code size of the system is approximately 2,000 lines (not including the GUI code). The current size of the keyword pattern base is 3,158 bytes, which can be increased if necessary.

4.2 Some Detailed Results

Next, we used our program to analyze an untrained data set called NET_TEST. There are two sources of NET_TEST. The first is [18], which consists of 225 game records on a CD. On average, there are 5 annotated positions in each game record. The second source consists of web sites. There are 249 positions in the web annotated by some of the best grandmasters in China, such as Y. C. Xu.

Table 2. Comparison of manually annotated answers and algorithm-generated results for WU1865. The horizontal axis represents the number of positions manually annotated. The vertical axis represents the number of positions generated by the algorithm, where E1 represents no element error and E2 represents inconsistent element error.

	0	1	2	3	4	5	6	7	8	9	Sum
0	85	10	9	0	0	0	0	1	0	0	105
1	4	98	10	0	0	0	1	0	0	0	113
2	9	10	548	5	0	1	2	1	1	0	577
3	0	0	10	82	1	1	4	0	0	0	98
4	0	0	0	2	97	0	2	0	0	0	101
5	1	0	0	2	1	29	3	2	0	1	39
6	0	0	0	0	5	1	186	9	0	0	201
7	0	0	0	1	1	0	9	393	4	2	410
8	0	0	0	0	0	0	0	2	61	0	63
9	0	0	0	0	0	0	0	0	1	36	37
E1	0	0	2	2	1	2	1	2	1	0	11
E2	1	0	19	16	23	7	28	13	3	0	110
Sum	100	118	598	110	129	41	236	423	71	39	1865

There are totally 1,418 annotated positions, 55 of which are not relevant to the measurement of the position. As a result, NET_TEST contains 1,363 annotated positions, comprised of 613 opening game positions, 627 middle game positions, and 123 endgame positions.

Before discussing the results of the experiment, we describe the properties of the test data. We discovered that there are three types of annotation in [17]

1. Grandmaster Wu made a conclusion to the specified position.
2. Grandmaster Wu discussed the advantages or disadvantages of the player's choice in a specific position.
3. In addition to 2, Grandmaster Wu discussed the advantages or disadvantages of several possible positions.

Our algorithm works well for points 1 and 2 because it is deterministic. However, in point 3, it is hard to determine which strategy is suitable for the critical position.

In our experiment, there are 351 no-keyword errors, 72 keyword-inconsistent errors, 353 scoring errors, and 100 slight errors.

In Table 3, most of the values are near the diagonal grids, and the number of significant errors is 253. Most errors occur in the column of class 4. The error rate in most rows and columns is less than or about $1/10$, except for val(2, 4) and val(5, 4). The performance of our algorithm is good and stable for most cases. It is also convenient to classify two error cases, E1 and E2, separately because they are the first candidates to be modified manually.

Table 3. Comparison of manually annotated answers and algorithm-generated results for NET_TEST. The horizontal axis represents the number of positions manually annotated. The vertical axis represents the number of positions generated by the algorithm, where E1 represents no element error and E2 represents inconsistent element error.

	0	1	2	3	4	5	6	7	8	9	Sum
0	7	0	0	2	0	1	1	0	0	0	11
1	0	28	3	2	3	1	1	0	1	0	39
2	0	0	190	13	74	4	13	1	1	1	297
3	0	1	1	64	7	1	3	2	0	1	80
4	0	0	0	1	55	3	3	2	0	0	64
5	1	1	2	11	70	69	12	3	2	1	172
6	0	1	0	2	16	3	52	0	0	0	74
7	0	1	7	14	37	2	10	99	0	1	171
8	0	0	0	0	4	2	0	0	20	0	26
9	0	0	1	0	1	1	0	0	0	3	6
E1	1	10	31	42	184	27	23	25	7	1	351
E2	0	2	9	15	29	4	10	2	1	0	72
Sum	9	44	244	160	480	118	128	134	32	8	1363

5   Concluding Remarks

We have proposed an automatic expert knowledge retrieval strategy that transforms human knowledge into information that can be easily implemented in computer programs. The experimental results show that the error rate of our strategy is low and errors in annotated game records are often detected. Using the program, a human expert can easily input Chinese Chess knowledge into the algorithm and confirm the computer's choices.

There are many books covering all phases of Chinese-Chess games. The current method of collecting data is to translate manually each game record into a computer file, which is very time consuming. Furthermore, the process is subject to human error. An automatic game-record-generating system can use either OCR (Optical Character Recognition) to obtain data from published works, or computer agents to find text data on the Web. It can then apply our automatic expert knowledge retrieval strategy to construct the information. In the future, we will incorporate machine learning techniques into the strategy to generate a new evaluation function based on the abstracted expert knowledge.

References

1. Chang, S.S., Kuo, L.P.: Si Tsan Chong Ti Wu Chu (The Mistakes in Playing Chinese Chess). Su Ron Qi Yi Publishing House (2001)
2. Chi, R.S.: Hsiang Chi Pu Chu Chu Yau (The Points of Opening Strategies in Chinese Chess). San Hai Wun Hua Publishing House, p. 330 (1990)

3. Chiang, C.S.: Lio Sin Pu Chu Sin Pien Tan Suo (Discovery of The New Variations of Popular Opening Strategies in Chinese Chess). Chen Du Xi Tai Publishing House (1996)
4. Cho, T.Y., Tsu, C.G.: Xiao Lie So Pao Run Min Ti Yu Publishing House (1990)
5. Condon, J.H., Belle, K.T.: In: Frey, P.W. (ed.) Chess Skill in Man and Machine, 2nd edn., pp. 201–210. Springer, New York (1983)
6. Fang, H.R., Hsu, T.-s, Hsu, S.C.: Construction of Chinese Chess Endgame Databases by Retrograde Analysis. In: Marsland, T., Frank, I. (eds.) CG 2001. LNCS, vol. 2063, pp. 96–114. Springer, Heidelberg (2002)
7. Gin, C.T., Yan, D.: Hsiang Chi Chong Chu Tsan Su Ja Si (Analysis of Middle Game Techniques in Chinese Chess). Su Ron Qi Yi Publishing House (1986)
8. Huang, S.L.: Hsiang Chi Kai Chu Tsan Li (Theory of Chinese Chess Opening Games). Shi Che Wun Wu Publishing House (1986)
9. Huang, S.L.: Shi Tsan Chong Chu Tsan Li (Theory of Chinese Chess Middle Games). Shi Che Wun Wu Publishing House (1986)
10. Hsu, T.-s., Liu, P.Y.: Verification of Endgame Databases. ICGA Journal 25(3), 132–144 (2002)
11. Iida, H.: Heuristic Theories on Game-Tree Search. Ph.D thesis. Tokyo University of Agriculture and Technology, Tokyo (1994)
12. Jansen, P.: Problematic Positions and Speculative Play. In: Marsland, T.A., Schaeffer, J. (eds.) Computers, Chess, and Cognition, pp. 169–182. Springer, New York (1990)
13. Levinson, R., Snyder, R.: Distance. Toward the Unification of Chess Knowledge. ICCA Journal 16(3), 228–229 (1993)
14. Lincke, T.R.: Strategies for the Automatic Construction of Opening Books. In: Marsland, T., Frank, I. (eds.) CG 2001. LNCS, vol. 2063, pp. 74–86. Springer, Heidelberg (2002)
15. Lincke, T.R.: Position-Value: Representation in Opening Books. In: Schaeffer, J., Müller, M., Björnsson, Y. (eds.) CG 2002. LNCS, vol. 2883, pp. 249–263. Springer, Heidelberg (2003)
16. Tu, J.M.: Hsiang Chi Tsan Chu Li Dian (Bible of Endgame Examples in Chinese Chess). San Hai Wun Hua Publishing House (1990)
17. Wu, G.L.: Hsiang Chi Pin Fa (Strategies of Chinese Chess Opening Games). A Computer Software published by Sohare Information Co. Ltd. (1998)
18. Wu, G.L.: 2001 Chung Kuo Hsiang Chi Ker Run Sai (Chinese Chess National Personal Contest in 2001). A Computer Software published by Sohare Information Co. Ltd. (2001)
19. Yen, S.J., Chen, J.C., Yang, T.N., Hsu, S.C.: Computer Chinese Chess. ICGA Journal 27(1), 3–18 (2004)

Automatic Strategy Verification for Hex*

Ryan B. Hayward, Broderick Arneson, and Philip Henderson

Department of Computing Science,
University of Alberta, Edmonton, Canada
{hayward,broderic,ph}@cs.ualberta.ca

Abstract. We present a concise and/or-tree notation for describing Hex strategies together with an easily implemented algorithm for verifying strategy correctness. To illustrate our algorithm, we use it to verify Jing Yang's 7×7 centre-opening strategy.

1 Introduction

Hex is the classic two-player board game invented by Piet Hein in 1942 and independently by John Nash around 1948 [1,2,6,7,8,9]. The game is named after the board, which consists of a parallelogram-shaped $m \times n$ array of hexagons, also called cells. Each player is assigned a set of stones and two opposing board sides; players alternately place a stone on an unoccupied cell; the first player to form a path connecting her[1] two sides with her stones wins the game. For example, Fig. 1 shows the start and end of a game on a 3×3 board. White succeeds in joining her two sides, so White wins this game. For more on Hex, see the recent survey by Hayward and Van Rijswijck [3] or the web page by Thomas Maarup [7].

Fig. 1. The start (left) and finish (right) of a Hex game on a 3×3 board

An intriguing aspect of the game of Hex is that for all $n \times n$ boards, although a winning first-player strategy is known to exist [1,2,9], explicit such strategies have been found only for small boards. While finding such strategies is routine on very small boards, the task quickly becomes challenging as board size increases. This is not surprising since, as Stefan Reisch has shown, determining the winner of arbitrary Hex positions is PSPACE-complete [11].

---

* Authors gratefully acknowledge the support of NSERC and the University of Alberta Games Group.
[1] For brevity we use 'she' and 'her' whenever 'she or he' and 'his or her' are meant.

H.J. van den Herik et al. (Eds.): CG 2006, LNCS 4630, pp. 112–121, 2007.
© Springer-Verlag Berlin Heidelberg 2007

Fig. 2. A winning first-player 3×3 Hex strategy. Fig. 1 shows one line of this strategy.

For 7×7, 8×8, and 9×9 boards, Jing Yang found strategies by hand [12,13,15,16]. Later, Hayward *et al.* found other 7×7 strategies by computer [4,5], while Noshita found 7×7 strategies and one 8×8 strategy similar to Yang's by hand [10]. For boards 10×10 or larger, no winning strategies are known.

As the search for winning strategies on larger boards continues, it is of interest to provide algorithms for verifying strategy correctness. Recently, Noshita described strategies in a manner that arguably facilitates human verification [10]. By contrast, in this paper we present a system that allows for computer verification. To demonstrate the utility of our system, we use it to confirm the correctness of Yang's original 7×7 strategy [13].

2   Excised Trees and Autotrees

The underlying feature of our verification system is the condensed tree notation we use to represent strategies.[2] Our notation allows the standard tree description of a strategy to be condensed in three ways. First, it permits the use of an "and" operation corresponding to the combinatorial sum of independent substrategies. Second, it permits the use of a macro descriptor for representing repeatedly occurring substrategies. Third, it allows all opponent moves to be excised from the tree by replacing each set of opponent responses with a single anonymous meta-response.

The first two of these three ideas are well known; for example, they were used by Yang in his description of his proofs [12,13,15,16]. The third idea, namely using excised trees, is new. In the rest of this section we illustrate the excision process and show that it does not hamper verification.

To begin, consider the first-player strategy tree in Fig. 2. The nodes at even depth indicate first-player moves; the nodes at odd depth indicate second-player moves; the game in Fig. 1 follows one root-to-leaf path through the tree. Notice that the first-player strategy described by the tree is *complete*: after each second-player move, there is a unique first-player response; after each first-player move,

---

[2] This notation could also be used for other two-player board games in which game pieces are fixed once they have been placed.

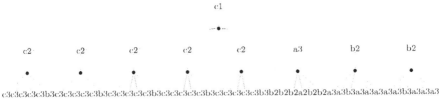

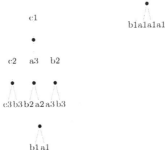

Fig. 3. The tree obtained from the strategy tree in Fig. 2 by replacing each set of opponent response nodes with a single "•" meta-node (top), and the excised tree obtained by then repeatedly merging identical subtrees (bottom)

there is every possible second-player response. Also, each leaf node establishes a first-player win, so this is a winning strategy for the first player.

Next, consider the two trees shown in Fig. 3. The top tree is obtained from the tree in Fig. 2 by excising nodes corresponding to second-player moves; each set of second-player moves is replaced with a single meta-node, indicated in our diagrams by a dot (•). The bottom tree is obtained from the top tree by repeatedly merging identical subtrees into a single subtree until, for each node, all subtrees are distinct. We refer to the bottom tree as an *excised tree*.

More generally, given any complete (but not necessarily winning) strategy tree, the following process, which we call *excision*, replaces the tree with an excised tree.

> *For each non-leaf first-player node, merge the children into a single meta-node. Next, as long as some second-player node has two identical subtrees, remove one of these subtrees.*

Excised trees represent equivalence classes of strategies, so some information is lost when a strategy tree is replaced with its excised tree. However, excision can be reversed in the following sense: for any excised tree E for a player, there is a set $\mathcal{S}$ of strategy trees such that E is the excised tree of every tree in $\mathcal{S}$. Furthermore, it is easy to construct all elements of $\mathcal{S}$ from E via the following process, which we call *restoration*:

> *At each meta-node m, for each possible opponent move to a cell c, select for the player's responding move any cell r that is the root of a subtree of m in which c does not appear.*

For example, consider the restoration process for the excised tree shown at the bottom of Fig. 3. Start with the top-most meta-node m^*, namely the child of $c1$. For this board position, the cell set of possible opponent moves is $\{a1, a2, a3, b1, b2, b3, c2, c3\}$. Consider the first such cell, $a1$. The cell sets of the subtrees of m^* are $\{c2, b3, c3\}$, $\{a3, b2, a2, b1, a1\}$, and $\{b2, a3, b3\}$. Since $a1$ is not in the first or third of these three cell sets, we can select the root of either the first or third subtree of m^*. Let us assume in this example that we always select the root of the first available subtree. Thus, as the response to $a1$ we select the root of the first subtree, namely $c2$. Continuing in this fashion, we select $c2$ as the response for opponent moves to $a2$, $a3$, $b1$, and $b2$, and we select $a3$ as the response for opponent moves to $b3$, $c2$, and $c3$. Having selected all responses to m^*, we continue in top-down order to process meta-nodes until all such nodes have been dealt with and the excised tree has been replaced with a complete strategy tree S' of S.

Notice that S' is different from the strategy tree S of Fig. 2 from which E was derived; for example, in the restoration process we never selected the root of the third subtree of m^* as a response to an opponent move. However, by repeating the restoration process once for each of the possible permutations of choices for r, we would construct all possible strategy trees associated with E, including S.

In the restoration process it will always be possible to find at least one value of r at each meta-node as long as the excised tree being restored was obtained from a complete strategy tree. This follows from Observation 1, which in turn follows from the fact that in Hex, stones never move once played.

With respect to a strategy, a π-move is a move made by player π. With respect to a strategy tree, a π-node is a node associated with a π-move, and a $\overline{\pi}$-node is a node associated with π's opponent.

Observation 1. *Let T be a complete Hex strategy tree for a player π, let p be a π-node of T that is not a leaf, let $S_1,\ldots,S_k$ be the subtrees of T rooted at the children of p, and for each S_j let P_j be the set of cells associated with the π-nodes of S_j. Then the combined intersection $I = P_1 \cap \ldots \cap P_k$ is empty.*

Proof. For each index j, let q_j be the cell associated with the root of S_j. T is complete, so $Q = \{q_1, \ldots, q_t\}$ corresponds to all possible opponent responses to p, namely all the unoccupied cells after the move p. Also, for each index j, q_j is occupied by an opponent's stone and so is not in P_j, and so is not in I. Thus I is empty.

The following is a corollary of the preceding observation.

Observation 2. *Let $E(T)$ be the excised tree obtained from a complete Hex strategy tree T for a player π, let m be a meta-node of $E(T)$ that is not a leaf, let $S_1,\ldots,S_k$ be the subtrees of $E(T)$ rooted at the children of m, and for each S_j let P_j be the set of cells associated with the π-nodes of S_j. Then the combined intersection $I = P_1 \cap \ldots \cap P_k$ is empty.*

We refer to the class of trees that we use in our verification system as "autotrees"; we use this term since such trees make explicit mention only of a player's *own* moves. Autotrees have the same form and function as excised trees; however, they may not have arisen via excision, and so we do not define them with respect to excision. An *autotree* is defined as follows: each node at one set of alternating levels is a special node called a *meta-node*; each node at the other set of alternating levels is labeled with a board cell.

We call an autotree *elusive* if it satisfies the conditions of Observation 2. Notice that restoration generates a complete strategy tree from an autotree if and only if the autotree is elusive.

As an initial step in our verification algorithm, we check whether the input autotree is elusive. The second and final step in our verification algorithm is to determine whether the strategies associated with the input autotree are winning. We call an autotree of a player *satisfying* if, for every leaf, the cells of the root-to-leaf path satisfy the conditions of a win, namely join the player's two sides on the Hex board. An elusive autotree represents a winning strategy if and only if the autotree is satisfying. This follows from the following theorems, which in turn follow by straightforward arguments from our definitions and the discussion to this point; we omit the details of the proofs.

Theorem 3. *For Hex, for any complete strategy tree there is a unique associated elusive excised tree, and for any elusive autotree there is a unique set of associated complete strategy trees. Furthermore, for any complete strategy tree S and the excised tree $E(S)$ derived from S, S is winning if and only if all strategy trees S' created via restoration from $E(S)$ are winning.*

Theorem 4. *An autotree represents a winning strategy if and only if the autotree is elusive and satisfying.*

3   And/or Autotrees with Leaf Patterns

To complete the description of our notation, we need only to describe how we add two features to autotrees: *and*-nodes and leaf patterns.

Notice that the children of a meta-node in an autotree correspond to an "or" decision in a strategy; depending on the opponent's move at the meta-node, the player will play the strategy corresponding to the first subtree, *or* the next subtree, *or* the next subtree, and so on; see the excised tree in Fig. 3. By contrast, in Hex as in many other games, a particular strategy often decomposes into two or more independent substrategies that each need to be followed.

Such "and" operations are easily incorporated into our notation by allowing each labeled node (namely, not a meta-node) of a modified autotree to have any number of children. We refer to autotrees that are modified in this way as *and/or autotrees* since, when interpreting them as strategies, the odd depth nodes (the meta-nodes) are *or*-nodes while the even depth nodes (with cell labels) are *and*-nodes.

Consider for example Fig. 4, which shows an *and/or* autotree for a winning 4×4 strategy. The root is an *and*-node, so we have to play all substrategies simultaneously; in this case, there is only one subtree so there is only one substrategy

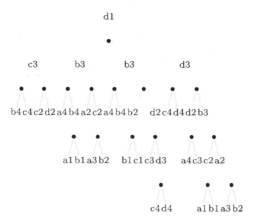

Fig. 4. An *and/or* autotree for a winning first-player 4×4 Hex strategy. Odd depth nodes (•) are "or"-nodes; even depth nodes (cell labels) are "and"-nodes. Fig. 5 shows one line of this strategy.

to follow. Suppose that the opponent's response to the player's initial move $d1$ is $b3$. Then the player can select any subtree not containing $b3$, say the first subtree; thus the player moves to $c3$, the root of the first subtree. This root is an *and*-node with two subtrees, so now the player has to follow these two substrategies simultaneously; the player must ensure that she reaches a leaf node in each of the subtrees of every *and*-node. For example, if the opponent's next move is at one of $\{b4, c4\}$, the player must immediately reply with the other of these two cells or risk not reaching a leaf of the $\{b4, c4\}$ subtree. Similarly, if the opponent's next move is at one of $\{c2, d2\}$, the player must immediately reply with the other of these two cells. If the opponent's next move is not in $\{b4, c4\}$ or $\{c2, d2\}$, the player can move anywhere. Fig. 5 illustrates another line of play of this strategy.

Finally, subtrees of *and/or* autotrees that correspond to isomorphic substrategies can be replaced with a special node corresponding to such substrategies. This is illustrated in Fig. 6, where two substrategy macros are used to simplify the tree of Fig. 4.

Modifying our verification algorithms to handle *and*- and *or*-nodes is straightforward. For *or*-nodes, the test for the elusive property is the same as with unmodified autotrees: check whether the combined intersection of all child nodes is

Fig. 5. The start (left) and finish (right) of one line of the strategy of Fig. 4

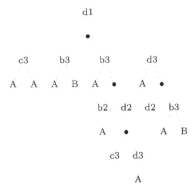

Fig. 6. An *and/or* autotree with two macro pattern nodes. This tree is equivalent to the tree in Fig. 4; pattern parameters have been omitted.

the empty set. For *and*-nodes, it is necessary to check whether the intersection of each pair of child nodes is empty. Another algorithmic approach one might take here is to expand the *and/or* autotree into the corresponding equivalent autotree; however, the resulting trees can be large,[3] so this approach would require significantly more space than our approach.

Testing the satisfying property on *and/or* autotrees involves checking every root-to-leaf path in the associated expanded autotree. For reasons of efficiency we do not want to generate the expanded autotree; we thus carry out this task in an implicit fashion. By using a simple indexing scheme for each root-to-leaf path in the *and/or* autotree, we can reconstruct the cell sets for each possible root-to-leaf path in the associated autotree. Each node stores the number of root-to-leaf paths it contains. We consider all such paths and verify that each satisfies the winning condition.

We implement the isomorphic substrategy feature in the simplest possible way, namely using macro substitution to generate the equivalent *and/or* autotree.

4   Verifying Yang's Proof

As a benchmark for testing our system, we used it to verify the first known winning 7×7 Hex strategy, namely Yang's original 7×7 center-opening strategy [14,13]. Yang described his strategy in an easily understood notation similar to that used in the C programming language; an applet that follows this strategy is available on his homepage[12]. The version of the strategy that we tested is from a preprint also available from his web page [14]. In Yang's notation, his strategy uses about 40 patterns (not counting pattern variations) comprising about six pages of text. A recursion tree indicating the hierarchy of his patterns is shown in Fig. 7.

---

[3] For example, an *and*-node with k subtrees of two nodes each corresponds in the expanded autotree to a node with 2^k subtrees.

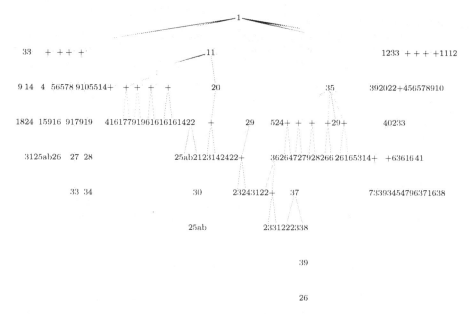

Fig. 7. Part of the recursion tree for Yang's proof. References to frequently occurring small patterns have been omitted. Labels indicate pattern numbers. Nodes labeled + are *and*-nodes; all other nodes are *or*-nodes.

```
( pattern8
  // called by: 1
  ((c6 BR) (d4 BR))
  (d6 e3 e4 e5 e6 f2 f3 f4 f5 f6 g1 g2 g3 g4 g5 g6)
  (c6 d4 BR)

  [(f3 [(pattern2ab (e3 e4) (d4 f3))]
       [(pattern2ab (g2 g3) (f3 BR))])
   (e5 [(d6) (e4)]
       [(pattern13 (e6 f4 f5 f6 g3 g4 g5 g6) (e5 BR))])
   (f2 [(pattern2ab (g1 g2) (f2 BR))]
       [(pattern9 (g5 g4 f5 f4 f3 e5 e4 e3) (BR f2 d4))])
   (e3 [(pattern17 (d6 e5 e6 f2 f3 f4 f5 g1 g2 g3 g4 g5) (c6 d4 e3 BR))]) ])
```

Fig. 8. Yang's Pattern 8 in our notation

We translated Yang's proof into our notation by hand, following his pattern naming convention. As an example of our notation, see Fig. 8. The first line gives the name of the pattern. The second line is a comment noting that the only pattern calling this pattern is Pattern 1. The third line gives the connections that are achieved by the pattern; in this case at least one of two connections is achieved, either between *c6* and the bottom right side of the board, or between *d4* and the bottom right side; this information is given only to aid in human debugging purposes and is not used by our algorithm. The fourth line lists the cells that

```
pattern1 connect: (TL BR)
   empty: (a1 a2 a3 a4 a5 a6 a7 b1 b2 b3 b4 b5 b6 b7 c1 c2 c3 c4 c5 c6 c7
          d1 d2 d3 d5 d6 d7 e1 e2 e3 e4 e5 e6 e7 f1 f2 f3 f4 f5 f6 f7 g1
          g2 g3 g4 g5 g6 g7)
  played: (TL d4 BR)
   stats: AND = 1480, OR = 2339, Leafs = 3514
   paths: 25574/25574
VALID pattern.
```

Fig. 9. Diagnostics returned after verifying Yang's proof

must be unoccupied at this point; the fifth line lists the cells that the player must already occupy. The subsequent lines describe the *and/or* autotree, where parentheses surround the subtrees of an *or*-node and square brackets surround the subtrees of an *and*-node.

In the process of verifying the description of Yang's proof, we found only one typographical error: in the description of Pattern 11 there is a call to Pattern 17 that should instead be a call to Pattern 19.

Our notation represents Yang's strategy in about 700 lines of text. The diagnostic message returned by our program after recursively verifying Yang's proof is shown in Fig. 9. The resulting tree had 1,480 *and*-nodes, 2,339 *or*-nodes, 3,514 leaves, and 25,574 implicit root-to-leaf paths. The verification took less than one second to execute on our computer, a single-processor Athlon64 3200+ with 1 gigabyte of memory.

5   Conclusions

We have introduced the notion of an *excised tree* as a compressed representation of a complete strategy tree from which all explicit opponent moves have been excised. We used excised trees in a simple algorithm that verified the correctness of Yang's original winning 7×7 Hex strategy.

One way in which our system could be improved would be to automate the process of translating strategies from other notations into our notation.

Another improvement concerns the number of paths that our algorithm checks in verifying the correctness of a strategy. Currently our system explicitly verifies that every possible cell set that a player might end up with contains a winning path. For example, for Yang's strategy this was a total of 25,574 cell sets that were checked. The problem with this approach is that the number of such cell sets, corresponding to the number of root-to-leaf paths in the complete strategy tree, increases exponentially in the board size.

Consider for example Martin Gardner's winning second-player strategy for the player with the longer sides on an $n \times n - 1$ board [2]. The strategy consists of the *and* of $f(n) = n \times (n-1)/2$ substrategies each consisting of the *or* of two moves. The associated excised tree thus has $2^{f(n)}$ root-to-leaf paths. Even for n as small as 14, $2^{f(n)} = 2^{91}$, and checking this many paths individually is currently computationally infeasible.

Thus, as board size increases, verification algorithms will be required that do not explicitly check the winning condition for each root-to-leaf path.

References

1. Gardner, M.: Mathematical Games. Scientific American 197, July, pp. 145–150, August, pp. 120–127, October, pp. 130–138 (1957)
2. Gardner, M.: The Scientific American Book of Mathematical Puzzles and Diversions, pp. 73–83. Simon and Schuster, New York (1959)
3. Hayward, R.B., van Rijswijck, J.: Hex and Combinatorics (formerly Notes on Hex). Discrete Mathematics 306, 2515–2528 (2006)
4. Hayward, R.B., Björnsson, Y., Johanson, M., Kan, M., Po, N., van Rijswijck, J.: Solving 7 × 7 Hex: Virtual Connections and Game-state Reduction. In: van den Herik, H.J., Iida, H., Heinz, E.A. (eds.) Advances in Computer Games (ACG10), Many Games, Many Challenges, pp. 261–278. Kluwer Academic Publishers, Boston (2004)
5. Hayward, R.B., Björnsson, Y., Johanson, M., Kan, M., Po, N., van Rijswijck, J.: Solving 7 × 7 Hex with Domination, Fill-in, and Virtual Connections. Theoretical Computer Science 349, 123–139 (2005)
6. Maarup, Th.: Hex – Everything You Always Wanted to Know About Hex but Were Afraid to Ask. Master's thesis, Department of Mathematics and Computer Science, University of Southern Denmark, Odense, Denmark (2005)
7. Maarup, Th.: Hex Webpage (2005), http://maarup.net/thomas/hex/
8. Nasar, S.: A Beautiful Mind. Touchstone, New York (1998)
9. Nash, J.: Some Games and Machines for Playing Them. Technical Report D-1164, Rand Corp. (1952)
10. Noshita, K.: Union-Connections and Straightforward Winning Strategies in Hex. ICGA Journal 28(1), 3–12 (2005)
11. Reisch, S.: Hex ist PSPACE-vollständig. Acta Informatica 15, 167–191 (1981)
12. Yang, J.: Jing Yang's Web Site (2003), http://www.ee.umanitoba.ca/~jingyang
13. Yang, J., Liao, S., Pawlak, M.: A Decomposition Method for Finding Solution in Game Hex 7x7. In: International Conference on Application and Development of Computer Games in the 21st Century, pp. 96–111 (November 2001)
14. Yang, J., Liao, S., Pawlak, M.: On a Decomposition Method for Finding Winning Strategy in Hex Game (2001),
    http://www.ee.umanitoba.ca/~jingyang/hexsol.pdf
15. Yang, J., Liao, S., Pawlak, M.: Another Solution for Hex 7x7. Technical report, University of Manitoba, Winnipeg, Canada (2002),
    http://www.ee.umanitoba.ca/~jingyang/TR.pdf
16. Yang, J., Liao, S., Pawlak, M.: New Winning and Losing Positions for 7x7 Hex. In: Schaeffer, J., Müller, M., Björnsson, Y. (eds.) CG 2002. LNCS, vol. 2883, pp. 230–248. Springer, Heidelberg (2003)

Feature Construction for Reinforcement Learning in Hearts

Nathan R. Sturtevant and Adam M. White

Department of Computing Science
University of Alberta, Edmonton, Canada
{nathanst,awhite}@cs.ualberta.ca

Abstract. Temporal difference (TD) learning has been used to learn strong evaluation functions in a variety of two-player games. TD-gammon illustrated how the combination of game tree search and learning methods can achieve grand-master level play in backgammon. In this work, we develop a player for the game of hearts, a 4-player game, based on stochastic linear regression and TD learning. Using a small set of basic game features we exhaustively combined features into a more expressive representation of the game state. We report initial results on learning with various combinations of features and training under self-play and against search-based players. Our simple learner was able to beat one of the best search-based hearts programs.

1 Introduction and Background

Learning algorithms have the potential to reduce the difficulty of building and tuning complex systems. But, there is often a significant amount of work required to tune each learning approach for specific problems and domains. We describe here the methods used to build a program to play the game of Hearts. This program is significantly stronger than a previously built expert-tuned program.

1.1 Hearts

Hearts is a trick-based card game, usually played with four players and a standard deck of 52 cards. Before play begins, 13 cards are dealt out to each player face-down. After all players have looked at their cards, the first player plays (leads) a card face-up on the table. The other players follow in order, if possible playing the same suit as was led. When all players have played, the player who played the highest card in the suit that was led wins or *takes* the trick. This player places the played cards face-down in his pile of taken tricks, and leads the next trick. This continues until all cards have been played.

The goal of Hearts is to take as few points as possible. A player takes points based on the cards in the tricks taken. Each card in the suit of hearts ($\heartsuit$) is worth one point, and the queen of spades (Q$\spadesuit$) is worth 13. If a player takes all 26 points, also called *shooting the moon*, they instead get 0 points, and the other players get 26 points each.

H.J. van den Herik et al. (Eds.): CG 2006, LNCS 4630, pp. 122–134, 2007.

There are many variations on the rules of Hearts for passing cards between players before the game starts, determining who leads first, determining when you can lead hearts, and providing extra points based on play. In this work we use a simple variation of the rules: there is no passing of cards and there are no limitations on when cards can be played.

Hearts is an imperfect information game because players cannot see their opponents' cards. Imperfect information in card games has been handled successfully using Monte-Carlo search, sampling possible opponent hands and solving them with perfect-information techniques. While there are known limitations of this approach [10], it has been quite effective in practice. The strongest bridge program, GIB [5], is based on Monte-Carlo search. This approach was also used for Hearts in Sturtevant's thesis work [12]. Instead of learning to play the imperfect-information game of Hearts, we build a program that learns to play the perfect-information version of the game. We then use this as the basis for a Monte-Carlo player which plays the full game.

1.2 Hearts-Related Research

There have been several studies on learning and search algorithms for the game of Hearts. Because Hearts is a multi-player game (more than two players) minimax search cannot be used. Instead max^n [7] is the general algorithm for playing multi-player games.

Perkins [9] developed two search-based approaches for multi-player imperfect information games based on max^n search. The first search method built a max^n tree based on move availability and value for each player. The second method maximized the max^n value of search trees generated by Monte-Carlo. The resultant players yielded a low to moderate level of play against human and rule-based based players.

One of the strongest computer Hearts programs [12], uses efficient multi-player pruning algorithms coupled with Monte-Carlo search and a hand-tuned evaluation function. When tested against a well-known commercial program[1], it beat the program by 1.8 points per hand on average, and 20 points per game. The game of Hearts is not 'solved': this program is still weaker than humans. We used this program as an expert to train against in some of our experiments for this paper. For the remainder of this paper we will refer to the full program as the Imperfect Information Expert (IIE) and the underlying solver as the Perfect Information Expert (PIE).

One of the first applications of temporal difference (TD) learning methods to Hearts was by [6]. These results, however, are quite weak. The resulting program was able to beat a random program, and in a small competition out-played other learned players. But, it was unable to beat Monte-Carlo based players.

Fujita *et al.* performd several studies on applying reinforcement learning (RL) algorithms to the game of Hearts. In their recent work [3], Fujita *et al.* modeled the game as a partially-observable Markov decision process (POMDP), where the

---

[1] Freeverse Software's 3D Hearts Deluxe, http://www.freeverse.com/hearts/

TDLearn(play history H, λ, $\Re$)
1   **for each** state $i = 0..n$ in H // $H[0]$ = TERMINAL; $H[n]$ = START
2   **if** TERMINAL($H[i]$)
3            $V(H[i]) = \Re$
4   **else**
5            $V(H[i]) = (1 - \lambda)V(H[i]) + \lambda V(H[i - 1])$

Fig. 1. Learning using TD(λ) given a history of game play

agent can observe the game state, but other dimensions of the state, such as the opponents' hands, are unobservable. Using a one step history, they constructed a new model of the opponents' action selection heuristic according to the previous action selection models. Although their learned player performed better than their previous players and also beat rule-based players, it is difficult to know the true strength of the learning algorithm and resultant player. This is due to the limited number of training games and the lack of validation of the final learned player. The work by Fujita *et al.* differs from ours in that they understate the feature selection problem, which was a crucial factor in the high level of play exhibited by our learning agents.

Finally, Fürnkranz *et al.* [4] have done initial work using RL techniques to employ an operational advice system to play hearts. A neural network was used with TD learning to learn a mapping from state abstraction and action selection advice to a number of move selection heuristics. This allowed their system to learn a policy using a very small set of features (15) and little training. However, this work is still in its preliminary stages and the resulting player exhibited only minor improvement over the greedy action selection of the operational advice system on which the learning system was built.

1.3   Reinforcement Learning

In reinforcement learning an agent interacts with an environment by selecting actions in various states to maximize the sum of scalar rewards given to it by the environment. In hearts, for example, a state consists of the cards held by and taken by each player in the game and negative reward is assigned each time you take a trick with points on it. The environment includes other players as well as the rules of the game.

The RL approach has strong links to the cognitive processes that occur in the brain and has proved effective in Robocup soccer [11], industrial elevator control [15] and backgammon [16]. These learning algorithms were able to obtain near optimal policies simply from trial and error interaction with the environment in high dimensional and sometimes continuous valued state and action spaces.

All of the above examples use TD learning [15]. In the simplest form of TD learning an agent stores a scalar value for the estimated reward of each state, $V(s)$. The agent selects the action that leads to the state with the highest utility. In the tabular case, the value function is represented as an array indexed by states. At each decision point or step of an episode the agent updates the value function according to the observed rewards. TD uses bootstrapping, where an agent updates its state value with the reward received on the current time step and the difference between the current state value and the state value on the previous time step. $TD(\lambda)$ updates all previously encountered state values according to an exponential decay, which is determined by λ. The pseudo-code for TD-learning used in this work is given in Figure 1, where H is the history of game play and $\Re$ is the reward (points taken) at the end of an episode.

1.4   Function Approximation

$TD(\lambda)$ is guaranteed to converge to an optimal solution in the tabular case, where a unique value can be stored for every state. Many interesting problems, such as Hearts, have a restrictively large of number states. Given a deck of 52 cards with each player being dealt 13 cards, there are $52!/13!^4 = 5.4 \times 10^{28}$ possible hands in the game. This is, in fact, only a lower bound on actual size of the state space because it does not consider the possible states that we could encounter while playing the game. Regardless, we cannot store this many states in memory, and even if we could, we do not have the time to visit and train in every state.

Therefore, we need some method to approximate the states of the game. In general this is done by choosing a set of features which approximate the states of the environment. A function approximator must map these features into a value function for each state in the environment. This generalizes learning over similar states and increases the speed of learning, but potentially introduces a generalization error as the features will not represent the state space exactly.

One of the simplest function approximators is the linear perceptron. A perceptron[2] computes a state value function as a weighted sum of its input features. Weights w are updated according to the formula:

$$w \leftarrow w + \alpha \cdot error_s \cdot \phi_s$$

given the current state s, current state output error $error_s$, which is provided by TD learning, and the features of the current state, ϕ_s.

Because the output of a perceptron is a linear function of its input, it can only learn linearly separable functions. Most interesting problems require a nonlinear mapping from features to state values. In many cases one would use a more complex function approximator, such as neural networks, radial basis functions, CMAC tile coding or kernel based regression. These methods can represent a variety of nonlinear functions.

An alternate approach, which we take, is to use a simple function approximator with more complicated set of features. We use linear function approximation for at least three reasons: (1) Linear regression is easy to implement, (2) the learning rate scales linearly with the number of features, and (3) the learned

---

[2] The computation performed by a perceptron with a real-value output is the same as in stochastic linear regression, so for the purpose of this work these terms are interchangeable.

network weights are easier to analyze. Our results demonstrate that a linear function approximation is able to learn and efficiently play the imperfect information game of Hearts.

Below we illustrate how additional features can allow a linear function approximator to solve a nonlinear task with a simple example [8]. Consider the task of learning the XOR function based on two inputs. A two-input perceptron cannot learn this function because XOR is not linearly separable. But, if we simply augment the network input with an extra input which is the logical AND of the first two inputs, the network can learn the optimal weights. In fact, all subsets of n points are always linearly separable in $n - 1$ dimensional space. Thus, given enough features, a non-linear problem can have a exact linear solution.

2   Learning to Play Hearts

Before describing our efforts for learning in Hearts, we examine the features of backgammon which make it an ideal domain for TD learning. In backgammon, pieces are always forced to move forward, except in the case of captures, so games cannot run forever. The set of moves available to each player are determined by a dice roll, which means that even from the same position the learning player may be forced to follow different lines of play in each game, unlike in a deterministic game such as chess where the exact same game can easily be repeated. Thus, self-play has worked well in backgammon.

Hearts has some similar properties. Each player makes exactly 13 moves in every game, which means we do not have to wait long to get exact rewards associated with play. Thus, we can quickly generate and play large numbers of training games. Additionally, cards are dealt randomly, so, like in backgammon, players are forced to explore different lines of play in each game. Another useful characteristic of Hearts is that even a weak player occasionally gets good cards. Thus, we are guaranteed to see examples of good play.

One key difference between Hearts and backgammon is that the value of board positions in backgammon tend to be independent. In Hearts, however, the value of any card is relative to what other cards have been played. For instance, it is a strong move to play the 10♡ when the 2-9♡ have already been played. But, if they have not been played, it is a weak move. This complicates the features needed to play the game.

2.1   Feature Generation

Given that we are using a simple function approximator, we need a rich set of features. In the game of Hearts, and card games in general, there are many very simple features which are readily accessible. For instance, we can have a single binary feature for each card that each player holds in their hand and for each card that they have taken (e.g. the first feature is true if Player 1 has the 2♡, the second is true if Player 1 has the 3♡, etc.). This would be a total of 104 features for each player and 416 total features for all players. This set of features fully specifies a game, so in theory it should be sufficient for learning, given a suitably powerful function approximator and learning algorithm.

However, consider a simple question like: "Does player 1 has the lowest heart left?" Answering this question based on such simple features is quite difficult, because the lowest heart is determined by the cards that have been played already. If we wanted to express this given the simple features described above, it would look something like: "[Player 1 has the 2♡] OR [Player 1 has the 3♡ AND Player 2 does not have the 2♡ AND Player 3 does not have the 2♡ AND Player 4 does not have the 2♡] OR [Player 1 has the 4♡ ...]". While this full combination *could* be learned by a non-linear function approximator, it is unlikely to be.

Another approach for automatically combining basic game features is GLEM [2]. GLEM measures the significance of groups of basic features, building mutually exclusive sets for which weights are learned. This approach is well-suited for many board games. But, as we demonstrated above, the features needed to play a card game well are quite complex and not easily learnable. Similar principles may be used to refine our approach, but we have yet to consider them in detail.

Our approach is to define a set of useful features, which we will call atomic features. These features are perfect-information features, so they depend on the cards other players hold. Then we built higher level features by combining these atomic features together. One set of atomic features used for learning can be found in Table 3.

Attempting to build manually all useful combinations of the atomic features would be tedious, error-prone, and time consuming. Instead, we take a more brute-force approach. Given a set of atomic features, we generate new features by exhaustively taking all pair-wise AND combinations of the atomic features. Obviously we could take this further by adding OR operations and negations. But, to limit feature growth we currently only consider the AND operator.

2.2   Learning Parameters

For all experiments described here we used TD-learning as follows. The value of λ was set to 0.75. We first generated and played out a game of Hearts using a single learning player and either three expert search players taken from Sturtevant's thesis work [12] (PIE and IIE), or three copies of our learned network for self-play. In most experiments we report results of playing the hand-tuned evaluation function (PIE) against the learned network, without Monte-Carlo. Thus, the expert player's search and evaluation function took full advantage of the perfect information provided (player's hands) and is a fair opponent for our perfect information learning player. Moves were selected using the max" algorithm with a lookahead depth of one to four, based on how many cards were left to play on the current trick. When backing up values in the search tree, we used our own network as the evaluation function for our opponents.

To simplify the learned network we only evaluated the game in states where there were no cards on the current trick. We did not learn in states where all the points had already been played. After playing a game we computed the reward for the learning player and then stepped backwards through the game, using

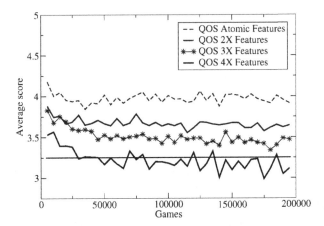

Fig. 2. Learning to not take the Q♠ using various combinations of atomic features. The break-even point is at 3.25.

TD-learning to update our target output and train our perceptron to predict the target output at each step. We did not attempt to train using more complicated methods such as TDLeaf [1].

2.3   Learning to Avoid the Q♠

Our first learning task was to predict whether we would take the Q♠. We trained the perceptron to return an output between 0 and 13, the value of the Q♠ in the game. In practice, we cut the output off with a lower bound of $0 + \epsilon$ and an upper bound of $13 - \epsilon$ so that the search algorithm could always distinguish between states where we expected to take the Q♠ versus states where we already had taken the Q♠, preferring those where we had not yet taken the queen. The perceptron learning rate was set to $1/(13 \times number\ active\ features)$.

We used 60 basic features, listed in Appendix A, as the atomic features for the game. Then, we built pair-wise, three-wise and four-wise combinations of these atomic features. The pair-wise combinations of these features results in 1,830 total features, three-wise combinations of the atomic features results in 34,280 total features, and four-wise combinations of features results in 523,685 total features. But, many of these features are mutually exclusive and can never occur. We initialized the feature weights randomly between $\pm 1/num\ features$.

The average score of the learning player during training is shown in Figure 2. This learning curve is performance relative to PIE. Except for the four-wise features, we did five training runs of 200,000 games against the expert program. Scores were then averaged between each run and over a 5,000 game window. With 13 points in each hand, evenly matched players would expect to take 3.25 points per hand. The horizontal line in the figure indicates the break-even point between the learned and expert programs.

These results demonstrate that the atomic features alone were insufficient for learning to avoid taking the Q♠. The pair-wise and three-wise features were

Table 1. Features predicting we can avoid the Q♠

Rank	Weight	Top Features - Predicting we will *not* take the Q♠			
1.	-0.103	One of J-2♠		Lead	Q♠ player has no other ♠
2.	-0.097	One of J-2♠	No ♡	Lead	Q♠ player has no other ♠
3.	-0.096	Two of J-2♠	We have K♠		Q♠ player has two other ♠.
4.	-0.093	One of J-2♠	No ♣	Lead	Q♠ player has no other ♠
5.	-0.090	One of J-2♠	No ♢	Lead	Q♠ player has no other ♠
148.	-0.040	One of J-2♠	We have Q♠		Lead player has no ♠

Table 2. Features predicting we will take the Q♠

Rank	Weight	Top Features - Predicting we will take the Q♠			
1.	0.125	We have Q♠	One of J-2♠		Lead
2.	0.123	We have Q♠	One of J-2♠		
3.	0.117	We have Q♠	No ♣	No ♡	Lead
4.	0.116	Only A/K/Q♠			Lead
5.	0.112	We have Q♠	No ♣	No ♡	No ♢

also insufficient, but are better than the atomic features alone. The four-wise combinations of features, however, are sufficient to beat the expert program.

The features which best predict avoiding the Q♠ are shown in Table 1. This player actually uses all atomic, pair-wise, three-wise, and four-wise features, so some of the best features in this table only have three atomic features. Weights are negative because they reduce our expected score in the game.

There are a few things to note about these features. First, we can see that the highest weighted feature is also one learned quickly by novice players: If the player with the Q♠ has no other spades and we can lead spades, we are guaranteed not to take the Q♠. In this case, leading a spade will force the Q♠ onto another player.

Because we have generated all four-wise combinations of features, and this feature only requires three atomic features to specify, we end up getting the same atomic features repeated multiple times with an extra atomic features added. The features ranked 2, 4, and 5 are copies of the first feature with one extra atomic feature added. The 148th ranked feature should seem odd, but we will explain it when looking at the features that predict that we will take the Q♠.

The features that best predict taking the Q♠ are found in Table 2. These features are slightly less intuitive. We might expect to find symmetric versions of the features from Table 1 in Table 2 (e.g., we have a single Q♠ and the player to lead has spades). This feature is among the top 300 (out of over 500,000) features, but not in the top five features.

What is interesting about Table 2 is the interactions between features. Again, we see that the atomic features which make up the 2nd ranked feature are a subset of the highest ranked feature. In fact, these two atomic features appear 259 times in the top 20,000 features. They appear 92 times as part of a feature

Table 3. Atomic Features Used to Learn the Q♠. Unless explicitly stated, all features refer to cards in our hand. The phrase "to start" refers to the initial cards dealt. "Exit" means we have a card guaranteed not to take a trick. "Short" means we have no cards in a suit. "Backers" are the J-2♠. "Leader" and "Q♠ player" refers to another player besides ourself.

we have A♠	we have K♠	we have Q♠
≥5 spades besides AKQ♠	0 spades besides AKQ♠	1 spades besides AKQ♠
2 spades besides AKQ♠	3 spades besides AKQ♠	4 spades besides AKQ♠
≥ 3 diamonds to start	0 diamonds to start	1 diamonds to start
2 diamonds to start	currently short diamonds	currently not short diamonds
opponent short diamonds	exit in diamonds	≥ 3 clubs to start
0 clubs to start	1 clubs to start	2 clubs to start
currently short clubs	currently not short clubs	opponent short clubs
exit in clubs	≥ 3 hearts to start	0 hearts to start
1 hearts to start	2 hearts to start	currently short hearts
currently not short hearts	opponent short hearts	exit in hearts
we have single Q♠	we have single A♠	we have single K♠
we have lead	Q♠ player has 0 backers	Q♠ player has 1 backers
Q♠ player has 2 backers	Q♠ player has ≥3 backers	Q♠ player has 0 shorts
Q♠ player has 1 shorts	Q♠ player has 2 shorts	Q♠ player has 3 shorts
Q♠ player has short diamonds	Q♠ player has short clubs	Q♠ player has short hearts
leader short spades	leader short diamonds	leader short clubs
leader short hearts	leader not short spades	leader not short diamonds
leader not short clubs	leader not short hearts	we have forced high spade
we have forced high diamond	we have forced high club	we have forced high heart

which decreases the chance of us taking the Q♠, while 167 times they increase the likelihood.

We can use this to explain what has been learned. Having the Q♠ with only one other spade in our hand means we are likely to take the Q♠ (feature 2 in Table 2). If we also have the lead (feature 1 in Table 2), we are even more likely to take the Q♠. But, if someone else has the lead, and they do not have spades (feature 148 in Table 1), we are much less likely to take the Q♠.

The ability to do this analysis is one of the benefits of putting the complexity into the features instead of the function approximator. If we rely on a more complicated function approximator to learn weights, it is very difficult to analyze the resulting network. Because we have simple weights on more complicated features it is much easier to analyze what has been learned.

2.4 Learning to Avoid Hearts

We used similar methods to predict how many hearts we would take within a game, and learned this independently of the Q♠. One important difference between the Q♠ and points taken from hearts is that the Q♠ is taken by one player all at once, while hearts are gradually taken throughout the course of the game. To handle this, we removed 14 Q♠ specific features and added 42 new atomic features to the 60 atomic features used for learning the Q♠. The new features were the number of points (0-13) taken by ourselves, the number of points taken (0-13) by all of our opponents combined, and the number of points (0-13) left to be taken in the game.

Given these atomic features, we then trained with the atomic (88), pair-wise (3,916) and three-wise (109,824) combinations of features. As before, we

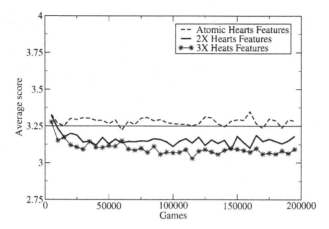

Fig. 3. Learning to not take the Hearts using various combinations of atomic features

present the results averaged over five training runs (200,000 games each) and then smoothed over a window of 5,000 games. The learning graph for this training is shown in Figure 3. An interesting feature of these curves is that, unlike learning the Q♠, we are already significantly beating the expert with the two-wise features set. It appears that learning to avoid taking hearts is a bit easier than avoiding the Q♠. However, when we go from two-wise to three-wise features, the increase in performance is much less. Because of this, we did not try all four-wise combinations of features.

2.5   Learning Both Hearts and the Q♠

Given two programs that separately learned partial components of the full game of Hearts, the final task is to combine them together. We did this by extracting the most useful features learned in each separate task, and then combined them to learn to play the full game. The final learning program used the top 2,000 features from learning to avoid hearts and the top 10,000 features used when learning to avoid the Q♠.

We tried training this program in two ways: first, by playing against PIE, and second, by self-play. During this training shooting the moon was disabled. The first results are plotted in Figure 4(a). Instead of plotting the learning curve, which is uninteresting for self-play, we plot the performance of the learned network against the expert program. We did this by playing games between the networks that were saved during training and the expert program. For two player types there are 16 ways to place those types into a four-player game, however two of these ways contain all learning players or all expert players. 100 games were played for each of the 14 possible arrangements for a total of 1400 hands played for each data point in the curve.

There are 26 points in the full game, so the break-even point, indicated by a horizontal line, falls at 6.5 points. Both the self-trained player and the

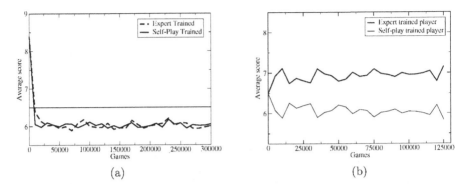

Fig. 4. (a) Performance of expert-trained and self-trained player against the expert. (b) Performance of self-trained and expert-trained programs against each other.

expert-trained player learn to beat the expert by the same rate, about 1 point per hand.

In Figure 4(b) we show the results from taking corresponding networks trained by self-play and expert-play and playing them in tournaments against each other. (Again, 1400 games for each player type.) Although both of these programs beat PIE program by the same margin, the program trained by self play managed to beat PIE by a large margin; again about 1 point per hand.

While we cannot provide a decisive explanation for why this occurs, we speculate that the player which only trains against the expert does not sufficiently explore different lines of play, and so does not learn to play well in certain situations of the game where the previous expert always made mistakes. The program trained by self-play, then, is able to exploit this weakness.

2.6   Imperfect Information Play

Given the learned perfect-information player, we then played it against IIE. For these tests, shooting the moon was enabled. Both programs used 20 Monte-Carlo models and analyzed the first trick of the game in each model (up to 4 ply). When just playing single hands, the learned player won 56.9% of the hands with an average score of 6.35 points per hand, while the previous expert program averaged 7.30 points per hand. When playing full games (repeated hands to 100 points), the learned player won 63.8% of the games with an average score of 69.8 points per game as opposed to 81.1 points per game for IIE.

3   Conclusions and Future Work

The work presented in this paper presents a significant step in learning to play the game of Hearts and in learning for multi-player games in general. We have shown that a search-based player with a learned evaluation function can learn to play Hearts using a simple linear function approximator and TD-learning with either expert-based training or self-play. Furthermore, our learned player beat one of the best-known programs in the world in the imperfect information version of Hearts.

There are several areas of interest for future research. First, we would like to extend this work to the imperfect information game to see if the Monte-Carlo based player can be beat. Next, there is a question of the search algorithm used for training and play. There are weaknesses in max" analysis that soft-max" [13] addresses; the paranoid algorithm [14] could be used as well.

Our ultimate goal is to play the full game of Hearts well at an expert level, which includes passing cards between players and learning to prevent other players from shooting the moon. It is unclear if just learning methods can achieve this or if other specialized algorithms may be used to tasks such as shooting the moon. But, this work is the first step in this direction.

Acknowledgments

We would like to thank Rich Sutton, Mark Ring, and Anna Koop for their feedback and suggestions regarding this work. The work was supported by Alberta's iCORE, the Alberta Ingenuity Center for Machine Learning, and NSERC.

References

1. Baxter, J., Trigdell, A., Weaver, L.: Knightcap: a Chess Program that Learns by Combining TD(λ) with Game-Tree Search. In: Proc. 15th International Conf. on Machine Learning, pp. 28–36. Morgan Kaufmann, San Francisco, CA (1998)
2. Buro, M.: From Simple Features to Sophisticated Evaluation Functions. In: van den Herik, J., Iida, H. (eds.) CG 1998. LNCS, vol. 1558, pp. 126–145. Springer, Heidelberg (1999)
3. Fujita, H., Ishii, S.: Model-based Reinforcement Learning for Partially Observable Games with Sampling-based State Estimation. In: Advances in Neural Information Processing Systems, Workshop on Game Theory, Machine Learning and Reasoning under Uncertainty (2005)
4. Fürnkranz, J., Pfahringer, B., Kaindl, H., Kramer, S.: Learning to Use Operational Advice. In: Proc. of the 14th European Conference on A.I. (2000)
5. Ginsberg, M.: GiB: Imperfect Information in a Computationally Challenging Game (2001)
6. Kuvayev, L.: Learning to Play Hearts. In: Proceedings of the 14th National Conference on Artificial Intelligence (AAAI-97) (1997)
7. Luckhardt, C., Irani, K.: An Algorithmic Solution of N-Person Games. In: AAAI-86, vol. 1, pp. 158–162 (1986)
8. Mitchell, T.M.: Machine Learning. McGraw-Hill, New York (1997)
9. Perkins, T.: Two Search Techniques for Imperfect Information Games and Application to Hearts. University of Massachusetts Technical Report, pp. 98–71 (1998)
10. Russell, S., Norvig, P.: Artificial Intelligence: A Modern Approach, 2nd edn. Prentice-Hall, Englewood Cliffs, NJ (2002)
11. Stone, P., Sutton, R.S.: Scaling Reinforcement Learning toward RoboCup Soccer. In: Proc. 18th ICML, pp. 537–544. Morgan Kaufmann, San Francisco,CA (2001)

12. Sturtevant, N.R.: Multi-Player Games: Algorithms and Approaches. PhD thesis, Computer Science Department, UCLA (2003)
13. Sturtevant, N.R., Bowling, M.H.: Robust Game Play against Unknown Opponents. In: AAMAS-2006, pp. 713–719 (2006)
14. Sturtevant, N.R., Korf, R.E.: On Pruning Techniques for Multi-Player Games. In: AAAI-2000 (2000)
15. Sutton, R., Barto, A.: Reinforcement Learning: An Introduction. MIT Press, Cambridge (1998)
16. Tesauro, G.: Temporal Difference Learning and TD-Gammon. Communications of the ACM 38(3), 58–68 (1995)

A Skat Player Based on Monte-Carlo Simulation

Sebastian Kupferschmid and Malte Helmert

Institut für Informatik
Albert-Ludwigs-Universität, Freiburg, Germany
{kupfersc,helmert}@informatik.uni-freiburg.de

Abstract. We apply Monte-Carlo simulation and alpha-beta search to
the card game of *Skat*, which is similar to Bridge, but sufficiently different to require some new algorithmic ideas besides the techniques
developed for Bridge. Our Skat-playing program, called DDS (Double
Dummy Solver), integrates well-known techniques such as *move ordering*
with two new search enhancements. *Quasi-symmetry reduction* generalizes symmetry reductions, disseminated by Ginsberg's Partition Search
algorithm, to search states which are "almost equivalent". *Adversarial
heuristics* generalize ideas from single-agent search algorithms like A* to
two-player games, leading to guaranteed lower and upper bounds for the
score of a game position. Combining these techniques with state-of-the-
art tree-search algorithms, our program determines the game-theoretical
value of a typical Skat hand (with perfect information) in 10 milliseconds.

1 Introduction

Although mostly unknown in the English-speaking world, the game of Skat is
the most popular card game in continental Europe, surpassed in world-wide
popularity only by Bridge and Poker. With about 30 million casual players and
about 40,000 people playing at a competitive level, Skat is mostly a German
phenomenon, although national associations exist in twenty countries on all six
inhabited continents. It is widely considered the most interesting card game for
three players.

Despite its popularity, Skat has not been studied extensively by the AI community. This is not due to a lack of challenge, as Skat is definitely a game of
skill: significant experience is required to reach tournament playing strength. So
far, all existing computer implementations play rather poorly. In this paper, we
explore how an existing approach for playing Bridge, *Monte-Carlo simulation*
using a fast solver for perfect information games, can be applied to the game of
Skat.

The paper is structured as follows. Section 2 briefly introduces the rules of
Skat. In Section 3, we review the idea of Monte-Carlo simulation for card games.
Section 4 describes the general architecture of DDS (Double Dummy Solver), is
followed by the central Section 5, which describes a fast algorithm for computing
the outcome of Skat games with perfect information. Section 6 presents empirical
results and Section 7 provides an estimation of the playing strength of the overall
system. Moreover, directions for future research are given.

H.J. van den Herik et al. (Eds.): CG 2006, LNCS 4630, pp. 135–147, 2007.

2 Skat

Skat is a three-player game played with 32 cards, a subset of the usual Bridge deck. At the beginning of a game, each player is dealt ten cards, which must not be shown or communicated to the other players. The remaining two cards, called the *skat*, are placed face down on the table. Like in Bridge, each hand is played in two stages, *bidding* and *card play*.

The bidding stage determines the alliances for this hand: the successful bidder, henceforth called the *declarer*, plays against the other two. Often, several players compete to become the declarer. In this event, the winner of the bidding process mostly depends on the number of jacks a player holds, and on their suits. Players may also improve their bids by declaring some special contracts (such as *hand*, *schneider*, and *schwarz* games), but these are nuances that we will not discuss further. We point out that, different to Bridge, bidding does *not* have a significant influence on the number of tricks needed to win the deal, with some minor exceptions.

We will not explain the bidding process further and refer to the official rules [7] for details. The declarer decides on the kind of game, for which there are six possibilities: (1) *grand* games, (2) *null* games, and (3-6) *suit* games for each of the four suits (♣, ♠, ♡, ◇).

Card play proceeds as in Bridge, except that the trumps and card ranks are different. In grand games, the four jacks are the only trumps. In suit games, the four jacks and the seven other cards of the selected suit are trumps. There are no trumps in null games. Non-trump cards are grouped into suits as in Bridge. Each card has an associated *point value* between 0 and 11, and the declarer must score more points than the opponents (i.e., at least 61 points) to win. Null games are an exception and follow *misère* rules: the declarer wins iff he scores no trick. Trumps, ranks, and point values of the cards are illustrated in Fig. 1.

Before declaring the game, the declarer may pick up the skat and then discard any two cards from his hand, face down. These cards count towards the declarer's score.

Ranks

Grand games	♣J, ♠J, ♡J, ◇J (trumps)
	A, 10, K, Q, 9, 8, 7 (non-trumps)
Suit games	♣J, ♠J, ♡J, ◇J, A, 10, K, Q, 9, 8, 7 (trumps)
	A, 10, K, Q, 9, 8, 7 (non-trumps)
Null games	A, K, Q, J, 10, 9, 8, 7

Point values

A: 11, 10: 10, K: 4, Q: 3, J: 2, 9: 0, 8: 0, 7: 0

Fig. 1. Ranks and point values of Skat cards. Higher ranking cards are listed further to the left.

3   Monte-Carlo Simulation

The main algorithmic problem when dealing with card games like Bridge or Skat is *uncertainty*. For perfect-information games like Chess, efficient algorithms exist that could be readily applied if it were not for the fact that the opponents' cards are hidden. In fact, the state space of these card games is comparatively small, and it is not too difficult to compute an optimal strategy with knowledge of the deal. However, taking randomness into account is much more challenging (cf. the work by Koller and Pfeffer [9]).

Monte-Carlo approaches, first proposed in this context by Levy [11] and later implemented by Ginsberg [6] in his Bridge-playing program GIB, reduce the problem to the perfect information case using the following strategy: whenever the computer player is asked to play a card, it generates a set of deals which are consistent with previous play. Each of these deals is then completely analyzed by a fast solver for perfect information games. In theory, this can be done with a traditional alpha-beta search engine. The results of these analyses are then used to vote on the card to play in the actual (uncertain) game.

The Monte-Carlo approach has two fundamental problems. The first problem is that the samples might not be representative of the real card distribution. This is not so much caused by the fact that only a limited number of deals are analyzed, because the law of large numbers guarantees that this statistical error can be made arbitrarily small. The real issue is that not all deals should be generated with equal probability, because different distributions are not equally plausible given the previous course of play.

To reflect this, we would need to take into account the *conditional probability* that an opponent will play a given card given a certain deal and previous play. For example, if the declarer starts the game with ♣A and the other players follow suit with ♣7 and ♣10, it is highly unlikely that the third player still holds a clubs card (except for ♣J, which is part of the trump suit, not the clubs suit). However, it is difficult to quantify information of that kind, both in theory and in practice.[1]

The second fundamental problem of the Monte-Carlo approach is that even if *all* possible deals are analyzed and the conditional probabilities are correct,[2] the algorithm does not play perfectly. Intuitively, the reason for this is the fact that the correct card to play may depend on information that the player cannot possibly know. Formally, this problem is discussed extensively by Frank and Basin [4].

Despite these fundamental limitations, Monte-Carlo-based approaches have been successful in the Computer Bridge world. Indeed, most current systems rely on sampling methods to some extent. We believe that they should be at least as effective for Computer Skat, and possibly more so, because the bidding phase of a Skat game allows for much less information gathering than in Bridge.

---

[1] *Mixed-Strategy Nash Equilibria* are the most commonly applied theoretical solution concept for such games [12].

[2] This condition requires an exact mental model of the opponents, and is thus not practically possible for human opposition.

4   General Architecture

Our Skat player consists of two parts, a bidding engine and a card play engine. The bidding engine is responsible for determining the highest bid that the player is willing to make and for deciding which two cards to discard in case it wins the bid. The card play engine is responsible for the actual card play after the game has been declared. At its core is a fast algorithm for solving Skat games with perfect information. We call this component DDS, the *double dummy solver*. The name is borrowed from Bridge terminology, even though the term *dummy* is Bridge-specific. DDS is explained in Section 5, while the rest of this section is dedicated to the bidding and card play engines.

4.1   Bidding Engine

In theory, it is possible to implement the bidding engine by Monte-Carlo sampling using the following strategy. First, select N random deals. Then, for each of the six kinds of games and each possible way of discarding two cards, query the double dummy solver to decide whether or not the game can be won. However, this requires $6 \cdot \binom{12}{2} \cdot N = 396N$ queries, which is prohibitively high even for a modest number of samples.

Typically, the choice of cards to discard is straightforward, as most candidates can be eliminated by simple rules of thumb. A mixed approach that computes Monte-Carlo samples for each kind of game and implements rules for discarding requires only $6N$ queries.

We have instead adopted a completely rule-based approach for both bidding and discard procedure. The rules were generated by the following learning algorithm. First, we used DDS to analyze 126,000 deals, where both the discarded cards and the kind of game were randomized. For each of the resulting hands of the computer player, the algorithm evaluated a number of hand-crafted features, e.g., number of jacks and length of each suit, and paired these with the outcome of the game (1 for win or 0 for loss). Then, a Least Mean Squares algorithm fitted a linear function from the feature space to the real numbers. The resulting *success estimator* was supposed to estimate the winning probability given the features of a hand. It was then used in place of DDS in the bidding stage of the game, so that instead of $396N$ calls to DDS, a corresponding number of computations of the success estimate was needed, which could be computed sufficiently fast.

Regrettably, the resulting bidding behavior leaves something to be desired. Although reasonable choices were made most of the time, some decisions were truly puzzling. While we do not discuss the bidding engine further, as this part contains no technical innovations and requires some domain knowledge for understanding the choice of features, we note that it is currently the Achilles heel of the Skat player.

4.2   Card Play Engine

As noted before, the card play engine is based on Monte-Carlo simulations using DDS. In Ginsberg's GIB, a score is calculated for each card and each sample deal. The algorithm plays the card with the highest average score. A similar scheme can be used for Skat. Yet, it is not immediately clear how the score should be measured. Counting the point total achieved by the computer player is not reasonable, since playing a card that reliably achieves a point total of 65 (and thus a win) is preferable to playing a card that leads to a total of 80 most of the time but rarely drops to 55 (and thus a loss).

However, simply counting whether or not the deal is won is also problematic, as there is no incentive for the algorithm to win the game by a score of 100:20, rather than, say, 63:57. Thus it will willingly give away points to the opponent as long as it does not see its victory endangered.[3] Because of statistical error and the inaccuracies of Monte-Carlo methods mentioned earlier, this can lead to situations where an easy victory is cast away lightly.

To avoid both problems, we follow a combined approach. Winning a hand is most important, so the set of winning cards of each sample deal is computed first. Those cards which win a maximal number of sample deals are further analyzed by computing the average point total across all sample deals. The card play engine finally selects a card which maximizes this value. This leads to card play that prefers safety to point accumulation, but accumulates points where safely possible.

To support this approach, DDS can run in a fast *qualitative mode*, in which it only determines whether a given card is a winner, and in a slower *quantitative mode*, in which it computes exact scores for cards. We describe these in the following section. Null games are always solved in qualitative mode as they end as soon as the declarer wins a trick. They are not computationally challenging and we do not describe them further, assuming grand or suit games in the following.

5   Double Dummy Solver

When running in qualitative mode, DDS uses a zero-window alpha-beta search to determine whether the score for the declarer when playing a given card is at least 61. When running in quantitative mode, DDS uses the MTD framework proposed by Plaat *et al.* [14]. Specifically, we employ the MTD(0) algorithm: First, we determine whether the declarer can achieve a single point. If the answer is positive, a new search determines whether he can achieve two points[4], three points, and so on, until the answer is negative. Although it is somewhat counter-intuitive that this procedure should be an improvement over standard alpha-beta search, the higher number of cut-offs, combined with the use of a transposition table for storing intermediate results, usually makes it much faster. Most recomputations of subtree values only require a transposition table lookup, reducing the

---

[3] This is a general problem for game- playing algorithms searching to the end of the game. Schaeffer reports similar "unreasonable" behavior in his checkers-playing program CHINOOK, which uses perfect endgame databases. The Chinook team went to some lengths to change this behavior CHINOOK.

[4] This is actually redundant, as scores of 1 or 119 are impossible.

```
def search(p):
  if p isa leaf position:
    return p.declarer_score ≥ 61
  else if p isa declarer node:
    for q in succ(p):
      if search(q) = true:
        return true
    return false
  else:
    for q in succ(p):
      if search(q) = false:
        return false
    return true
```

Fig. 2. Basic search algorithm

cost of re-search. In our experiments, zero-window search outperformed standard alpha-beta by an order of magnitude when using a transposition table.

Therefore, regardless of mode, in the following we can assume that we are conducting a zero-window search. For clarity of presentation, we will only consider the most common search window $[60, 61]$, although other search bounds do occur in quantitative mode. The basic search algorithm, without any search enhancements, is shown in Fig. 2. In the rest of this section, we discuss four enhancements to the basic search algorithm which lead to significant speed-ups, concluding our presentation with a refined search algorithm that includes all enhancements.

5.1 Transposition Table

The first and obvious enhancement to Fig. 2 is the use of a transposition table. However, using a transposition table efficiently in this setting requires some care. At every stage of the game, the current position can be adequately represented by the player to move, the remaining cards, the cards in the current trick (if any), and the *running score*, i.e., the number of points won by the declarer in previous tricks. The running score is an important part of the position because it influences the evaluation of the position (win/loss and exact point value). However, it has *no* effect on the optimal strategy for the rest of the game, the *subgame* rooted at this position in game theory terminology.

Therefore, to keep the transposition table small and to allow as many lookups as possible, it is desirable not to consider the running score a part of the position information. This means that it is not sufficient to store win/loss values in the transposition table, not even for the qualitative solver: a subgame in which the declarer can achieve 40 points can be either a win or a loss, depending on the running score as this subgame is reached. Thus, the transposition table must always store exact point values or bounds on exact values, not boolean results.

Using transposition tables in this way, we have a simple cut-off criterion for search nodes: a subgame is not searched further if (1) the transposition table shows that the total of the running score and the lower bound on the future score is at least 61, or (2) the total of the running score and the upper bound on the future score is at most 60.

5.2 Move Ordering

It is commonly known that alpha-beta search performance is drastically influenced by the order in which the different move alternatives are considered [13].

Typical implementations of alpha-beta search use three heuristics for finding a good ordering, i.e., one where an optimal move is considered early: *transposition table moves*, *history heuristic*, and *killer heuristic*. All these techniques are roughly based on the idea that a move which is good in a certain context is often good in other contexts. We have implemented the first of these techniques. If playing a certain card leads to a cut-off in some subgame, this card is always considered first when this subgame is later reexamined with different search bounds.

The remaining cards are ordered with the aim of reducing branching. If there is currently no card on the table, we prefer playing a suit of which the other players hold at least one card (so that they must follow suit), but only few cards (so that their choice is limited). More to the point, for each card we multiply the number of allowed answers for the other two players, preferring cards which minimize this value. Within a suit, cards of higher rank are preferred.

In our experiments, this ordering heuristic reduces the average number of investigated search nodes by a factor of 3.45, while reducing the average search time by a factor of 3.02, compared to the original, arbitrary move ordering. We analyze the impact of move ordering in more detail in Section 6, together with the effect of the other search enhancements introduced now.

5.3 Quasi-symmetry Reduction

Ginsberg reports that his Bridge-playing program GIB [6] is accelerated by an order of magnitude by replacing alpha-beta search with his *Partition Search* algorithm [5]. Partition Search aims at increasing the number of subgames that can be solved by transposition table lookups. It does so by not storing single game positions but *equivalence classes* of game positions in the transposition table. This is very effective in Bridge as the number of equivalent positions can be expected to be high.

Many of the equivalences of Bridge positions are due to the fact that only the *relative* rank of cards is important for determining optimal play; the *absolute* rank is irrelevant. We say that two cards held by the same player are *rank-equivalent* iff they are in the same suit and no card on the table or in another player's hand is ranked between them.

Unfortunately, unlike Bridge, it is usually the case in Skat games that *all* rank-equivalent cards must be considered because of different point values. In a Bridge game, a player that holds both ♠K and ♠Q need never play the king before the queen (or vice versa). The same is not true of a Skat position. For example, if the player can win the trick playing the king, eventually winning the deal by a margin of 61:59 points, playing the queen instead might lose the deal. Especially if the difference in value between the two cards is greater than one, for example in the case of ♠10 and ♠K, the lines of play that begin with these cards often look completely different.

However, it can be proven that if two cards are rank-equivalent, then the difference between the values of the subgames started by playing either of these cards is bounded by their difference in point value [10]. Thus, if we can prove that playing ♠10 results in a declarer score of 72 and ♠K is no longer in play, then playing ♠Q results in a declarer score in the interval $[65, 79]$, since in this situation ♠10 and ♠Q are rank-equivalent, and the difference in point value is $10 - 3 = 7$.

We use rank-equivalence for a technique we call *quasi-symmetry reduction*, which decreases the branching factor of interior nodes of the search tree. Whenever the search algorithm considers playing a card c which is rank-equivalent to a previously considered card c', we fetch the transposition table entry for the position reached by playing c' and check if there is any hope in playing c instead of c'. For example, if the transposition table shows that playing ♠Q at some declarer node yields at most 57 points, then playing ♠K can yield at most 58 points, so that the move need not be considered.

In our experiments, exploiting quasi-symmetries significantly reduces the number of search nodes. However, much of this gain is countered by an increased cost per node for rank-equivalence checking and transposition table lookups. The most efficient version of the algorithm, which is the one we report on, only exploits quasi-symmetries for cards of which the point values differ by at most one.

In our experiments, quasi-symmetry reduction reduces the average number of search nodes by a factor of 2.38 and average running time by a factor of 2.03.

5.4   Adversarial Heuristics

As a final search enhancement, the double dummy solver uses a forward pruning technique which we will now describe. In Section 5.1, we explained that whenever a position is re-explored during search, the search algorithm fetches a lower bound L and upper bound U on the declarer score in this subgame from the transposition table. If M is the running declarer score, then the subgame is not searched further if $M + L \geq 61$ or $M + U \leq 60$. Our forward pruning technique extends this early termination check to positions which are *not* present in the transposition table. To this end, we must compute (preferably narrow) bounds L and U for arbitrary subgames.

How can such bounds be calculated? In single-agent search problems, lower bounds on the actual search cost are typically computed by *relaxing* the problem at hand, i.e., by increasing the set of allowed moves. For example, the minimal weighted matching heuristic for Sokoban [8] can be interpreted as the length of an optimal solution to a relaxed problem where boxes may be moved to adjacent empty squares regardless of the position of the man. The Manhattan heuristic for the $n^2 - 1$ puzzle can be similarly understood as the length of an optimal solution to a relaxed problem where tiles may always be moved to adjacent positions, even if these are occupied.

When extending these ideas to an adversarial search context, care must be taken to reflect correctly the role of the MAX and MIN players. For any given subgame, we can compute an *upper bound* to the score of the MAX player by *extending* the set of possible moves for MAX and/or *reducing* the set of possible moves for MIN. Conversely, a *lower bound* can be computed by extending the set of possible moves for MIN and/or reducing the set of possible moves for MAX. Any such modification leads to correct bounds that can be exploited during search without compromising the validity of the search in any way, unlike common forward pruning techniques such as *null-move pruning* in Chess or Buro's *ProbCut* [3] in Othello. We call bounds derived in such a way *adversarial heuristics* because of their similarity to heuristics used in single-agent (non-adversarial) search.

The key to good adversarial heuristics is modifying the sets of allowed moves in such a way that the resulting bounds are reasonably narrow, but cheap to compute. For lower bounds on the declarer score in Skat, the following two modifications of the game rules satisfy this criterion.

1. The declarer may only play cards that are guaranteed to win the trick. If this means that he has no legal moves, the opponents may claim the remaining points.

The rationale between this modification is that the optimal strategy for the opponents becomes difficult to compute once they are able to control the game. We eliminate this expensive computation by requiring the declarer to force the game.

2. In addition to normal moves, an opponent may swap the point values of two cards in his hand before playing a card, provided that the two cards are in the same suit.

This modification eliminates a strategic dilemma for the opponents. In some situations, it is difficult to decide whether they should play a card of minimal *point value* or minimal *rank*. For example, consider a diamonds game where the declarer plays ♣J and is thus guaranteed to win the trick. The first opponent holds two trumps, ♡J and ♢A. In some situations, it is preferable to play ♡J, only losing two points to the declarer. In other situations, it is better to play ♢A, losing eleven points to the declarer but keeping the higher-ranked card in order to win a trick later. In the modified game, the best reply is obvious: swap the point values of the ace and jack, so that the ace is worth two points and the jack is worth eleven points, then play the ace.

We point out that swapping the point values of cards is rarely needed, because rank ordering and point value ordering are consistent for all non-trump suits, and in the case of grand games even for the trump suit. Thus, the second modification does not usually have a large impact on the quality of the bounds.

```
def search(p):
  if p isa leaf position:
    return p.declarer_score ≥ 61
  else if p isa declarer node:
    if p in transposition_table:
      (L, U) := transposition_table(p)
    else:
      (L, U) := adversarial_heuristics(p)
    if M + L ≥ 61:
      return true
    if M + U ≤ 60:
      return false
    for q in order_moves(succ(p)):
      if q ~ q' for q' considered earlier:
        (L',U') := transposition_table(q')
        if M + U' + δ(q,q') ≤ 60:
          continue
      if search(q) = true:
        return true
    return false
  else:
    ... {analogous to declarer node case}
```

Fig. 3. Search algorithm with all search enhancements; $\delta(q, q')$ denotes the point difference between the two cards being played. We omit some details like updating the transposition table, which are handled in the standard way.

An experienced Skat player will notice that computing an optimal strategy in the modified game is almost trivial, except for situations where an opponent need not follow suit. Indeed, the value of a game position in the modified game can be computed in time $O(N)$, where N is the number of remaining cards. For details on how this can be done, we refer to the first author's master's thesis [10]. Similar ideas can be applied to obtain *upper bounds* on the declarer score. However, this case is slightly more complicated, so we refer to [10] again for details.

In our experiments, using adversarial heuristics reduces the average number of search nodes by a factor of 1.80 and the average run time by a factor of 1.58. The complete search algorithm, including all enhancements, is shown in Fig. 3.

6   Experiments

In Section 5, we described the implementation of DDS, focusing on three central search enhancements: move ordering, quasi-symmetry reduction, and adversarial heuristics. In this section, we provide a more detailed empirical analysis of the performance gains offered by these features.

In practice, it is not sufficient to consider the effectiveness of a search enhancement in isolation. It is quite possible that a given enhancement leads to a drastic improvement in run time by itself, but offers no gain when implemented together with other features. For this reason, we evaluated all possible combinations of move ordering (MO), quasi-symmetry reduction (QR) and adversarial heuristics (AH), including the empty set. All configurations used a transposition table.

Table 1. Mean value, standard deviation and median of node count and running time for 100,000 randomly generated Skat games

Features			Nodes × 1,000			Run time in seconds		
MO	QR	AH	Mean	St.dev.	Median	Mean	St.dev.	Median
			2,772	8,853	181	0.84	3.11	0.04
×			804	3,669	29	0.28	1.42	0.01
	×		1,163	3,457	102	0.41	1.36	0.03
		×	1,538	5,223	57	0.53	2.02	0.02
×	×		317	1,499	17	0.12	0.65	0.01
×		×	626	3,182	12	0.20	1.10	0.01
	×	×	658	2,152	34	0.25	0.87	0.01
×	×	×	244	1,300	7	0.11	0.60	0.01

Table 1 shows that the three enhancements work well in combination. Although the speedups are not completely orthogonal, each of the three features is a useful addition to all configurations without it. The results are based on 100,000 randomly generated suit games. Grand games are easier, null games much easier to solve. The high standard deviations and low medians show that the results are far from being normally distributed. Most deals are solved very quickly, but occasional outliers heavily influence the average case performance.

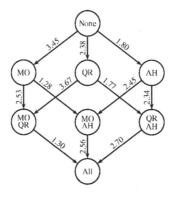

Fig. 4. Node count reductions for the various search enhancements

Fig. 4 depicts a lattice illustrating the usefulness of adding each feature to any configuration. The arrow labels show the reduction of average node count achieved when going from one configuration to another. The figure shows that there are some diminishing returns, but the general picture is quite positive.

7   Results and Future Directions

We have shown that by enhancing a state-of-the-art game-playing algorithm with a number of suitable search enhancements, it is possible to build a fast double dummy solver for the game of Skat (called DDS). Of course, in practice we are not just interested in the performance of DDS, but also in the quality of play of the overall system.

This is somewhat harder to quantify because such experiments are difficult to automate. We played 18 games against human and machine opposition. Against two human players of moderate strength, the system ended on a close second place. Post-mortem analysis revealed flawless card play but improvable bidding behavior. Against two computer players[5] the system played very convincingly, winning every single game when it played as a declarer and all games but one when it played in the opposing side [10].

The first logical next step for future work is to improve the bidding engine. In theory, the general approach of learning rules from a set of features and self-play data seems reasonable. However, our choice of features and learning algorithm might not be the best possible. Alternatively, hand-crafted rules could be used, but this is tedious and requires expert domain knowledge.

A second possible direction for further study is the investigation of alternative search algorithms, such as proof number search [1] or B^* [2]. The classical drawback of these approaches is their high memory consumption, but in Skat games, all visited positions can easily be kept in memory.

Third, it would be interesting to apply our search enhancements to other games. (1) Quasi-symmetry reduction is a potentially useful technique for all games where points are accumulated in the course of play, which includes, but is not limited to, all trick-based card games. (2) Adversarial heuristics can be usefully applied in a similar way to classical heuristics whenever a complete solution of a game is feasible. (3) We see a good potential for other games amenable to Monte-Carlo approaches, for games with a strong threat structure such as Go-Moku and Hex, and for Amazon subgames.

References

1. Allis, L.V., van der Meulen, M., van den Herik, H.J.: Proof-Number Search. *Artificial Intelligence* 66(1), 91–124 (1994)
2. Berliner, H.: Search and Knowledge. In: Proceedings of the International Joint Conference on Artificial Intelligence (IJCAI-77), pp. 975–979 (1977)
3. Buro, M.: ProbCut: An Effective Selective Extension of the Alpha-Beta Algorithm. ICCA Journal 18(2), 71–76 (1995)
4. Frank, I., Basin, D.A.: Search in Games with Incomplete Information: A Case Study using Bridge Card Play. Artificial Intelligence 100(1–2), 87–123 (1998)

---

[5] Played by XSKAT 3.4; cf. http://www.xskat.de/, a program with rule-based card play.

5. Ginsberg, M.L.: Partition Search. In: Proceedings of the Thirteenth National Conference on Artificial Intelligence and Eighth Innovative Applications of Artificial Intelligence Conference (AAAI/IAAI-96), Portland, Oregon, pp. 228–233 (1996)
6. Ginsberg, M.L.: GIB: Steps toward an Expert-Level Bridge-Playing Program. In: Dean, T. (ed.) Proceedings of the Sixteenth International Joint Conference on Artificial Intelligence (IJCAI-99), Stockholm, Sweden, pp. 584–593 (1999)
7. International Skat Players Association: International skat and tournament order 1999, (2003), http://www.skatcanada.ca/canada/forms/rules-2003.pdf
8. Junghanns, A., Schaeffer, J.: Sokoban: Enhancing General Single-Agent Search Methods using Domain Knowledge. Artificial Intelligence 129(1–2), 210–251 (2001)
9. Koller, D., Pfeffer, A.: Representations and Solutions for Game-Theoretic Problems. Artificial Intelligence 94(1–2), 167–215 (1997)
10. Kupferschmid, S.: Entwicklung eines Double Dummy Skat Solvers – mit einer Anwendung für verdeckte Skatspiele. Master's thesis, University of Freiburg (2003)
11. Levy, D.N.L.: The Million Pound Bridge Program. In: Levy, D.N.L., Beal, D.F. (eds.) Heuristic Programming in Artificial Intelligence — The First Computer Olympiad, Ellis Horwood, pp. 95–103 (1989)
12. Osborne, M.J., Rubinstein, A.: A Course in Game Theory. MIT Press, Cambridge (1994)
13. Pearl, J.: Heuristics — Intelligent Search Strategies for Computer Problem Solving. Addison-Wesley, London, UK (1984)
14. Plaat, A., Schaeffer, J., Pijls, W., de Bruin, A.: Best-First Fixed-Depth Minimax Algorithms. Artificial Intelligence 87(1–2), 255–293 (1996)

A Retrograde Approximation Algorithm for One-Player Can't Stop

James Glenn[1], Haw-ren Fang[2], and Clyde P. Kruskal[3]

[1] Department of Computer Science
Loyola College in Maryland, Baltimore, MD, USA
jglenn@cs.loyola.edu
[2] Department of Computer Science and Engineering,
University of Minnesota, Minneapolis, MN, USA
hrfang@cs.umn.edu
[3] Department of Computer Science
University of Maryland, College Park, MD, USA
kruskal@cs.umd.edu

Abstract. A one-player, finite, probabilistic game with perfect information can be presented as a bipartite graph. For one-player Can't Stop, the graph is cyclic and the challenge is to determine the game-theoretical values of the positions in the cycles. In this contribution we prove the existence and uniqueness of the solution to one-player Can't Stop, and give an efficient approximation algorithm to solve it by incorporating Newton's method with retrograde analysis. We give results of applying this method to small versions of one-player Can't Stop.

1 Introduction

Retrograde analysis has been well developed and successfully applied to deterministic, finite, two-player zero-sum games with perfect information, such as Awari [2], checkers [3], and chess [4]. For some probabilistic games, such as Yahtzee and Can't Stop[1], retrograde algorithms are less practical due to the high complexity. Therefore, we studied the simplified one-player game instead.

A one-player probabilistic game can be represented as a bipartite graph, in which one set of nodes corresponds to deterministic events while the other corresponds to random events. For one-player Yahtzee, the graph representation is acyclic, which simplifies algorithm design and allows the game to be solved easily [1] [5]. In some games, the graph representation is cyclic, which causes difficulty in designing a bottom-up retrograde algorithm. We are particularly interested in one-player Can't Stop, developed an approximation algorithm to solve this game

---

[1] Can't Stop was designed by Sid Sackson and marketed by Parker Brothers. It is currently out of print but will be republished by Face 2 Face Games in 2006. The rules can be found at http://en.wikipedia.org/wiki/Can't_Stop and http://www.boardgamegeek.com/game/41. Also see Appendix for a short description.

H.J. van den Herik et al. (Eds.): CG 2006, LNCS 4630, pp. 148–159, 2007.
© Springer-Verlag Berlin Heidelberg 2007

by incorporating Newton's method with retrograde analysis, and give results of applying the method to small versions of the game.

The organization of this paper is as follows. Section 2 formulates the problem. In Section 3 we prove that one-player Can't Stop has a unique solution, and give an efficient retrograde algorithm to solve it. Section 4 presents the indexing scheme. Section 5 summarizes the results of the experimental tests. A conclusion is provided in Section 6. A short description of Can't Stop game rules is provided in Appendix.

2   Problem Formulation

A one-player, finite and probabilistic game with perfect information can be represented as a directed, bipartite *game graph* $G = (U, V, E)$, where U and V are two disjoint sets of vertices and E is the set of edges. (An edge (x, y) must have either $x \in U$ and $y \in V$ or $x \in V$ and $y \in U$.) The graph may be cyclic. A *position* is a vertex $w \in U \cup V$.

In Yahtzee a turn consists of a dice roll followed by a move. In Can't Stop a turn consists of a sequence of partial turns, each of which is a dice roll followed by a move. A *roll position* is a vertex $u \in U$. A *move position* is a vertex $v \in V$.

For each non-terminal roll position u, a *dice roll* is a random event. The weight $0 < p((u, v)) \leq 1$ indicates the probability that the game in roll position u will change into move position v. So,

$$\sum_{\forall v \text{ with } (u,v) \in E} p((u, v)) = 1 \,. \tag{1}$$

A *move* (v, u) (from a move position to a roll position) is a deterministic choice.

A *partial turn* (u_1, u_2) *(from roll position u_1 to roll position u_2)* consists of a dice roll followed by a move. It is represented by a pair of edges $((u_1, v), (v, u_2))$ in G. A *turn* is a sequence of partial turns $(u_0, u_1), (u_1, u_2), \ldots, (u_{k-1}, u_k)$. As noted above, in Yahtzee, a turn consists of exactly one partial turn, and in Can't Stop a turn may consist of many partial turns.

We associate each vertex with a number, as the expected cost (or penalty) of playing the optimal strategy starting from that vertex. The information is stored in a *cost database*, which is presented as a function $f : (U \cup V) \rightarrow R$.

The goal of one-player Can't Stop is to play so as to minimize the expected number of turns to finish the game. Therefore, $f(u)$ is the expected number of remaining turns to finish the game starting at roll position u, in optimal play.

The cost function f satisfies that for all non-terminal roll positions $u \in U$,

$$f(u) = g(u) + \sum_{\forall v \text{ with } (u,v) \in E} p((u, v))f(v) \,, \tag{2}$$

where $g(u)$ is the step cost (or step penalty) at u.

In the game of one-player Can't Stop, $g(u)$ indicates whether it is the first partial turn for a turn. More precisely,

$$g(u) = \begin{cases} 1, \text{ if } u \text{ is the starting position for a turn,} \\ 0, \text{ otherwise.} \end{cases} \tag{3}$$

For the optimal playing strategy, we minimize the cost (or penalty)[2]. Therefore, for all non-terminal move positions $v \in V$,

$$f(v) = \min_{\forall u \text{ with } (v,u) \in E} f(u). \tag{4}$$

For all positions $w \in U \cup V$, $f(w)$ is also called the *position value* of w.

A terminal vertex indicates the end of a game. We assume all terminal vertices, denoted by z, are roll positions (in U) with $f(z) = g(z)$. For one-player Can't Stop, a terminal vertex z is reached when the player completes three columns, and therefore no additional rolling of dice is required (i.e., $f(z) = g(z) = 0$).

A cost database f satisfying both conditions (2) and (4) is called a *solution*. A game is *solved* if a solution is obtained. Unless otherwise noted, all the game graphs in this paper stand for finite, one-player, probabilistic games with perfect information.

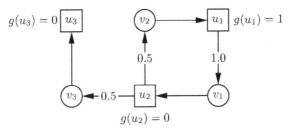

Fig. 1. An example of game graph $G = (U, V, E)$

We illustrate an example in Figure 1, where $g(u_1) = 1$, $g(u_2) = g(u_3) = 0$, $p((u_1, v_1)) = 1$, and $p((u_2, v_2)) = p((u_2, v_3)) = 0.5$. This example simulates the last stage of a game of one-player Can't Stop. A turn begins at position u_1. At position u_2, the first player has 50% probability to complete three columns, and the other 50% probability to fall back to u_1. By (2) and (4),

$$\begin{aligned} f(v_1) &= f(u_2) = \tfrac{1}{2}f(v_2) + \tfrac{1}{2}f(v_3), \\ f(v_2) &= f(u_1) = f(v_1) + 1, \\ f(v_3) &= f(u_3) = 0. \end{aligned} \tag{5}$$

The unique solution is $f(u_1) = f(v_2) = 2$ and $f(u_2) = f(v_1) = 1$.

---

[2] If the goal of a game is to maximize some value, we can multiply it by -1 and minimize.

The problem of solving a one-player probabilistic game is formulated as follows. Suppose we are given a game graph $G = (U, V, E)$ of a one-player, finite, probabilistic game with perfect information, and its step cost function $g : U \to R$. First, we investigate the existence and uniqueness of the solution (i.e., the cost database $f : (U \cup V) \to R$ that satisfies both conditions (2) and (4)). Second, we design an efficient algorithm to construct the cost database, assuming a solution exists.

3   Retrograde Analysis for One-Player Probabilistic Games

A retrograde algorithm typically consists of three phases: initialization phase, propagation phase, and the final phase. In the initialization phase, the terminal vertices are associated with their position values. In the propagation phase, the information is propagated iteratively back to its predecessors until no propagation is possible. The final phase deals with the undetermined vertices.

Subsection 3.1 gives an algorithm to construct the cost database of an acyclic game graph. In Subsection 3.2, we prove that one-player Can't Stop has a unique solution. In Subsection 3.3, we give an approximation algorithm to construct the cost database for a game graph with cycles.

3.1   Game Graph Is Acyclic

For games with acyclic game graphs, such as one-player Yahtzee, the bottom-up propagation procedure is clear. Algorithm 1 gives the pseudocode to construct the cost database for an acyclic game graph.

Consider Alg. 1. Assuming all terminal vertices are in U, the set S_2 is initially empty and (†) is not required. However, it is useful for the reduced graph $\hat{G}$ in Algs. 2 and 3. We call a vertex *determined* if its position value is known. By (2) and (4), a non-terminal vertex cannot be determined until all its children are determined. The sets S_1 and S_2 store all determined but not yet propagated vertices. A vertex is removed from them after it is propagated. The acyclic property ensures that all vertices are determined at the end of the propagation phase. Therefore, a final phase is not required. The optimal playing strategy is clear: given $v \in V$, always make the move (v, u) with the minimum $f(u)$.

Note that in Alg. 1, an edge (u, v) can be visited as many times as the outdegree of u because of (*) and (**). The efficiency can be improved as follows. We associate each vertex with a number of undetermined children, and decrease the value by one whenever a child is determined. A vertex is determined after the number is decreased down to zero. As a result, each edge is visited only once and the algorithm is linear. This is called the *children counting* strategy. For games like Yahtzee, the *level* of each vertex (the longest distance to the terminal vertices) is known *a priori*. Therefore, we can compute the position values level by level. Each edge is visited only once without counting the children.

Lemma 1. *If a game graph is acyclic, its solution exists and is unique.*

Require: $G = (U, V, E)$ is acyclic.
Ensure: Program terminates with (2) and (4) satisfied.   ▷ Lemma 1
  $\forall u \in U,\ f(u) \leftarrow g(u)$, the step cost.   ▷ Initialization Phase
  $\forall v \in V,\ f(v) \leftarrow \infty$.
  $S_1 \leftarrow \{\text{terminal positions in } U\}$
  $S_2 \leftarrow \{\text{terminal positions in } V\}$
  $\forall u \in S_1 \cup S_2$, set $f(u)$ to be its value.   ▷ (†)
  repeat   ▷ Propagation Phase
    for all $u \in S_1$ **do**
      for all $(v, u) \in E$ **do**
        $f(v) \leftarrow \min\{f(v), f(u)\}$
        if all children of v are determined **then**   ▷ (*)
          $S_2 \leftarrow S_2 \cup \{v\}$
        end if
      end for
    end for
    $S_1 \leftarrow \emptyset$
    for all $v \in S_2$ **do**
      for all $(u, v) \in E$ **do**
        $f(u) \leftarrow f(u) + p((u, v)) f(v)$
        if all children of u are determined **then**   ▷ (**)
          $S_1 \leftarrow S_1 \cup \{u\}$
        end if
      end for
    end for
    $S_2 \leftarrow \emptyset$
  until $S_1 \cup S_2 = \emptyset$

Algorithm 1. Construct cost database f for acyclic game graph $G = (U, V, E)$

Proof. In an acyclic graph, the level for each vertex is well-defined. In Alg. 1, the position values are uniquely determined level by level. Hence, the solution exists and is unique.   □

3.2   Game Graph Is Cyclic

If a game graph is cyclic, a solution may not exist. Even if it exists, it may not be unique. We give a condition under which a solution exists and is unique in Lemma 2. The proof uses the Fixed Point Theorem[3]. With Lemma 2, we prove that the game graph of one-player Can't Stop has a unique solution in Theorem 2.

Theorem 1 (Fixed Point Theorem). *If a continuous function $f : R \longrightarrow R$ satisfies $f(x) \in [a, b]$ for all $x \in [a, b]$, then f has a fixed point in $[a, b]$ (i.e., $f(c) = c$ for some $c \in [a, b]$).*

---

[3] See, for example, `http://mathworld.wolfram.com/FixedPointTheorem.html`

Lemma 2. *A cyclic game graph $G = (U, V, E)$ has a solution if,*

1. *For all $u \in U$, $g(u) \geq 0$.*
2. *For each non-terminal vertex u, there is a path from u to a terminal vertex.*
3. *There exists some $w \in U$ such that the graph is acyclic after removing the outgoing edges of w.*

In addition, if the vertex w in condition 3 satisfies $g(w) > 0$, then the solution is unique with all position values non-negative.

Proof. Let $\hat{G} = (U, V, \hat{E})$ be the graph obtained by removing all of the outgoing edges from w in G (i.e., $\hat{E} = \{(u, v) : (u \in U - \{w\}) \wedge (v \in U \cup V) \wedge ((u, v) \in E)\}$). By condition 3, $\hat{G} = (U, V, \hat{E})$ is acyclic. All the terminal vertices other than w in $\hat{G}$ are also terminal in G. Let x be the estimated position value of w. We can construct a database for $\hat{G}$ by Alg. 1. However, we propagate in terms of x (i.e., treat x as a variable during the propagation), though we know the value of x. For example, assuming $x = 6$, we write $\min\{\frac{1}{2}x, \frac{1}{3}x + 2\} = \frac{1}{2}x$ instead of 3. We use $\hat{f}(x, y)$ to denote the position value of $y \in U \cup V$ of $\hat{G}$ in terms of x. At the end of Alg. 1, we compute $\hat{f}(x, w)$ with edges in $E - \hat{E}$ in terms of x by (2). The values of $\hat{f}(x, y)$ for all $y \in U \cup V$ constitute a solution to G, if and only if $\hat{f}(x, w)$ equals x in value. The main theme of this proof is to discuss the existence and uniqueness of x satisfying $\hat{f}(x, w) = x$.

Iteratively applying (2) and (4), all $\hat{f}(x, y)$ for $y \in U \cup V$ are in the form $ax + b$, where $0 \leq a \leq 1$. We are particularly concerned with $\hat{f}(x, w)$. Let $\hat{f}(x, w) = a(x)x + b(x)$, where $a(x)$ and $b(x)$ are real functions of x. By (2) and (4), it is not hard to see that $a(x)$ is non-increasing, $b(x)$ is non-decreasing, and both $a(x)$ and $b(x)$ are piecewise constant. Hence $\hat{f}(x, w)$ is piecewise linear, continuous and non-decreasing in terms of x. By condition 1, $\hat{f}(0, w) = b(0) \geq g(w) \geq 0$. By condition 2, $a(x) < 1$ for x large enough. Since $a(x)$ is non-increasing and $a(x) < 1$ for x large enough, $f(x) < x$ for x large enough. By Theorem 1, there exists $x \geq 0$ such that $\hat{f}(x, w) = x$.

By condition 1, $\hat{f}(0, w) \geq g(w)$. Assuming $g(w) > 0$, then $\hat{f}(w, 0) > 0$. Moreover, $\hat{f}(x, w) = a(x)x + b(x)$ is piecewise linear and continuous with $0 \leq a(x) \leq 1$ for $x \in R$. Therefore, $x \leq 0$ implies $\hat{f}(x, w) > x$. Let $x_0 > 0$ be the smallest solution to $\hat{f}(x, w) = x$. Since $\hat{f}(0, w) > 0$ and $a(x)$ is non-increasing, $a(x_0) < 1$ and therefore $\hat{f}(x, w) < x$ for $x > x_0$. We may conclude that the additional condition $g(w) > 0$ guarantees the solution $x_0 > 0$ to $\hat{f}(x, w) = x$ is unique. Hence the game graph G has a unique solution with all position values non-negative.   □

Consider the strongly connected components of the game graph of one-player Can't Stop. Each strongly connected component consists of all the positions with a certain placement of the squares and various placements of the at most three markers. The roll position with no marker is the *anchor* of the component. When left without a legal move, the game goes back to the anchor, and results in a cycle. The outgoing edges of each non-terminal component lead to the anchors in the supporting components. The terminal components are those in which the

player has won three columns. Each terminal component has only one vertex
with position value 0.

Theorem 2. *The game graph of one-player Can't Stop has a unique solution.*

Proof. The proof is by finite induction. We split the graph into strongly connected components, and consider the components in bottom-up order.

Given a non-terminal component with the anchors in its supporting components having position values non-negative and uniquely determined, we consider the subgraph induced by the component and the anchors in its supporting components. This subgraph satisfies all the conditions in Lemma 2, where the terminal positions are the anchors in the supporting components. Therefore, it has a unique solution with all position values non-negative. By induction, the solution to the game graph of one-player Can't Stop exists and is unique. □

3.3   Retrograde Approximation Algorithms

If we apply Alg. 1 to a game graph with cycles, then the vertices in the cycles cannot be determined. A naive algorithm to solve the game is described as follows. Given a cyclic game graph $G = (U, V, E)$, we prune some edges so the resulting $\hat{G} = (U, V, \hat{E})$ is acyclic, and then solve $\hat{G}$ by Alg. 1. The solution to $\hat{G}$ is treated as the initial estimation for G, denoted by a cost function $\hat{f}$. We approximate the solution to G by recursively updating $\hat{f}$ using (2) and (4). If $\hat{f}$ converges, it converges to a solution to G. The pseudocode is given in Alg. 2.

An example is illustrated by solving $G = (U, V, E)$ in Figure 1. We remove (u_1, v_1) to obtain the acyclic graph $\hat{G}$. The newly terminal vertex is u_1. Let $\hat{f}(u_1) = 1$, which is a reasonable initial guess since $f(u_1) \geq g(u_1) = 1$ in G. The solution for $\hat{G}$ is $\hat{f}(u_1) = \hat{f}(v_2) = 1$ and $\hat{f}(u_2) = \hat{f}(v_1) = \frac{1}{2}$. The update is repeated with $\hat{f}(u_1) = \frac{3}{2}, \frac{7}{4}, \ldots, \frac{2^{n+1}-1}{2^n}, \ldots$, which converges to 2. Hence $\hat{f}$ converges to the solution to G. Let e_n be the difference between $\hat{f}(u_1)$ at the nth step and the converged value; then $\frac{e_{n+1}}{e_n} = \frac{1}{2}$. In other words, it converges linearly.

Consider Alg. 2. For one-player Can't Stop, it is natural to prune the outgoing edges of the anchors and obtain the acyclic $\hat{G}$. In (†), we assign an estimated value to each vertex terminal in $\hat{G}$ but not terminal in G (i.e., the newly terminal positions). For efficiency, we do not have to recompute the whole (*) and (**). Updating with the recent cost changes of the children is sufficient.

If the conditions in Lemma 2 are satisfied (e.g., a strongly connected component of one-player Can't Stop), $\hat{G}$ can be obtained by pruning the outgoing edges of the anchor w. In this case, Alg. 2 corresponds to the steepest descent method without line search in numerical optimization, so linear convergence is expected.

The proof of Lemma 2 reveals that if we solve $f(x, w) = x$ using Newton's method[4], then quadratic convergence can be expected. In other words, if we use

---

[4] See, for example, `http://mathworld.wolfram.com/NewtonsMethod.html`

Ensure: If $\hat{f}$ converges, it converges to a solution to $G = (U, V, E)$.

    Obtain an acyclic graph $\hat{G} = (U, V, \hat{E})$, where $\hat{E} \subset E$.   ▷ Estimation Phase

    Compute the solution $\hat{f}$ to $\hat{G}$ by Algorithm 1.   ▷ (†)

    Use $\hat{f}$ as the initial guess for G.

    $S_1 \leftarrow \{\text{terminal positions of } \hat{G} \text{ in } U\}$

    $S_2 \leftarrow \{\text{terminal positions of } \hat{G} \text{ in } V\}$

    repeat   ▷ Approximation Phase

        for all $u \in S_1$ **do**

            for all $(v, u) \in E$ **do**

                $\hat{f}(v) \leftarrow \min_{\forall w} \text{ with } _{(v,w)\in E} \hat{f}(w)$ ▷ (*)

                $S_2 \leftarrow S_2 \cup \{v\}$

            end for

        end for

        $S_1 \leftarrow \emptyset$

        for all $v \in S_2$ **do**

            for all $(u, v) \in E$ **do**

                $\hat{f}(u) \leftarrow g(u) + \sum_{\forall w} \text{ with } _{(u,w)\in E} p((u, w)) \hat{f}(w)$ ▷ (**)

                $S_1 \leftarrow S_1 \cup \{u\}$

            end for

        end for

        $S_2 \leftarrow \emptyset$

    until $\hat{f}$ converges.

Algorithm 2. A naive algorithm to solve a cyclic game graph $G = (U, V, E)$

e_n to denote the difference between the estimation and the solution at the nth step, $\frac{e_{n+1}}{e_n^2} \approx c$ for some constant c when the estimate is close enough to the solution[5]. An example is illustrated with the game graph in Figure 1 as follows. We treat u_1 as w in Lemma 2, and let x be the initial estimate of the position value of u_1. Then $\hat{f}(x, v_2) = x$, $\hat{f}(x, v_1) = \hat{f}(x, u_2) = \frac{1}{2}x$, and $\hat{f}(x, u_1) = \frac{1}{2}x + 1$. Solving $\frac{1}{2}x + 1 = x$, we obtain $x = 2$, which is the exact position value of u_2. In this small example we obtain the solution by one iteration. In practice, multiple iterations are expected to reach the solution. The pseudocode is given in Alg. 3.

    Consider Alg. 3. In the estimation phase, the better the initial estimate of position value of w (denoted by x), the fewer steps are needed to reach the solution.

4   Indexing Scheme

In practice, a game graph $G = (U, V, E)$ can be too big to fit in physical memory. For one-player Can't Stop, we partition the graph into strongly connected components. Before constructing a cost database for a component, we have all

---

[5] In our case $\hat{f}(x, w)$ is piecewise linear. Hence Newton's method can reach the solution in a finite number of steps. In practice, however, rounding errors may create minor inaccuracy.

Require: $G = (U, V, E)$ satisfies the conditions in Lemma 2.
Ensure: $\hat{f}$ converges to a solution to G in the rate of Newton's method.
   Let x denote the estimate for position value of w in Lemma 2.   ▷ Estimation Phase
   Obtain the acyclic graph $\hat{G} = (U, V, \hat{E})$ by removing the outgoing edges of w.
   repeat   ▷ Approximation Phase
      Solve $\hat{G}$ (in terms of x) with the current estimate x for w by Algorithm 1.
      Compute $\hat{f}(x, w)$ with $E - \hat{E}$ by (2) in terms of x. Denote the result by $ax + b$.
      $x \leftarrow \frac{b}{1-a}.$ ▷ (The solution to $ax + b = x$ is $x = \frac{b}{1-a}$.)
   until $\hat{f}(x, w) = x$ in value.

Algorithm 3. An efficient algorithm to solve a game graph with one anchor

its supporting databases constructed. The construction is in bottom-up order, until the game is solved.

4.1 Indexing Scheme for Can't Stop

Consider one-player Can't Stop. Let x_i denote the number of steps from the square to the top at column 'i'. Each strongly connected component of the game graph consists of all the positions with some particular $(x_2, x_3, \ldots, x_{12})$, where $0 \leq x_i \leq 2i - 1$ for $i = 2, 3, \ldots, 7$ and $0 \leq x_i \leq 27 - 2i$ for $i = 7, 8, \ldots, 12$ and at most three of the x_i are zero. $(x_2', x_3', \ldots, x_{12}')$ is a supporting component of $(x_2, x_3, \ldots, x_{12})$ if and only if $x_i' \leq x_i$ for $i = 2, 3, \ldots, 12$ and $(x_2', x_3', \ldots, x_{12}') \neq (x_2, x_3, \ldots, x_{12})$.

A terminal component has three zero squares, and contains only one position in the game graph G (win three columns, the end of a game). Each position in a non-terminal component $(x_2, x_3, \ldots, x_{12})$ is $(y_2, y_3, \ldots, y_{12})$ where each $y_i \leq x_i$ and for at most three i, $y_i < x_i$. All positions are indexed using

$$\sum_{c=2}^{12} y_c \prod_{d=2}^{c-1} l(d) \tag{6}$$

as a hash value, where $l(d)$ denotes the length of column d. When the cost database cannot fit in main memory, it is possible to reorder the columns (e.g., consider a position to be $(x_2, x_{12}, x_3, x_{11}, \ldots)$ if that provides for better access patterns). The cost databases for the anchor positions and the non-anchor positions are maintained separately, since the position values of the non-anchor positions in a component are used only when computing the position value for the anchor in that component.

In practice, we discard the cost database for the non-anchor positions and reconstruct them using the cost database for the anchor positions as necessary (as when simulating perfect play). If storage is abundant and speed is important, it would also be possible to use a file named $x_2 x_3 \ldots x_{12}.ijk$ to store all the position values, where $i < j < k$ are the columns where $x_i \neq y_i$, $x_j \neq y_j$, and $x_k \neq y_k$. The offset of a position would be $y_i + y_j x_i + y_k x_i x_j$ in the file. The naming and indexing convention is the same for positions with two markers or fewer. The cost database for component $(x_2, x_3, \ldots, x_{12})$ consists of all files $x_2 x_3 \ldots x_{12}*$. The largest component is $(3, 5, 7, 9, 11, 13, 11, 9, 7, 5, 3)$, which contains the position of the beginning of the game.

4.2   Algorithms

When the game graph is split, we can construct the cost databases component-by-component in the bottom-up order. Algorithm 2 is applied to the subgraph consisting of the component, its outgoing edges, and the positions that the outgoing edges connect to. For each supporting component, there is only one position the parent component connects to, the anchor (i.e., the position with no markers). In Alg. 2, we may propagate the information from the supporting databases to the component, so the supporting databases are not required in the propagation phase.

5   Experiments

As proof of concept, we have solved simple versions of one-player Can't Stop. These simpler versions use 3-, 4-, and 5-sided dice instead of 6-sided dice and may have shorter columns than the official version. Let (n, k) Can't Stop denote the one-player game played with four n-sided dice with the shortest column k spaces long. Columns 2 and $2n$ are the shortest columns and column $n + 1$ is the longest. Adjacent columns always differ in length by 2 spaces. The official version is then $(6, 3)$ Can't Stop.

For $n = 2, 3, 4$ and $k = 1, 2, 3$ (and also $n = 5, k = 1$) we have implemented Alg. 3 in Java and solved (n, k) Can't Stop. We used an initial estimate of 1.0 for the position value of each vertex. Table 1 shows, for each version of the game, the size of the game graph, the time it took the computer to solve the game, and the average number of turns needed to win the game when using the optimal strategy. The size of the game graph is given as the number of anchor vertices (i.e., vertices representing the beginning of a turn with no markers placed), and the total number of vertices in all of the anchor vertices' strongly connected components (which includes vertices representing the middle of a turn when the markers have been placed on the board). Symmetry allows us to ignore about half of the anchor vertices in our implementation since the position represented by $(x_2, x_3, \ldots, x_{12})$ is equivalent to $(x_{12}, x_{11}, \ldots, x_2)$.

Note that for fixed n, the time to solve the game is roughly proportional to the number of vertices. When n increases there is also an additional cost due to the increased number of outgoing edges from each vertex in U. For $n = 3$ there are 15 neighbors of each vertex (representing the 15 different outcomes of rolling four 3-sided dice); for $n = 4$ there are 35 neighbors.

The average position value also affects the running time. For larger values of k or n, the average position value is higher; higher position values will require more iterations in Alg. 3 to converge. Table 2 shows the number of iterations required for convergence when solving $(4, 3)$ Can't Stop.

Table 1. Results of solving simple versions of Can't Stop

(n, k)	Anchor vertices	Total vertices	Time	Optimal Turns
$(2, 1)$	15	225	0.166s	1.298
$(2, 2)$	44	1,936	0.405s	1.347
$(2, 3)$	95	9,025	0.601s	1.400
$(3, 1)$	308	64,372	1.70s	1.480
$(3, 2)$	1,432	787,600	5.05s	1.722
$(3, 3)$	4,378	4,934,006	23.3s	1.890
$(4, 1)$	12,913	20,802,843	4m50s	2.187
$(4, 2)$	83,456	289,091,584	58m50s	2.454
$(4, 3)$	333,069	2,104,663,011	6h7m	2.700
$(5, 1)$	921,174	7,105,015,062	2d20h	2.791

Table 2. For Can't Stop $(4, 3)$, # of iterations required for ranges of position values

Position value	States	Mean iter.	Position value	States	Mean iter.
1.0 - 1.1	50,044	3.31	1.9 - 2.0	6,326	8.27
1.1 - 1.2	21,147	3.41	2.0 - 2.1	8,096	8.32
1.2 - 1.3	8,842	3.73	2.1 - 2.2	8,797	8.61
1.3 - 1.4	13,535	4.32	2.2 - 2.3	7,598	8.90
1.4 - 1.5	9,617	5.00	2.3 - 2.4	5,210	9.18
1.5 - 1.6	5,829	5.88	2.4 - 2.5	2,574	9.43
1.6 - 1.7	3,524	6.70	2.5 - 2.6	684	9.61
1.7 - 1.8	3,157	7.51	2.6 - 2.7	75	9.76
1.8 - 1.9	4,321	8.10	Total	159,376	5.13

6   Conclusion

We used a bipartite graph to abstract a one-player probabilistic game with the goal to maximize some expected game value or to minimize the expected cost. We investigated the game of one-player Can't Stop, and proved that its optimal solution exists and is unique. To obtain the optimal solution, we developed a new approximation algorithm that converges quadratically, by incorporating Newton's method with retrograde analysis.

We successfully constructed the databases of the simplified models with 3-sided and 4-sided dice. The optimal solution of one-player Can't Stop can be used as the approximate solution of two-player Can't Stop. Two-player Can't Stop can be presented as a four-partite graph. Given a position, a function f is defined as the probability that the first player wins the game. The goal is to build a database representing f. In addition to building the optimal databases of one-player Can't Stop, we plan to tackle two-player Can't Stop in the future.

Acknowledgments

The authors thank David Slater for pointing out that the algorithm can be improved by taking advantage of symmetry.

References

1. Glenn, J.: An Optimal Strategy for Yahtzee. Technical Report CS-TR-0002, Loyola College in Maryland, 4501 N. Charles St, Baltimore MD 21210, USA (May 2006)
2. Romein, J.W., Bal, H.E.: Solving the Game of Awari using Parallel Retrograde Analysis. IEEE Computer 36(10), 26–33 (2003)
3. Schaeffer, J., Björnsson, Y., Burch, N., Lake, R., Lu, P., Sutphen, S.: Building the Checkers 10-Piece Endgame Databases. In: van den Herik, H.J., Iida, H., Heinz, E.A. (eds.) 10th Advances in Computer Games (ACG10), Many Games, Many Challenges, pp. 193–210. Kluwer Academic Publishers, Boston, USA (2004)
4. Thompson, K.: 6-Piece Endgames. ICCA Journal 19(4), 215–226 (1996)
5. Woodward, P.: Yahtzee: The Solution. Chance 16(1), 18–22 (2003)

Appendix: Can't Stop Rules

We summarize the game rules of Can't Stop, largely taken from Wikipedia[6]: The game equipment consists of four dice, a board, a set of eleven markers for each player, and three neutral markers. The board consists of eleven columns of spaces, one column for each of the numbers 2 through 12. The columns (respectively) have 3, 5, 7, 9, 11, 13, 11, 9, 7, 5 and 3 spaces each. The object of the game is to move your markers up the columns, and be the first player to complete three columns.

On a player's turn he[7] rolls all four dice. He then divides the four dice into two pairs, each of which has an associated total. (For example, if he rolled 1 - 3 - 3 - 4 he could make a 4 and a 7, or a 5 and a 6.) If the neutral markers are off of the board then they are brought on to the board on the columns that correspond to these totals. If the neutral markers are already on the board in one or both of these columns then they are advanced one space upward. If the neutral markers are on the board, but only in columns that cannot be made with any pair of the current four dice, then the turn is over and the player gains nothing.

After moving the markers the player chooses whether or not to roll again. If he stops, then he puts markers of his color in the locations of the current neutral markers. If on a later turn he restarts this column, he starts building from the place he previously claimed. If he does not stop then he must be able to advance at least one of the neutral markers on his next roll, or all progress on this turn is lost.

When a player reaches the top space of a column, that column is won, and no further play in that column is allowed. The first player to complete three columns wins the game.

---

[6] http://en.wikipedia.org/wiki/Can't_Stop
[7] We use 'he' when both 'she' and 'he' are possible.

Improving Depth-First PN-Search: $1 + \varepsilon$ Trick

Jakub Pawlewicz and Łukasz Lew

Institute of Informatics,
Warsaw University, Warsaw, Poland
{pan,lew}@mimuw.edu.pl

Abstract. Various efficient game problem solvers are based on PN-Search. Especially depth-first versions of PN-Search like DF-PN or PDS – contrary to other known techniques – are able to solve really hard problems. However, the performance of DF-PN and PDS decreases drastically when the search space significantly exceeds the available memory. A straightforward enhancement trick to overcome this problem is presented. Experiments on Atari Go and Lines of Action show great practical value of the proposed enhancement.

1 Introduction

In many two-person zero-sum games with perfect information we frequently encounter game positions there often with a non-trivial but forced win for one of the players. To compute such a winning strategy a large tree search is performed. The most profitable algorithms that successfully perform the task of finding a solution are the Proof-number search algorithms [1]. They may search as deep as 20 plies and even more deeply. However, the usage of the basic PN-Search is limited to rather short runs because of the high memory requirements. A straightforward improvement is the PN^2 algorithm which was thoroughly investigated in [3]. Yet, it is still a best-first algorithm and therefore needs memory to work with. So, the length of a single run is still limited. As a further improvement, several depth-first versions of PN-Search have appeared to overcome the memory requirements problem. They were successful in many fields. Seo [7] successfully applied it to many difficult problems in his Tsume-Shogi solver using PN*. The PDS algorithm – an extension of PN* was developed by Nagai [5]. Another successful algorithm is DF-PN [6] which is a straightforward transformation of PN-Search to a depth-first algorithm.

Yet even, all these methods lose their effectiveness on very hard problems when the search lasts so long that the number of positions to explore significantly exceeds the available memory. The methods spend most of the time repeatedly re-producing trees stored in the transposition table but overwritten by a search in other branches of a game tree. So, we must admit that this kind of performance leak does not only occur in alpha-beta search.

From these results, there is room for improvement of these methods. For instance, Winands et al. [9] presented a method called PDS-PN used in the LOA program MiA. This variation was created by taking the best of PN^2 and

H.J. van den Herik et al. (Eds.): CG 2006, LNCS 4630, pp. 160–171, 2007.
© Springer-Verlag Berlin Heidelberg 2007

PDS. Kishimoto and Müller [4] successfully applied DF-PN in their tsume-Go problems solver. They enhanced DF-PN by additional threshold increments.

This paper presents a more general approach of threshold increments, which reduces the number of tree reproductions during the search and results in a quite efficient practical enhancement. The enhancement is applicable both to DF-PN and PDS and possibly to other variants of tree-traversing algorithms.

A deeper understanding of DF-PN is needed to understand the merits of the enhancement. Section 2 precisely describes the transformation of the PN-Search algorithm to a depth-first search algorithm. Section 3 provides a further insight into DF-PN search and shows its weak point along with a remedy. The same section presents also an application of our enhancement to PDS. Section 4 presents results of experiments. The last section concludes.

2   A Depth-First Transformation of PN-Search

Section 2.1 briefly describes PN-Search. Section 2.2 describes DF-PN as a depth-first transformation of PN-Search.

2.1   PN-Search

For detailed description of PN-Search we refer to [1]. We recall only a few selected properties essential for an analysis in the later sections.

The algorithm maintains a tree in which each node represents a game position. With each node we associate two numbers: the *proof-number* (PN) and the *disproof-number* (DN).

The PN(DN) of a node v is the minimum number of leaves in the subtree rooted at v valued *unknown* such that if they change theirs value to true(false) then the value of v would also change to true(false).

In other words, the PN(DN) is a lower bound on the number of leaves to expand in order to prove(disprove) v.

PN and DN for a proved node are set to 0 and $+\infty$ and for a disproved node are set to $+\infty$ and 0. In an unsolved leaf node we set both PN and DN to 1. In an unsolved internal node PN and DN can be calculated recursively as shown in Fig. 1. A square denotes an OR node and a circle denotes an AND node.

The algorithm iteratively selects a leaf and expands it. To minimize the total number of expanded nodes in the search it chooses a leaf to expand in a such way that proving(disproving) it decreases the root's PN(DN) by 1. Such a leaf is called the *most proving node* (MPN).

It can be easily shown that an MPN always exists. This leaf may be found by traversing downwards through the tree and choosing a child only on the basis of the PNs and DNs of the node's children. In a node of type OR(AND) we choose a child with the minimum PN(DN). That is the child with the same PN(DN) as its parent.

After expanding the MPN, PNs and DNs are updated by going back on the path up to the root. The algorithm stops when it determines a game value of

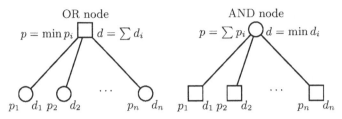

Fig. 1. In an OR node PN is a minimum of children's PNs while DN is a sum of children's DNs. In an AND node PN is a sum of children's PNs while DN is a minimum of children's DNs.

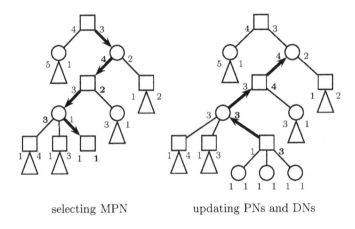

selecting MPN   updating PNs and DNs

Fig. 2. Selecting MPN and updating PNs and DNs

the root, i.e., if one of the root's numbers will be infinity while the other will drop to zero. An example of selecting an MPN and updating PNs and DNs is shown in Fig. 2.

2.2 DF-PN

As in Fig. 2 we see that updating PNs and DNs can be stopped before we reach the root. It happens for instance when updating a node does change neither PN nor DN. In that case we may begin the search of the next MPN from the last updated node instead of from the root.

In fact we can shorten the way up even more. If the next MPN is in a subtree rooted in the node currently visited we can suspend update of the parent and further ancestors. Observe that we need valid values of PNs and DNs of the ancestors only while traversing downwards and selecting the MPN. Then, in general, we can suspend updating the ancestors as long as the MPN resides in the subtree of the node.

To take advantage of the above observation we introduce PN and DN thresholds. The thresholds are stored only for nodes along the path from the root to the current node. Let p and d be the PN and DN of node v. We want to define the thresholds pt and dt for node v in such a way that if $p < pt$ and $d < dt$ then there exists an MPN in the subtree rooted at v and conversely if there exists an MPN in the subtree rooted at v then $p < pt$ and $d < dt$.

We will now determine the rules for setting the thresholds. For the root we set the thresholds to $+\infty$. Clearly, the condition $p < +\infty$ and $d < +\infty$ holds only if the tree is not solved.

Now we take a closer look on what happens in an internal OR node (Fig. 3). Let p and d be the node's PN and DN respectively. Let pt and dt be

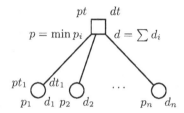

Fig. 3. Visiting the first child in an OR node and setting the thresholds

its thresholds. Assume the node has n children. Assume the i-th child's PN and DN equal p_i and d_i. Without loss of generality we assume $p_1 \leq p_2 \leq \ldots \leq p_n$.

The subtree where the MPN resides is rooted at the child with the minimum PN. Since p_1 is the smallest value, MPN lies in the leftmost subtree, so we are going to visit the first child. We have to set the thresholds pt_1 and dt_1 for this child such that if the PN or DN reaches its threshold (i.e., if $p_1 \geq pt_1$ or $d_1 \geq dt_1$) then MPN must lie outside this child's subtree.

Now, we set constraints for pt_1 to deduce the actual value. When p_1 exceeds p_2, then the second child will have the minimum PN, and MPN will no longer lie in the first child's subtree. Hence $pt_1 \leq p_2 + 1$. When p_1 reaches pt, but does not exceed p_2, then the PN p in the parent will also reach its threshold pt, and by pt threshold definition, MPN will no longer be a descendant of the parent and thus it will not lie in the child's subtree. Hence, the second constraint is $pt_1 \leq pt$. These two constraints give us the formula $pt_1 = \min(pt, p_2 + 1)$.

Consequently, when d_1 increases such that d reaches dt, then again MPN will be outside the current subtree. Now we calculate how d changes when d_1 changes to d_1'. Let d' denote DN of the parent, after DN of the first child has changed from d_1 to d_1'. Then we have $d' = d + (d_1' - d_1)$. We are interested whether $d' \geq dt$. Replacing d' and rewriting the inequality we may obtain the answer. That is: if $d_1' \geq dt - d + d_1$ then $d' \geq dt$. Therefore we have a formula for the DN threshold $dt_1 = dt - d + d_1$.

In summary, we arrive at the following formulas for an OR node for the first child's thresholds:

$$pt_1 = \min(pt, p_2 + 1)\,, \tag{1}$$

$$dt_1 = dt - d + d_1\,. \tag{2}$$

It remains to show that for the above thresholds the inequalities $p_1 < pt_1$ and $d_1 < dt_1$ hold if and only if MPN is in the first child's subtree. We have already seen that if $p_1 \geq pt_1$ or $d_1 \geq dt_1$ then MPN is not in the child's subtree. So assume $p_1 < pt_1$ and $d_1 < dt_1$. Then $p < pt$ and $d < dt$, thus MPN lies in the parent's subtree. Moreover $p_1 < p_2 + 1$. This is the same as $p_1 \leq p_2$. Thus the first child has the smallest PN among all children. Therefore MPN lies in the first child's subtree.

Similarly, we arrive at the formulas for an AND node for the first child's thresholds (assuming $d_1 \leq d_2 \leq \ldots \leq d_n$):

$$pt_1 = pt - p + p_1\,, \tag{3}$$

$$dt_1 = \min(dt, d_2 + 1)\,. \tag{4}$$

A main advantage is that using thresholds we can suspend updates as long as it is possible.

A second important advantage of this approach is the possibility of switching from (1) maintaining a whole search tree to (2) using a transposition table to store positions' (nodes') PN and DN.

If in such implementation the algorithm visits a node, we may try to retrieve its PN and DN from the transposition table searching for an early cut. In case of failure we initialize the node's PN and DN the same way as we do it for a leaf allowing the further search to reevaluate them. The resulting algorithm called DF-PN [6] is a depth-first algorithm, and it can be implemented in a recursive fashion.

The main property of DF-PN is the following. If all nodes can be stored in a transposition table, then the nodes are expanded in the same order as in the standard PN-Search. Of course, in DF-PN, we are not forced to store all nodes, and usually we store only a fraction of them, with a slight loss of efficiency.

3   Enhancement

Section 3.1 shows a usual scenario for which DF-PN has a poor performance. Section 3.2 shows our enhancement (i.e., the $1 + \varepsilon$ trick) applied to DF-PN. Section 3.3 describes an analogous improvement of PDS.

3.1   Weak Point of DF-PN

We remind that in case of a failure in retrieving the children's values from a transposition table, DF-PN initializes the children's PNs and DNs to 1 and re-search their values. Usually in an OR(AND) node with a large subtree all the children's PNs (DNs) are very similar, because DF-PN searches the child with the smallest PN (DN) and returns as soon as it exceeds the second smallest number.

Let us consider the following typical situation during a run of DF-PN. Assume we are in an OR node with at least two children. Assume further that the threshold is big and search lasts so long that most of the investigated nodes do not fit in the transposition table.

After some time, the searched tree will be so large, that the algorithm will not be able to store most of the searched nodes. Now DF-PN will make a recursive tree search for the first child with a PN threshold fixed to the second child's PN plus one. Eventually, search will return but due to the weak threshold the new PN will be only slightly greater than before the recursion. For an example see Fig. 4.

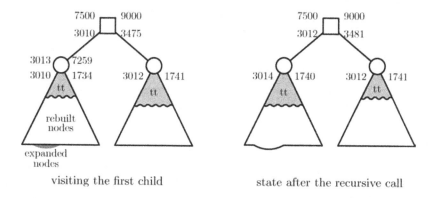

visiting the first child   state after the recursive call

Fig. 4. Example of a single recursive call in an OR node during a run of DF-PN. The numbers are potential values during the search. Gray part of a tree marked as *tt* denotes nodes stored in a transposition table. The left picture shows what happens during the recursive call for the first child. The state after the call is shown in the right picture.

Then the control goes back to the parent level and we call DF-PN for its second child, again setting the PN threshold to its sibling's PN plus one. After expensive reconstruction of the second child's tree, its PN increases insignificantly and we will have to switch again to its sibling. However, most information from the previous search in the first child has been lost due to insufficient memory.

We see that for each successive recursive call, we have to rebuild almost the whole child's tree. Usually the number of recursive calls in the parent node is *linear* to the parent's PN threshold.

3.2   The $1 + \varepsilon$ Trick

The above example shows that when we are in an OR node, setting the PN threshold to the number one larger than p_2 can lead to quite a large number of visits in a single child, causing multiple reconstructions of a tree rooted in that child. To be more effective we should spend more time in a single node, doing some search in advance.

We have a constraint $pt_1 \le p_2 + 1$ for the child's PN threshold in an OR node. We can relax that constraint somewhat to a small multiplicity of p_2, for example to $1 + \varepsilon$, where ε is a small real number greater than zero. Thus we change the constraint to $pt_1 \le \lceil p_2(1 + \varepsilon) \rceil$ and the old formula (1) transforms into the new formula (5) for the child's PN threshold in an OR node

$$pt_1 = \min(pt, \lceil p_2(1 + \varepsilon) \rceil). \tag{5}$$

So, after each recursive call the child's PN increases by a constant factor rather by a constant addend. More precisely after the call either the parent's DN threshold is reached or the child's PN increases by at least $1 + \varepsilon$ times. Therefore, a single child can be called at most $\log_{1+\varepsilon} pt = O(\log pt)$ times before reaching the parent's PN threshold. As a consequence the described trick has a nice property of reducing the number of recursive calls from linear to *logarithmic* in the parent's PN threshold.

This enhancement not only improves the way a transposition table is used, but also reduces the overhead of multiple replaying the same sequences of moves. Here we may observe that using the presented trick we lose the property of visiting nodes in the same order as in PN-Search.

3.3   Application to PDS

The $1 + \varepsilon$ trick is also applicable to PDS. In PDS we use two thresholds pt and dt like in DF-PN. The main difference is their meaning. The algorithm remains in a node until both thresholds are reached or the node is solved. PDS introduces the notion of proof-like and disproof-like nodes. When making a recursive call at a child with PN p_1 and DN d_1, it sets the thresholds $pt_1 = p_1 + 1$ and $dt_1 = d_1$ if the node is proof-like, and $pt_1 = p_1$ and $dt_1 = d_1 + 1$ if the node is disproof-like. PDS uses a straightforward heuristic to decide whether the node is proof-like or disproof-like. We refer the reader to [5] for the details.

The similar weakness as in DF-PN, described in 3.1, harms PDS. Here we apply our technique by setting the thresholds: $pt_1 = \lceil p_1(1 + \varepsilon) \rceil$, $dt_1 = d_1$ for a proof-like child, and $pt_1 = p_1$, $dt_1 = \lceil d_1(1 + \varepsilon) \rceil$ for a disproof-like child.

4   Experiments

Below we examine the practical efficiency of DF-PN and PDS with and without the presented enhancement. We focus on the solving times for various set-ups. First, Subsection 4.1 describes the background for the performed experiments. Then the results of these experiments are described. In Subsection 4.2 we test the influence of the size of a transposition table. In Subsection 4.3 we compare the algorithms under tournament conditions. In Subsection 4.4 we explore capabilities to solve hard problems.

4.1   Experimental Environment

We choose two games for our experiments. The first one is Atari Go, the capture game of Go. In Subsection 4.2 we take as a starting position a 6×6 board with a crosscut in the centre. The second game is Lines of Action. For more information we refer to Winands' web page [8]. The rules and testing positions, used in Subsection 4.3 and 4.4, were taken from that web page.

Our implementations of four search methods in the Atari Go and LOA games do not have any game-specific enhancements. For a transposition table, TwoBig scheme [2] is used.

For the accelerated version of DF-PN with the $1 + \varepsilon$ trick, ε was set empirically to 1/4. With larger values of ε, enhanced DF-PN tends to over-explore significantly some nodes. With smaller values, enhanced DF-PN is usually slower.

In PDS, spending more time in one child is a common behavior because the ending condition requires to exceed both thresholds simultaneously. Usage of $1 + \varepsilon$ trick in PDS makes this deep exploring behavior even more exhaustive, which often leads to over-exploring. Therefore ε should be much smaller in enhanced version of PDS. We found 1/16 as the best ε value.

All experiments were performed on 3GHz Pentium 4 with 1GB RAM under Linux.

4.2   The Size of a Transposition Table, Tested on Atari Go

We run all four methods with different transposition table sizes. The results are shown in Fig. 5 and exact times for the sizes between 2^{12} and 2^{22} nodes are shown in Table 1.

Table 1. Solving times in seconds of Atari Go 6×6 with a crosscut

Algorithm	2^{12}	2^{13}	2^{14}	2^{15}	2^{16}	2^{17}	2^{18}	2^{19}	2^{20}	2^{21}	2^{22}
					TT size in nodes						
DF-PN with $1 + \varepsilon$ trick	672	236	95	98	55	70	32	77	38	54	68
DF-PN	–	–	3403	279	189	187	104	168	111	113	106
PDS with $1 + \varepsilon$ trick	–	4895	3246	1017	468	440	527	350	400	235	269
PDS	–	–	–	4057	1448	1124	776	494	353	261	195

Obviously enhanced DF-PN is the fastest method and plain DF-PN is the second fastest.

Setting the size greater than 2^{20} does not noticeably affect the times. In that range there is no remarkable difference between plain and enhanced PDS. For sizes smaller than 2^{20}, enhanced PDS becomes faster than plain PDS.

A noticeable drop of performance can be observed when the size is below 2^{16}. Within a 2 hour time limit the following results are te be reported: (1) PDS is unable to solve the problem for transposition table with a size of 2^{14} nodes, (2) DF-PN with a size of 2^{13} nodes and (3) the enhanced PDS with a size of 2^{12} nodes.

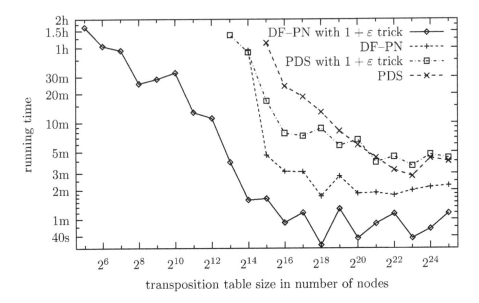

Fig. 5. Solving times of Atari Go 6 × 6 with a crosscut

The enhanced versions are much better for really small transposition tables. For sizes 2^{14} and smaller, enhanced DF-PN is far better than any other method. Enhanced DF-PN is able to solve the problem almost not using a transposition table at all. With a memory of 256 nodes it needed 1535 seconds and with a memory of 32 nodes it needed 5905 seconds. Of course there is a substantial information stored in the local variables in each recursive call.

4.3 Efficiency Under Tournament Conditions, Tested on a Set of Easy LOA Positions

We have already seen that DF-PN with the $1 + \varepsilon$ trick performs excellently when the search space significantly exceeds the size of a transposition table. In this Subsection we check the practical value of the methods on the set of 488 LOA positions. The purpose of this test is to evaluate the efficiency of solving the positions with tournament time constraints, as it is desired in the best computer programs.

The size of a transposition table is set to 2^{20} nodes and should fit into the memory of most computers. The results are shown in Fig. 6. The figure was created by measuring solving time for each method for every position from the set. Then for every time limit we can easily find the number of solved positions with time not exceeding the limit. Exact numbers of solved positions for selected time limits are shown in Table 2.

Here enhanced DF-PN is clearly the most efficient method, plain DF-PN is the second best and both PDS versions are the least efficient. The difference between enhanced PDS and plain PDS is unnoticeable.

Table 2. Numbers of solved positions from the set `tscg2002a.zip` [8] for selected time limits

Algorithm	Time limit									
	0.5s	1s	2s	5s	10s	20s	30s	1m	2m	5m
DF-PN with $1 + \varepsilon$ trick	214	278	343	410	438	457	468	478	482	486
DF-PN	184	246	304	379	419	449	457	471	479	486
PDS with $1 + \varepsilon$ trick	100	154	217	299	358	414	430	450	471	481
PDS	102	144	214	300	365	409	425	451	466	480

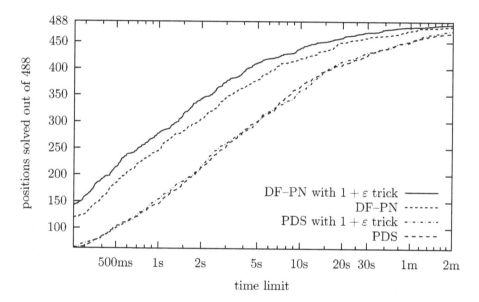

Fig. 6. Numbers of solved easy LOA positions from the set `tscg2002a.zip` [8] for given time limit

4.4 Efficiency of Solving Hard Problems, Tested on a Set of Hard LOA Positions

This experiment is aimed at checking our ability of solving harder problems in reasonable time. The 286 test positions were taken from [8]. Again we define the size of a transposition table to be 2^{20} nodes. The results are shown in Fig. 7 and the exact numbers of solved positions for selected time limits are shown in Table 3.

Again, as in the previous test, the enhanced DF-PN is the most efficient method, plain DF-PN is the second best, and both PDS versions are the least efficient. The difference between enhanced PDS and plain PDS is now noticeable. For each time limit greater than 90 seconds enhanced PDS solves more positions than plain PDS. It shows the advantage of enhanced PDS over plain PDS for harder positions.

To illustrate the speed differences in numbers for each two methods we calculated a geometric mean of the ratios of solving times (see Table 4). The geometric

Table 3. Numbers of solved positions from the set `tscg2002b.zip` [8] for selected time limits.

Algorithm	Time limit										
	5s	10s	20s	30s	1m	2m	3m	5m	10m	20m	30m
DF-PN with $1 + \varepsilon$ trick	44	107	158	182	228	258	272	280	285	286	286
DF-PN	25	64	123	154	198	241	258	273	282	284	285
PDS with $1 + \varepsilon$ trick	1	18	52	80	138	186	212	228	257	272	278
PDS	3	22	56	83	134	179	206	224	252	265	274

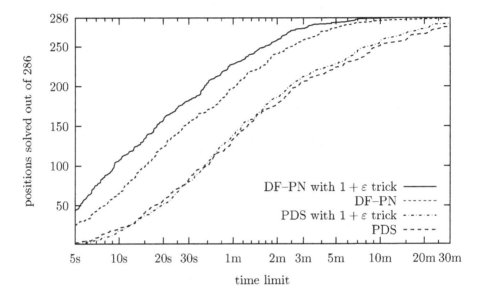

Fig. 7. Numbers of solved hard LOA positions from the hard set `tscg2002b.zip` [8] for given time limit

Table 4. Overall comparison for positions from the set `tscg2002b.zip`. The number r in the row A and the column B says A is r times faster than B in average.

	DF-PN	enhanced DF-PN	PDS	enhanced PDS
DF-PN	1.00	0.63	2.83	2.64
enhanced DF-PN	1.58	1.00	4.46	4.17
PDS	0.35	0.22	1.00	0.93
enhanced PDS	0.38	0.24	1.07	1.00

mean is more appropriate for averaging ratios than the arithmetic mean because of the following property: if A is r_1 times faster than B, B is r_2 times faster than C, and A is r_3 times faster than C then $r_3 = r_1 r_2$.

5   Conclusions and Future Work

The $1 + \varepsilon$ trick has been introduced to enhance DF-PN. We have shown that the trick can also be used in PDS. The experiments showed a noticeable speedup when the search space significantly exceeds the size of a transposition table.

The trick is particularly well-suited to DF-PN, since the experiments have shown a large advantage of enhanced DF-PN over the other methods. The AtariGo experiment has shown that this method performs extremely well in low memory conditions. Moreover, DF-PN with the $1 + \varepsilon$ trick was the most efficient method in the "real-life" experiment on the LOA positions so it should be valuable in practice. The $1 + \varepsilon$ trick is possibly applicable to other threshold-based depth-first versions of PN-Search.

The nice property of the enhancement is that it can be added to existing implementations with little effort. Its value has to be confirmed in other games different from Atari Go and LOA. We notice that in other games and in other depth-first variants of PN-Search, the value of ε can be different, and it should be further investigated.

We hope that the presented enhancement becomes attractive for a brute-force search for solving games. However, there are still some weak points in PN-based methods and more work for improvements is required to make depth-first PN-Search even more useful.

References

1. Allis, L.V., van der Meulen, M., van den Herik, H.J.: Proof-Number Search. Artificial Intelligence 66, 91–124 (1994)
2. Breuker, D.M., Uiterwijk, J.W.H.M., van der Herik, H.J.: Replacement Schemes and Two-Level Tables. ICCA J. 19(3), 175–180 (1996)
3. Breuker, D.M., Uiterwijk, J.W.H.M., van der Herik, H.J.: The PN²-Search Algorithm. In: van den Herik, H.J., Monien, B. (eds.) 9th Advances in Computer Games (ACG9), pp. 115–132. Department of Computer Science, Universiteit Maastricht, Maastricht, The Netherlands (2001)
4. Kishimoto, A., Müller, M.: Search Versus Knowledge for Solving Life and Death Problems in Go. In: Twentieth National Conference on Artificial Intelligence (AAAI-05), pp. 1374–1379 (2005)
5. Nagai, A.: A New AND/OR Tree Search Algorithm using Proof Number and Disproof Number. In: Proceeding of Complex Games Lab Workshop, pp. 40–45, Tsukuba, ETL (November 1998)
6. Nagai, A.: Df–pn Algorithm for Searching AND/OR Trees and its Applications. Ph.d. thesis, The University of Tokyo, Tokyo, Japan (2002)
7. Seo, M., Iida, H., Uiterwijk, J.W.H.M.: The PN*–Search Algorithm: Application to Tsume-Shogi. Artificial Intelligence 129(1–2), 253–277 (2001)
8. Winands, M.H.M.: Mark's LOA Homepage (2007), http://www.cs.unimaas.nl/m.winands/loa/
9. Winands, M.H.M., Uiterwijk, J.W.H.M., van den Herik, H.J.: An Effective Two-Level Proof-Number Search Algorithm. Theoretical Computer Science 313(3), 511–525 (2004)

Search Versus Knowledge Revisited Again

Aleksander Sadikov and Ivan Bratko

Faculty of Computer and Information Science,
University of Ljubljana, Ljubljana, Slovenia
{aleksander.sadikov,bratko}@fri.uni-lj.si

Abstract. The questions focusing on diminishing returns for additional search effort have been a burning issue in computer chess. Despite a lot of research in this field, there are still some open questions, e.g., what happens at search depths beyond 12 plies, and what is the effect of the program's knowledge on diminishing returns? The paper presents an experiment that attempts to answer these questions. The results (a) confirm that diminishing returns in chess exist, and more importantly (b) show that the amount of knowledge a program has influences when diminishing returns will start to manifest themselves.

1 Introduction

The phenomenon of diminishing returns for additional search effort in chess has been a burning issue in the game-playing scientific community. The consensus was that diminishing returns, as in all the other games observed, do exist in chess as well. However, for a very long time the experimental results did not back up this opinion convincingly. This even drove research in another direction, namely, to find reasons what it is that masks the manifestation of diminishing returns in chess [7,8]. Finally, Heinz [5,6] introduced rigorous statistical methods to self-play experiments, and with that (a) pointed out that all previous experiments lacked statistical significance, and (b) with his new self-play experiment that featured much longer matches between successive versions of FRITZ 6 statistically proved the existence of diminishing returns in chess [3,4]. Later, Haworth [2] confirmed and slightly improved Heinz's statistical analysis.

There are, however, still a few questions that in our opinion should be asked. What happens at search depths beyond 12 plies? And what is the effect of knowledge the program has, on diminishing returns? Thus, our goal was to investigate the effect of search depth on the quality of play, and *not* be limited in search depth. We studied this effect for evaluation functions with various levels of knowledge. In this context, we were especially interested in the phenomena of diminishing returns for additional search effort.

How is it possible not to be limited in search depth when doing such experiments? After all, as Heinz pointed out, one has to play a huge number of games to obtain statistically significant results. We had to come up with a completely different strategy. We made very deep searches (unlimited for all practical purposes) possible by concentrating on chess endgames with a limited number of

H.J. van den Herik et al. (Eds.): CG 2006, LNCS 4630, pp. 172–180, 2007.

pieces. The number of pieces defines the space of all possible legal positions — and for four-piece endgames it is no problem to store all legal positions in an array in computer memory. Searching then simply involves performing repetitive 2-ply (or even 1-ply) minimax calls on the fields in this array. The other idea we came up with was to simulate heuristic functions of various quality (knowledge contents) by corrupting the perfect evaluation function in the form of endgame tablebase in a controlled manner. This enables us to have at our disposal programs of comparable knowledge. As a bonus, with this approach we sidestep a problem that Junghanns et al. [7,8] complained about, namely the problem of not having an oracle for chess to compare the decisions of their programs with. Actually, that is exactly how we arrived at the idea to simulate heuristic functions in the first place.

We do sacrifice something in return, though: (a) we limit ourselves only to chess endgames and (b) the heuristics we use, while real-valued, are somewhat different from the usual ones used in chess-playing programs.

The paper is organized as follows. Section 2 explains in detail our experimental design, while Sect. 3 presents the results and comments on them. Some conclusions are given in the last section.

2   Experimental Design

This section describes the experimental setup. First, the evaluation function is characterized in Section 2.1. Second, the heuristic used is explained in greater detail by Section 2.2. Finally, Section 2.4 describes the variables measured in the experiment.

2.1   The Evaluation Function

As already mentioned in the introduction, we have at our disposal an absolutely correct evaluation function for an endgame — we chose the King-Bishop-Bishop-King (KBBK) endgame. The tablebase tells us how many moves are needed to reach mate in the case that both players play optimally, and is measured in moves. It is in the form of a database that consists of all possible legal positions and their evaluations, and can be generated with a tablebase construction algorithm based on retrograde analysis [13,14]. There are two special cases: value 0 means Black is mated and value 255 means that Blackhas a draw (either the position is a stalemate or Black can capture one of the Bishops).

It should be noted that length-to-mate as an evaluation function requires slight modification when used with minimax. The reason is that length-to-mate decreases along the best-play line, whereas minimax backed-up values preserve the value along the best-play line. To account for this peculiarity of length-to-mate, in the experiments we used the rule $1 + \max(v_i)$ instead of $\max(v_i)$.[1]

---

[1] A constant 1 is added to reflect an additional ply when going one layer deeper in the minimax tree.

As in our experiments all the lines in the game tree search were of equal length, this modification merely amounts to adding a constant to all the position values at the same level.

It may be argued that distance-to-mate is a rather artificial evaluation function from the usual point of view in chess where the task is to win, and not necessarily to win in the quickest way. Therefore, in chess, an evaluation function is usually interpreted as an indicator of the probability to win. However, distance-to-mate can in fact also be interpreted as such an indicator. Although virtually all the positions in KBBK are won for White, an imperfect player may have difficulties in actually mating in 50 moves.[2] Such an imperfect player will have much better chances to win in a position where mate is possible in two moves, than in a position that requires 10 or more moves to mate — this reasoning corresponds with our notion that positions that take more moves to reach mate are harder. Realistic evaluation functions for chess-playing programs also typically prefer shorter paths to mate (win) and should thus also have some correlation with our distance to mate.

2.2 Simulation of Heuristic Errors

For the purpose of simulating the imperfection of heuristic evaluation functions, we corrupted the ideal evaluation function in a controlled manner. Our method of doing this is as follows. We take a position value and add to it a certain amount of Gaussian noise, described with the formula:

$$P(x) = \frac{1}{\sigma\sqrt{2\pi}}e^{-(x-\mu)^2/(2\sigma^2)} \ . \tag{1}$$

The formula gives the probability $P(x)dx$ that given the correct evaluation μ and standard deviation σ, the random errors, $x \in \mathbf{R}$, will take on a value in the range $[x, x + dx]$. The error of the new evaluation is $\mu - x$. We do this for all positions in the database, including the positions where Black is already mated (special value 0). The corruption is symmetrical, meaning that there is practically equal chance that the new evaluation will be optimistic or pessimistic. We allow x to take on a negative value — in this way we are able to preserve symmetry for positions that have true values close or equal to 0.

The level of corruption is regulated by the parameter σ, the standard deviation, which controls the dispersion of the corrupted values x around the correct values μ. The standard deviation is measured in moves. For example, if σ equals 1.0, this means that approximately two thirds of corrupted evaluations are within 1.0 move around the true evaluation and over 95% of corrupted evaluations are within 2.0 moves (two standard deviations) around the true evaluation.

---

[2] The number of moves to mate allowed by the rules of chess is 50 for the KBBK endgame.

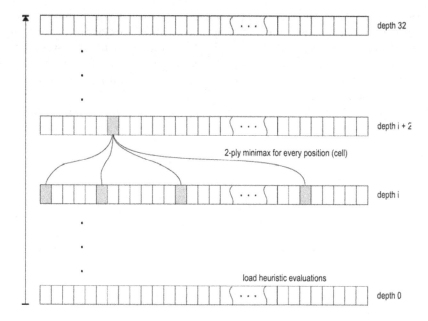

Fig. 1. An efficient way of computing backed-up values

2.3   Search Engine

The search engine we used is the standard fixed-depth minimax search. The only built-in knowledge it has is the ability to detect fatal errors for White (stalemates and losing a piece). The true value for stalemate or for losing a piece would correspond to a distance to mate larger than 50 because of the draw by 50-move rule. Such a high value is virtually impossible to corrupt with noise to the extent when it becomes comparable to other values. Therefore, the search engine was assumed reliably to detect stalemates and losses of a piece. In contrast, the ability to detect mate is not given.[3] If mate is encountered during the search, it is evaluated using the corrupted evaluation function just like every other position. This evaluation does not necessarily match the true value for mate, which is zero.

We can quickly search to high search depths of 32 plies, and beyond if desired, by exploiting the fact that the KBBK endgame only has a comparatively small number of positions (approximately 4 million under some symmetries) which we can all store in a sort of transposition table. Each unique position is given its own unique index. This index specifies its place in the transposition table, which is big enough to hold all the positions. The table thus holds only position evaluations: the actual position is uniquely determined by its location (index) in the table. We start at depth 0 by loading the values from a

---

[3] Even if the ability to detect mate is given, the observations are qualitatively the same as in the paper.

(corrupted) database, then move on to depth 2, perform a 2-ply minimax search and use the results of the previous depth as evaluations of the leaves, store results of depth 2 search, move on to depth 4 and so forth. The procedure is demonstrated in Figure 1. This back-up procedure, similar to endgame tablebase construction algorithm, is much more efficient than forward search. For example, a classic search engine would require 4 million 32-ply minimax searches to calculate the evaluations of all positions of depth 32. Instead, our method requires sixteen times 4 million 2-ply minimax searches to accomplish the very same task.

2.4   Variables Observed

Our goal is to vary the search depth and the quality of heuristic function and observe how well a program would play under these conditions. Therefore White was guided by a corrupted evaluation function. Additionally, White was allowed to use a simple mechanism to avoid repeating the same position over and over again. The mechanism kept a list of all positions that already occurred in the game and if the position was to be repeated a different move was selected — the next best move according to the evaluation function. In contrast, Black was always playing optimally.

We measured the quality of play as the average number of moves above what an optimal white player (playing with a non-corrupted database) would need. This statistic is computed as White's average performance loss (APL), that is, the difference between the number of moves needed by White for all sample positions and the number of moves needed for all sample positions using optimal play, divided by the number of positions in the sample (10,000):

$$\text{APL} = \frac{\sum_{pos \in \mathcal{S}} h(pos) - \sum_{pos \in \mathcal{S}} o(pos)}{|\mathcal{S}|} \; . \tag{2}$$

$\mathcal{S}$ is a sample of 10,000 legal KBBK positions[4] of difficulty greater than 10 (moves-to-mate), $h(pos)$ returns the number of plies that a computer player using a heuristic evaluation function (with some initial corruption level σ) needs to mate the ideal opponent starting from position pos, and $o(pos)$ is the number of plies a perfect player needs to mate the ideal opponent from the starting position pos. APL is measured in plies. The better the quality of play, the lower the value of APL, and vice versa.

The results so obtained vary somewhat because of the random process of simulating heuristic errors. To obtain a more stable result we repeated the experiments several (ten) times for every combination of search depth and magnitude of heuristic errors and averaged the results. The sample $\mathcal{S}$ was the same for all settings.

---

[4] It would be completely possible to increase the sample to 100,000 positions or even to all legal KBBK positions, but we deem that a waste of resources as this would have only a miniscule effect on the results.

3   Experimental Results and Discussion

The experimental results are presented in Figs. 2 and 3. The latter is a zoomed in detail of the first one to show more precisely what happens at higher search depths. The x-axis represents search depth in plies and the y-axis the APL statistic also measured in plies. The curves in the figures represent evaluation functions of different heuristic quality. They are marked in the legend by their corresponding level of noise σ. Each curve is in fact an averaged curve of ten evaluation functions with the same σ.

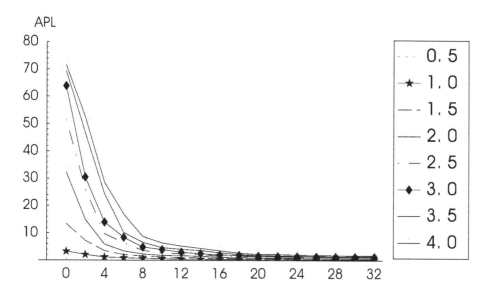

Fig. 2. Quality of play using corrupted evaluation functions

3.1   Knowledge Versus Search

The experimental results show, as expected, that the quality of play gradually increases (decreasing APL) with deeper searches for all levels of heuristic errors. The interesting part, though, is the shape of the curves. It represents the relationship between search effort and the quality of knowledge.

Figures 2 and 3 show that better knowledge is much more useful with shallower search, and that at very high search depths it almost loses its meaning. There is very little difference in quality of play between a program with evaluation function with σ of 4.0 and an almost perfect-knowledge program with σ of 0.5 at search depths of about 30 plies. However, the results also indicate that if we would like to improve the play at extreme search depths it could only be done by increasing knowledge.

It is interesting that this result seems to clash with the model proposed by Junghanns and Schaeffer [7]. They write: "... as one moves to higher performance

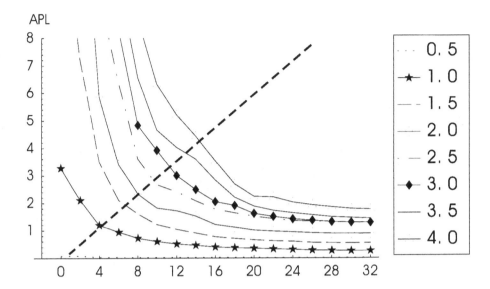

Fig. 3. The turning point for diminishing returns

levels, the slope of the isocurves increase. *This implies that for shallow search depths, more knowledge is required to move to a higher isocurve than for deeper search depths.*" The latter statement is not confirmed by our experiments.

3.2 Diminishing Returns

What do our results have to tell about the diminishing returns? First of all, they confirm their existence in the game of chess. True, the results were obtained only for a subset of chess, yet we see no reason for not believing that by adding more pieces on the board our results would still hold. Perhaps, some time in the future this can also be proven with better computers and larger tablebases.

A more interesting observation stemming from Fig. 3 is, however, that the point at which diminishing returns start to manifest themselves is dependant on the quality of the heuristic function. All of the curves in the figure exhibit a sort of turning point, yet, as indicated by the dashed line in the figure, this turning point comes later for worse evaluation functions. The dashed line shows how this turning point is dependant on the level of knowledge — while it intersects with the $\sigma = 1.0$ curve already at about ply 4, it intersects with the $\sigma = 4.0$ curve only somewhere around ply 14. The worse the quality of heuristic function, the later diminishing returns start to manifest themselves.

Interestingly, this speaks in favor of the argument that it is the quality of knowledge that is to fault that it took so long for diminishing returns to be proven as argued by Junghanns et al. [7,8]. The programs used for experiments, with the current level of knowledge, most likely simply did not reach the search depth at which the curve relating knowledge and search would turn.

If the hypothesized relation between the quality of knowledge and search effort as given first by Michie [10] and then by Berliner et al. [1] holds, then our result comes as no surprise. To attain equal quality of play the program with less knowledge must search deeper than a more knowledgeable one. Therefore diminishing returns for additional search effort come later for programs with less knowledge. This result thus in turn backs up the validity of the hypothesized relation.

4   Conclusions

The most interesting results gained by our alternative approach at investigating the effect of search depth and heuristic quality on the level of play are: (a) confirming the existence of diminishing returns for additional search effort in chess, and especially (b) showing that the point at which diminishing returns start to show themselves depends on the amount of knowledge the program has.

Perhaps leaving aside statistical considerations raised by Heinz [5], though we do agree with them, we should point out another issue that surely had an effect on the results of previous self-play experiments. The quality of knowledge the programs had has to be taken into account — and those programs were far less knowledgeable than the programs nowadays.

Here, an interesting question arises. The contemporary programs have reached or even surpassed the level of human world champion. Yet, one wonders, how good their knowledge actually is? How do they compare to the so-called God's algorithm, the tablebase-guided optimal play that requires no search beyond ply one? One only has to take one look at Marc Bourzutschky's and Yakov Konoval's recently constructed 7-man tablebases and one particular KQBNKQB position that is a win in 330 moves and it all becomes perfectly clear to him or her. As Tim Krabbé put it in his online diary, item 311 [9]: "As usual, the play is weird, incomprehensible, and beautiful." Neither humans nor today's strongest computers (without tablebases, relying only on their evaluation function) have any idea how to win such a position (against a perfect opponent). The point we are trying to make is that even the best contemporary programs are far from perfection in their knowledge. So why would we expect them to show diminishing returns?

We would also like to add that similar results (though at the moment only preliminary — just one or two runs instead of ten as for the KBBK endgame) were obtained for King-Rook-King and King-Queen-King-Rook chess endgames. Some of these results, although for the purpose of other research, can be seen in [11,12].

The generalization of our results on the complete chess game, as opposed to just the endgames analyzed, is an interesting question. As we see it, it all boils down to two issues: (a) the distance to mate as an evaluation function, and (b) what is the qualitative difference between a chess endgame and chess itself; in other words what, if anything, changes by putting more pieces on the board?

References

1. Berliner, H., Goetsch, G., Campbell, M., Ebeling, C.: Measuring the Performance Potential of Chess Programs. Artificial Intelligence 43(1), 7–21 (1990)
2. Haworth, G.: Self Play: Statistical Significance. ICGA Journal 26(2), 115–118 (2003)
3. Heinz, E.A.: A New Self-Play Experiment in Computer Chess. Technical Report No. 608 (MIT-LCS-TM-608), Laboratory for Computer Science, Massachussetts Institute of Technology, USA (2000)
4. Heinz, E.A.: New Self-Play Results in Computer Chess. In: Marsland, T., Frank, I. (eds.) CG 2001. LNCS, vol. 2063, pp. 267–282. Springer, Heidelberg (2002)
5. Heinz, E.A.: Self-Play Experiments in Computer Chess Revisited. In: van den Herik, H.J., Monien, B. (eds.) 9th Advances in Computer Games (ACG9), pp. 73–91. Department of Computer Science, Universiteit Maastricht, Maastricht, The Netherlands (2001)
6. Heinz, E.A.: Follow-up on Self-play, Deep Search, and Diminishing Returns. ICGA Journal 26(2), 75–80 (2003)
7. Junghanns, A., Schaeffer, J.: Search Versus Knowledge in Game-Playing Programs Revisited. In: Proceedings of the 15th International Joint Conference on Artificial Intelligence, pp. 692–697. Morgan Kaufmann, San Francisco (1997)
8. Junghanns, A., Schaeffer, J., Brockington, M., Björnsson, Y., Marsland, T.A.: Diminishing Returns for Additional Search in Chess. In: van den Herik, H.J., Uiterwijk, J.W.H.M. (eds.) 8th Advances in Computer Chess (ACC8), pp. 53–67. Department of Computer Science, University of Maastricht, Maastricht, The Netherlands (1997)
9. Krabbé, T.: Open Chess Diary (2006), http://www.xs4all.nl/~timkr/chess2/diary.htm
10. Michie, D.: A Theory of Advice. Machine Intelligence 8, 151–170 (1977)
11. Sadikov, A.: Propagation of Heuristic Evaluation Errors in Game Graphs. PhD thesis, University of Ljubljana, Faculty of Computer and Information Science (2005)
12. Sadikov, A., Bratko, I., Kononenko, I.: Bias and Pathology in Minimax Search. Theoretical Computer Science 349(2), 268–281 (2005)
13. Thompson, K.: Retrograde Analysis of Certain Endgames. ICCA Journal 9(3), 131–139 (1986)
14. Thompson, K.: 6-Piece Endgames. ICCA Journal 19(4), 215–226 (1996)

Counting the Number of Three-Player Partizan Cold Games

Alessandro Cincotti

Research Unit for Computers and Games, School of Information Science,
Japan Advanced Institute of Science and Technology, Ishikawa, Japan
cincotti@jaist.ac.jp

Abstract. We give upper and lower bounds on $S_3[n]$ equal to the number of three-player partizan cold games born by day n. In particular, we give an upper bound of $O(S_2[n]^3)$ and a lower bound of $\Omega(S_2[n])$ where $S_2[n]$ is the number of surreal numbers born by day n.

1 Introduction

Games represent a conflict of interests between two or more parties and, as a consequence, they are a good framework to study complex problem-solving strategies. Typically, a real-world economical, social or political conflict involves more than two parties and a winning strategy is often the result of coalitions. For this reason, it is important to determine the winning strategy of a player in the worst scenario, i.e., assuming that all his/her opponents are allied against him/her.

It is therefore, a challenging and fascinating problem to extend the field of combinatorial game theory [1,3] so as to allow more than two players. Past effort to classify impartial three-player combinatorial games (the theories of Li [5] and Straffin [8]) have made various restrictive assumptions about the rationality of one's opponents and the formation and behavior of coalitions. Loeb [6] introduces the notion of a stable winning coalition in a multi-player game as a new system of classification of games. Differently, Propp [7] adopts in his work an agnostic attitude toward such issues, and seeks only to understand in what circumstances one player has a winning strategy against the combined forces of the other two.

Cincotti [2] presents a theory to classify three-player partizan games adopting the same attitude. Such a theory represents a possible extension of Conway's theory of partizan games [3,4] and it is therefore both a theory of games and a theory of numbers.

In order to understand the mathematical structure of three-player partizan games, counting the number of cold games born by day n is a crucial point. We recall that the number of surreal numbers born by day n is $S_2[n] = 2^{n+1} - 1$. Moreover, a lower and upper bound of two-player games is given by Wolfe and Fraser in [9].

H.J. van den Herik et al. (Eds.): CG 2006, LNCS 4630, pp. 181–189, 2007.

2   Three-Player Partizan Games

For the sake of self-containment we recall in this section the main results concerning three-player partizan games obtained in the previous work [2].

2.1   Basic Definitions

Definition 1. *If L, C, R are any three sets of numbers previously defined and*

1. *no element of L is $\geq_L$ any element of $C \cup R$, and*
2. *no element of C is $\geq_C$ any element of $L \cup R$, and*
3. *no element of R is $\geq_R$ any element of $L \cup C$,*

then $\{L|C|R\}$ is a number. All numbers are constructed in this way.

This definition for numbers is based on the definition and comparison operators for games given in the following two definitions.

Definition 2. *If L, C, R are any three sets of games previously defined then $\{L|C|R\}$ is a game. All games are constructed in this way.*

We introduce three different relations $(\geq_L, \geq_C, \geq_R)$ that represent the subjective point of view of every player which is independent from the point of view of the other players.

Definition 3. *We say that*

1. *$x \geq_L y$ iff ($y \geq_L$ no x^C and $y \geq_L$ no x^R and no $y^L \geq_L x$),*
2. *$x \geq_C y$ iff ($y \geq_C$ no x^L and $y \geq_C$ no x^R and no $y^C \geq_C x$),*
3. *$x \geq_R y$ iff ($y \geq_R$ no x^L and $y \geq_R$ no x^C and no $y^R \geq_R x$).*

Numbers are totally ordered with respect to $\geq_L$, $\geq_C$, and $\geq_R$ but games are just partially-ordered, e.g., there exist games x and y for which we have neither $x \geq_L y$ nor $y \geq_L x$.

Definition 4. *We say that*

1. *$x =_L y$ if and only if ($x \geq_L y$ and $x \leq_L y$),*
2. *$x =_C y$ if and only if ($x \geq_C y$ and $x \leq_C y$),*
3. *$x =_R y$ if and only if ($x \geq_R y$ and $x \leq_R y$),*
4. *$x = y$ if and only if ($x =_L y$, $x =_C y$, and $x =_R y$),*
5. *$x + y = \{x^L + y, x + y^L | x^C + y, x + y^C | x^R + y, x + y^R\}$.*

Moreover, it is possible to classify numbers in 11 classes as shown in Table 1. The entries '?' are unrestricted and indicate that we can have different outcomes. For further details, see [2].

Table 1. Classification of numbers

Short notation	Class	Left starts	Center starts	Right starts
$= 0$	$=_L 0, =_C 0, =_R 0$	Right wins	Left wins	Center wins
$>_L 0$	$>_L 0, <_C 0, <_R 0$	Left wins	Left wins	Left wins
$>_C 0$	$<_L 0, >_C 0, <_R 0$	Center wins	Center wins	Center wins
$>_R 0$	$<_L 0, <_C 0, >_R 0$	Right wins	Right wins	Right wins
$=_{LC} 0$	$=_L 0, =_C 0, <_R 0$	Center wins	Left wins	Center wins
$=_{LR} 0$	$=_L 0, <_C 0, =_R 0$	Right wins	Left wins	Left wins
$=_{CR} 0$	$<_L 0, =_C 0, =_R 0$	Right wins	Right wins	Center wins
$<_{CR} 0$	$=_L 0, <_C 0, <_R 0$	?	Left wins	Left wins
$<_{LR} 0$	$<_L 0, =_C 0, <_R 0$	Center wins	?	Center wins
$<_{LC} 0$	$<_L 0, <_C 0, =_R 0$	Right wins	Right wins	?
< 0	$<_L 0, <_C 0, <_R 0$	?	?	?

2.2   Examples of Numbers

According to the construction procedure, every number has the form $\{L|C|R\}$, where $L, C,$ and R are three sets of earlier constructed numbers. At day zero, we have only the empty set $\emptyset$ therefore the earliest constructed number could only be $\{L|C|R\}$ with $L = C = R = \emptyset$, or in the simplified notation $\{ \ | \ | \ \}$. We denote it by 0.

The first day we have only three new numbers which we call $1_L = \{0| \ | \ \}$, $1_C = \{ \ |0| \ \}$, and $1_R = \{ \ | \ |0\}$. We observe that $\{0|0| \ \}$, $\{0| \ |0\}$, $\{ \ |0|0\}$, and $\{0|0|0\}$ are not numbers. Table 2 shows the numbers created the second day.

Note 1. In [2] the list of numbers created the second day was incomplete because we can create 24 (not 18) different numbers.

Table 2. Numbers created the second day

$\{1_L| \ | \ \}$	$\{ \ |1_C| \ \}$	$\{ \ | \ |1_R\}$	$\{1_C|1_L| \ \}$
$\{0, 1_C, 1_R| \ | \ \}$	$\{ \ |0, 1_L, 1_R| \ \}$	$\{ \ | \ |0, 1_L, 1_C\}$	$\{1_R| \ |1_L\}$
$\{0, 1_C| \ | \ \}$	$\{ \ |0, 1_L| \ \}$	$\{ \ | \ |0, 1_L\}$	$\{ \ |1_R|1_C\}$
$\{0, 1_R| \ | \ \}$	$\{ \ |0, 1_R| \ \}$	$\{ \ | \ |0, 1_C\}$	$\{1_C, 1_R| \ | \ \}$
$\{0|1_L| \ \}$	$\{1_C|0| \ \}$	$\{1_R| \ |0\}$	$\{ \ |1_L, 1_R| \ \}$
$\{0| \ |1_L\}$	$\{ \ |0|1_C\}$	$\{ \ |1_R|0\}$	$\{ \ | \ |1_L, 1_C\}$

3   Counting the Numbers

How many numbers will be created by day n?

Definition 5. *Let S_2 and S_3 be respectively Conway's surreal numbers and the new set of numbers previously defined. We define three different maps $\pi : S_3 \to S_2$ as follows:*

1. $\pi_L(\{x^L|x^C|x^R\}) = \{\pi_L(x^L)|\pi_L(x^C), \pi_L(x^R)\}$
2. $\pi_C(\{x^L|x^C|x^R\}) = \{\pi_C(x^C)|\pi_C(x^L), \pi_C(x^R)\}$
3. $\pi_R(\{x^L|x^C|x^R\}) = \{\pi_R(x^R)|\pi_R(x^L), \pi_R(x^C)\}$

Theorem 1. *For any* $x, y \in S_3$

1. $x \leq_L y \iff \pi_L(x) \leq \pi_L(y)$
2. $x \leq_C y \iff \pi_C(x) \leq \pi_C(y)$
3. $x \leq_R y \iff \pi_R(x) \leq \pi_R(y)$

Proof. 1. If $x \leq_L y$ then $\nexists x^L \geq_L y$ and, by the inductive hypothesis

$$\nexists \pi_L(x^L) \geq \pi_L(y) \Rightarrow \nexists \pi_L(x)^L \geq \pi_L(y). \tag{1}$$

Moreover, $\nexists y^C \leq_L x$, $\nexists y^R \leq_L x$, and by the inductive hypothesis

$$\left. \begin{array}{l} \nexists \pi_L(y^C) \leq \pi_L(x) \\ \nexists \pi_L(y^R) \leq \pi_L(x) \end{array} \right\} \Rightarrow \nexists \pi_L(y)^R \leq \pi_L(x). \tag{2}$$

Conversely, if $\pi_L(x) \leq \pi_L(y)$ then

$$\nexists \pi_L(x)^L \geq \pi_L(y) \Rightarrow \nexists \pi_L(x^L) \geq \pi_L(y) \tag{3}$$

and by the inductive hypothesis $\nexists x^L \geq_L y$.
Also,

$$\nexists \pi_L(y)^R \leq \pi_L(x) \Rightarrow \left\{ \begin{array}{l} \nexists \pi_L(y^C) \leq \pi_L(x) \\ \nexists \pi_L(y^R) \leq \pi_L(x) \end{array} \right. \tag{4}$$

and by the inductive hypothesis $\nexists y^C \leq_L x$ and $\nexists y^R \leq_L x$.
2. Analogous to *1*.
3. Analogous to *1*.   □

We have two corollaries of the above theorem.

Corollary 1. *If* $x \in S_3$ *is a number then* $\pi_L(x)$, $\pi_C(x)$, *and* $\pi_R(x)$ *are numbers.*

Corollary 2. *Let* $x, y \in S_3$ *be two numbers. Then* $x = y$ *if and only if* $\pi_L(x) = \pi_L(y)$, $\pi_C(x) = \pi_C(y)$, *and* $\pi_R(x) = \pi_R(y)$.

It follows that to every number $x \in S_3$ there corresponds a unique triple $(\pi_L(x), \pi_C(x), \pi_R(x))$ of surreal numbers. Table 3 shows all numbers born by day 2 and their corresponding triples of surreal numbers.

Theorem 2. *Let* $x = \{x^L|x^C|x^R\} \in S_3[n]$ *be a number born by day n. Then* $\pi_L(x)$, $\pi_C(x)$, *and* $\pi_R(x) \in S_2[n]$.

Proof. By the hypothesis x^L, x^C, and $x^R \in S_3[n-1]$ and by the inductive hypothesis $\pi_L(x^L)$, $\pi_L(x^C)$, and $\pi_L(x^R) \in S_2[n-1]$ therefore $\pi_L(x) \in S_2[n]$. Analogously, $\pi_C(x)$ and $\pi_R(x)$ are numbers born by day n.   □

Unfortunately, the above theorem is not reversible.

Table 3. Numbers born by day 2 and their corresponding triples of surreal numbers

	Day 0	Day 1	Day 2	
$=$	$\{\ |\ |\ \}$ (0,0,0)			
$>_L$		$\{0|\ |\ \}$ (1,-1,-1)	$\{1_L|\ |\ \}$	(2,-2,-2)
$>_L$			$\{0,1_C,1_R|\ |\ \}$	(1,-2,-2)
$>_L$			$\{0,1_C|\ |\ \}$	(1,-1,-2)
$>_L$			$\{0,1_R|\ |\ \}$	(1,-2,-1)
$>_L$			$\{0|1_L|\ \}$	(1/2,-1/2,-2)
$>_L$			$\{0|\ |1_L\}$	(1/2,-2,-1/2)
$>_C$		$\{\ |0|\ \}$ (-1,1,-1)	$\{\ |1_C|\ \}$	(-2,2,-2)
$>_C$			$\{\ |0,1_L,1_R|\ \}$	(-2,1,-2)
$>_C$			$\{\ |0,1_L|\ \}$	(-1,1,-2)
$>_C$			$\{\ |0,1_R|\ \}$	(-2,1,-1)
$>_C$			$\{1_C|0|\ \}$	(-1/2,1/2,-2)
$>_C$			$\{\ |0|1_C\}$	(-2,1/2,-1/2)
$>_R$		$\{\ |\ |0\}$ (-1,-1,1)	$\{\ |\ |1_R\}$	(-2,-2,2)
$>_R$			$\{\ |\ |0,1_L,1_C\}$	(-2,-2,1)
$>_R$			$\{\ |\ |0,1_L\}$	(-1,-2,1)
$>_R$			$\{\ |\ |0,1_C\}$	(-2,-1,1)
$>_R$			$\{1_R|\ |0\}$	(-1/2,-2,1/2)
$>_R$			$\{\ |1_R|0\}$	(-2,-1/2,1/2)
$=_{LC}$			$\{1_C|1_L|\ \}$	(0,0,-2)
$=_{LR}$			$\{1_R|\ |1_L\}$	(0,-2,0)
$=_{CR}$			$\{\ |1_R|1_C\}$	(-2,0,0)
$<_{CR}$			$\{1_C,1_R|\ |\ \}$	(0,-2,-2)
$<_{LR}$			$\{\ |1_L,1_R|\ \}$	(-2,0,-2)
$<_{LC}$			$\{\ |\ |1_L,1_C\}$	(-2,-2,0)

Example 1. Let's consider $1_L + 1_C + 1_R = \{\{\ |1_R|1_C\}|\{1_R|\ |1_L\}|\{1_C|1_L|\ \}\}$. We observe that $\pi_L(x) = \pi_C(x) = \pi_R(x) = -1$ therefore they all belong to $S_2[1]$ but $1_L + 1_C + 1_R \notin S_3[1]$ because it will be created only the third day.

It follows that a rough upper bound on $S_3[n]$ is given by the number of distinct triples of surreal numbers born by day n, i.e., $(S_2[n])^3$. Moreover, a simple lower bound is given by $S_2[n]$.

Theorem 3. *For any* $x, y \in S_3$

1. $\pi_L(x + y) = \pi_L(x) + \pi_L(y)$
2. $\pi_C(x + y) = \pi_C(x) + \pi_C(y)$
3. $\pi_R(x + y) = \pi_R(x) + \pi_R(y)$

Proof. 1.

$$\pi_L(x + y) = \pi_L(\{x^L + y, x + y^L | x^C + y, x + y^C | x^R + y, x + y^R\}) \quad (5)$$
$$= \{\pi_L(x^L + y), \pi_L(x + y^L)|$$
$$\pi_L(x^C + y), \pi_L(x + y^C), \pi_L(x^R + y), \pi_L(x + y^R)\}$$

$$= \{\pi_L(x^L) + \pi_L(y), \pi_L(x) + \pi_L(y^L)|$$
$$\pi_L(x^C) + \pi_L(y), \pi_L(x) + \pi_L(y^C),$$
$$\pi_L(x^R) + \pi_L(y), \pi_L(x) + \pi_L(y^R)\}$$
$$= \pi_L(x) + \pi_L(y)$$

2. Analogous to *1*.
3. Analogous to *1*.   □

Theorem 4. *Let* $x \in S_3$ *be a number. Then*

1. $\pi_L(x) + \pi_C(x) \leq 0$,
2. $\pi_L(x) + \pi_R(x) \leq 0$,
3. $\pi_C(x) + \pi_R(x) \leq 0$.

Proof.   *1.* We observe that

$$\pi_L(x^L) + \pi_C(x) < \pi_L(x^L) + \pi_C(x^L) \tag{6}$$

and by the inductive hypothesis

$$\pi_L(x^L) + \pi_C(x^L) \leq 0. \tag{7}$$

Analogously,
$$\pi_L(x) + \pi_C(x^C) < \pi_L(x^C) + \pi_C(x^C) \leq 0 \tag{8}$$

therefore no left option of $\pi_L(x) + \pi_C(x)$ is ≥ 0.
2. Analogous to *1*.
3. Analogous to *1*.   □

Theorem 5. *Let* $x \in S_3[n]$ *and* $y \in S_3[m]$ *be two numbers. Then* $x + y \in S_3[n + m]$.

Proof. We recall that $x + y = \{x^L + y, x + y^L | x^C + y, x + y^C | x^R + y, x + y^R\}$. By the hypothesis, x^L, x^C, and x^R belong to $S_3[n-1]$ and y^L, y^C, y^R belong to $S_3[m-1]$. By the inductive hypothesis, $x^L + y$, $x + y^L$, $x^C + y$, $x + y^C$, $x^R + y$, $x + y^R$ belong to $S_3[n+m-1]$ therefore $x + y \in S_3[n+m]$.   □

3.1 Lower and Upper Bound

Below we give a more accurate upper and lower bound for each class. We start recalling four statements.

1. The number of surreal numbers born by day n is

$$S_2[n] = 2^{n+1} - 1 \tag{9}$$

2. The number of positive (negative) surreal numbers born by day n is

$$\frac{1}{2}(S_2[n] - 1) \tag{10}$$

3. The number of positive (negative) dyadic fraction born by day n is

$$\frac{1}{2}(S_2[n] - 2n - 1) \tag{11}$$

4. The following equality holds

$$1^2 + 2^2 + \ldots + n^2 = \frac{n(n+1)(2n+1)}{6} \tag{12}$$

Definition 6. *We define*

$$(i+1)_L = \{i_L | \ | \} \tag{13}$$
$$(j+1)_C = \{ \ |j_C| \ \} \tag{14}$$
$$(k+1)_R = \{ \ | \ |k_R\} \tag{15}$$

where $i, j, k \in \mathbf{N}$ and $0_L = 0_C = 0_R = 0$.

Definition 7. *We define*

$$\left(\frac{2p+1}{2^{q+1}}\right)_{LC} = \left\{ \left(\frac{p}{2^q}\right)_{LC} \left| \left(\frac{p+1}{2^q}\right)_{LC} \right| \right\} \tag{16}$$

$$\left(\frac{2p+1}{2^{q+1}}\right)_{LR} = \left\{ \left(\frac{p}{2^q}\right)_{LR} \left| \ \right| \left(\frac{p+1}{2^q}\right)_{LR} \right\} \tag{17}$$

where $p, q \in \mathbf{N}$.

Note 2. If $\left(\frac{p}{2^q}\right) \in \mathbf{N}$ then $\left(\frac{p}{2^q}\right)_{LC} = \left(\frac{p}{2^q}\right)_{LR} = \left(\frac{p}{2^q}\right)_{L}$. Analogously, if $\left(\frac{p+1}{2^q}\right) \in$ $\mathbf{N}$ then $\left(\frac{p+1}{2^q}\right)_{LC} = \left(\frac{p+1}{2^q}\right)_{LR} = \left(\frac{p+1}{2^q}\right)_{L}$.

Theorem 6. *If $x = \left(\frac{2p+1}{2^{q+1}}\right)_{LC}$ is born the bth day then $\pi_R(x) = -b$.*

Proof. By definition

$$\pi_R\left(\left(\frac{2p+1}{2^{q+1}}\right)_{LC}\right) = \left\{ \left| \pi_R\left(\left(\frac{p}{2^q}\right)_{LC}\right), \pi_R\left(\left(\frac{p+1}{2^q}\right)_{LC}\right) \right\} \tag{18}$$

We observe that either $\left(\frac{p}{2^q}\right)_{LC}$ or $\left(\frac{p+1}{2^q}\right)_{LC}$ must be born the $(b-1)^{th}$ day therefore by the inductive hypothesis we have $\pi_R(x) = \{ \ | - (b-1)\} = -b$. □

Below we make eight observations

1. The first class contains only the number 0.
2. In the class $>_L 0$, $\pi_L(x)$ is positive, $\pi_C(x)$ and $\pi_R(x)$ are negative therefore we have an upper bound of $\frac{1}{8}(S_2[n] - 1)^3$. Using Theorem 4 and the equality 12 we can refine this value obtaining $\frac{1}{24}(S_2[n]^2 - 1)S_2[n]$.

In contrast, we can express every triple $(i, -j, -k)$ by the number $\{(i-1)_L, (k-1)_C, (j-1)_R| \ | \}$ where $i-j \leq 0$, $i-k \leq 0$ and $i,j,k \in \mathbf{Z}^+$. Moreover, for every positive dyadic fraction we can create two different numbers $\left(\frac{2p+1}{2^{q+1}}\right)_{LC}$ and $\left(\frac{2p+1}{2^{q+1}}\right)_{LR}$ corresponding respectively to $\left(\frac{2p+1}{2^{q+1}}, -\frac{2p+1}{2^{q+1}}, -b\right)$ and $\left(\frac{2p+1}{2^{q+1}}, -b, -\frac{2p+1}{2^{q+1}},\right)$ where b is the day $\left(\frac{2p+1}{2^{q+1}}\right)$ was born. Summing up, we have a lower bound of $\frac{1}{6}n(n+1)(2n+1) + S_2[n] - 2n - 1$.

3. The classes $>_C 0$ and $>_R 0$ are analogous to the class $>_L 0$.
4. If $x =_{LC} 0$ then $x^R = \emptyset$ therefore $\pi_R(x) = \{ \ |\pi_R(x^L), \pi_R(x^C)\} \in \mathbf{Z}^-$. If $x \in S_3[n]$ then by Theorem 2, $-n \leq \pi_R(x)$ therefore n is an upper bound. Moreover, we observe that the number $\{\{ \ |0|1_C\}|\{0| \ |1_L\}| \ \}$ corresponding to $(0, 0, -1)$ belongs to $S_3[3]$ therefore the lower bound is n with $n > 2$.
5. The classes $=_{LR} 0$ and $=_{CR} 0$ are analogous to the class $=_{LC} 0$.
6. In the class $<_{CR} 0$, $\pi_L(x) = 0$, $\pi_C(x)$ and $\pi_R(x)$ are negative therefore we have an upper bound of $\frac{1}{4}(S_2[n] - 1)^2$.
   In contrast, we can express every triple $(0, -j, -k)$ by the number $\{(k-1)_C, (j-1)_R| \ | \}$ where $j,k \in \mathbf{Z}^+$ with $j, k \geq 2$. Moreover, for every positive dyadic fraction we can create the number $\left(\frac{2p+1}{2^{q+1}}\right)_{LC} + \left(\frac{2p+1}{2^{q+1}}\right)_{RL}$ corresponding to $\left(0, -b-\frac{2p+1}{2^{q+1}}, -b+\frac{2p+1}{2^{q+1}}\right)$ where b is the day $\left(\frac{2p+1}{2^{q+1}}\right)$ was born. Summing up, we have a lower bound of $(n-1)^2 + \frac{1}{2}(S_2[\lfloor n/2 \rfloor] - 2\lfloor n/2 \rfloor - 1)$.
7. The classes $<_{LR} 0$ and $<_{LC} 0$ are analogous to the class $<_{LR} 0$.
8. In the last class, we have an upper bound of $\frac{1}{8}(S_2[n] - 1)^3$ because $\pi_L(x)$, $\pi_C(x)$, and $\pi_R(x)$ are all negative.
   We recall that
   (a) If $x <_{CR} 0 \in S_3[n-2]$ and $y = \{ \ |1_R|1_C\} \in S_3[2]$ then $x+y < 0 \in S_3[n]$.
   (b) If $x <_{LR} 0 \in S_3[n-2]$ and $y = \{1_R| \ |1_L\} \in S_3[2]$ then $x+y < 0 \in S_3[n]$.
   (c) If $x <_{LC} 0 \in S_3[n-2]$ and $y = \{1_C|1_L| \ \} \in S_3[2]$ then $x+y < 0 \in S_3[n]$.
   To be sure that the sets of numbers given by (a), (b), and (c) are disjoint, we do not consider the numbers $x <_{CR} 0 \in S_3[n - 2]$ corresponding to $(0, -j, -k)$ where either $j = 2$ or $k = 2$. Analogously, we do not consider the numbers $x <_{LR} 0 \in S_3[n - 2]$ corresponding to $(-j, 0, -k)$ where either $j = 2$ or $k = 2$ and the numbers $x <_{LC} 0 \in S_3[n - 2]$ corresponding to $(-j, -k, 0)$ where either $j = 2$ or $k = 2$. Therefore, we have a lower bound of $\frac{3}{2}S_2[\lfloor n/2 \rfloor - 1] + 3n^2 - 24n - 3\lfloor n/2 \rfloor + \frac{99}{2}$.

Summarizing, we have an upper bound of

$$\frac{1}{4}S_2[n]^3 + \frac{3}{8}S_2[n]^2 - \frac{5}{4}S_2[n] + 3n + \frac{13}{8} = O(S_2[n]^3) \tag{19}$$

and a lower bound of

$$3S_2[n] + \frac{3}{2}S_2[\lfloor n/2 \rfloor] + \frac{3}{2}S_2[\lfloor n/2 \rfloor - 1] + \tag{20}$$

$$n^3 + \frac{15}{2}n^2 - \frac{65}{2}n - 6\lfloor n/2 \rfloor + 49 = \Omega(S_2[n])$$

Table 4 shows the results so far obtained but to establish the exact value of $S_3[n]$ as well as the canonical form of a three-player game is still an open problem.

Table 4. Results obtained so far

x	Lower bound	Upper bound
$= 0$	1	1
$>_L 0$	$S_2[n] + \frac{1}{3}n^3 + \frac{1}{2}n^2 - \frac{11}{6}n - 1$	$\frac{1}{24}(S_2[n]^2 - 1)S_2[n]$
$>_C 0$	$S_2[n] + \frac{1}{3}n^3 + \frac{1}{2}n^2 - \frac{11}{6}n - 1$	$\frac{1}{24}(S_2[n]^2 - 1)S_2[n]$
$>_R 0$	$S_2[n] + \frac{1}{3}n^3 + \frac{1}{2}n^2 - \frac{11}{6}n - 1$	$\frac{1}{24}(S_2[n]^2 - 1)S_2[n]$
$=_{LC} 0$	$n, n > 2$	n
$=_{LR} 0$	$n, n > 2$	n
$=_{CR} 0$	$n, n > 2$	n
$<_{CR} 0$	$\frac{1}{2}S_2[\lfloor n/2 \rfloor] + n^2 - 2n - \lfloor n/2 \rfloor + \frac{1}{2}, n > 1$	$\frac{1}{4}(S_2[n] - 1)^2$
$<_{LR} 0$	$\frac{1}{2}S_2[\lfloor n/2 \rfloor] + n^2 - 2n - \lfloor n/2 \rfloor + \frac{1}{2}, n > 1$	$\frac{1}{4}(S_2[n] - 1)^2$
$<_{LC} 0$	$\frac{1}{2}S_2[\lfloor n/2 \rfloor] + n^2 - 2n - \lfloor n/2 \rfloor + \frac{1}{2}, n > 1$	$\frac{1}{4}(S_2[n] - 1)^2$
< 0	$\frac{3}{2}S_2[\lfloor n/2 \rfloor - 1] + 3n^2 - 24n - 3\lfloor n/2 \rfloor + \frac{99}{2}, n > 3$	$\frac{1}{8}(S_2[n] - 1)^3$

Acknowledgments

I would like to thank the anonymous referees for their useful comments.

References

1. Berlekamp, E.R., Conway, J.H., Guy, R.K.: Winning Ways For Your Mathematical Plays. Academic Press, San Diego (1982)
2. Cincotti, A.: Three-Player Partizan Games. Theoretical Computer Science 332, 367–389 (2005)
3. Conway, J.H.: On Numbers and Games. Academic Press, San Diego (1976)
4. Knuth, D.: Surreal Numbers. Addison-Wesley, London, UK (1974)
5. Li, S.Y.R.: N-Person Nim and N-Person Moore's Games. Game Theory 7, 31–36 (1978)
6. Loeb, D.E.: Stable Winning Coalitions. In: Nowakowski, R.J. (ed.) Games of No Chance, vol. 29, pp. 451–471. MSRI Publ. Cambridge University Press (1994)
7. Propp, J.: Three-player Impartial games. Theoretical Computer Science 233, 263–278 (2000)
8. Straffin Jr., P.D.: Three-Person Winner-Take-All Games with Mc-Carthy's Revenge Rule. College Journal of Mathetmatics 16, 386–394 (1985)
9. Wolfe, D., Fraser, W.: Counting the Number of Games. Theoretical Computer Science 313, 527–532 (2004)

LUMINES Strategies

Greg Aloupis, Jean Cardinal, Sébastien Collette*, and Stefan Langerman**

Département d'Informatique,
Université Libre de Bruxelles, Brussels, Belgium
{greg.aloupis,jcardin,secollet,stefan.langerman}@ulb.ac.be

Abstract. We analyze a new popular video-game called LUMINES, which was developed by SONY for the PSP platform. It involves a sequence of bichromatic 2×2 blocks that fall in a grid and must be shifted or rotated by the player before they land. Patterns of monochromatic 2×2 blocks in the terrain are regularly deleted. The primary goal is to contain the terrain within a fixed height and, if possible, clear the grid.

We deal with questions such as: (1) Can the game be played indefinitely? and (2) Can all terrains be eliminated? We examine how answers to these questions depend on the size of the grid and the rate at which blocks are deleted.

1 Introduction

LUMINES[1] is a popular puzzle video game for the PSP[2] platform, originally released in December 2004 in Japan. The original concept was proposed by Tetsuya Mizuguchi and half a million units of this game have been sold within the first year of its release. In this paper, we consider a model that follows the properties of the game as closely as possible. The game is described below.

In a two-dimensional grid game, a sequence of 2×2 blocks falls and the user may rotate and shift each block before it reaches the existing terrain. From this instant, the user may no longer shift or rotate that block, and if only one of the two columns of the block is supported by the terrain just below it, the other 2x1 column continues to fall without intervention until it is also supported. Once a new terrain has been formed, the next block falls.

Each of the four square cells that form a block is colored either black or white, and any cell that belongs to a 2×2 monochromatic square in the terrain is marked. Note that some of those 2×2 monochromatic squares could overlap. At regular intervals, a vertical line sweeps across the terrain and deletes marked cells. If other cells existed above the deleted ones in the terrain, then they collapse so that a new terrain is formed. An example is illustrated in Figure 1.

The main goal of the game is to manipulate the falling blocks so that the terrain remains within the grid. A secondary goal is to clear the terrain entirely

---

* Aspirant du F.N.R.S.
** Chercheur qualifié du F.N.R.S.
[1] LUMINES is a trademark of Bandai.
[2] PSP is a trademark of Sony Computer Entertainment Inc.

H.J. van den Herik et al. (Eds.): CG 2006, LNCS 4630, pp. 190–199, 2007.

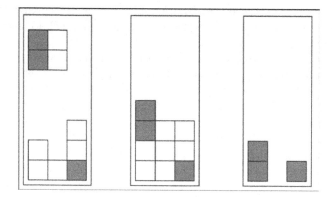

Fig. 1. Example of a block falling: both 2×1 columns fall until supported, the two overlapping squares are deleted, and the the remainder of the terrain collapses

as much as possible, as this is worth many points. Variations that the game offers include trying to build a target terrain from a given one.

The actual game is played on a grid with 16 columns and 10 rows. While one block is falling, the player can see what the next three blocks will be. The rates at which the sweep-line appears and moves depend on the level of the game.

There are four kinds of blocks (see Figure 2), and six shapes overall:

- (1-2) monochromatic (black and white),
- (3) H-blocks, consisting of two black cells above two white cells,
- (4-5) L-blocks, with 3 black and one white cell (and vice versa),
- (6) X-blocks, no two cells of the same color in the same row or column.

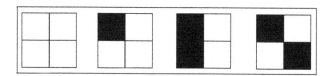

Fig. 2. The four block types in LUMINES: monochromatic, L-block, H-block, X-block

In this work we analyze the game and discuss how to play as best as possible. Related works include analyses of similar board games, where blocks fall and must be arranged by the player. TETRIS, the best known falling-block game invented by Alexey Pajitnov in 1985, has been extensively studied. Burgiel [4] proved that there exists a sequence of tetrominoes that will fill up the board no matter how the player places them. This improves a previous result of Brzutowski [3] who showed that there is no winning strategy if the computer uses the information of how the previous tetrominoes are placed. Similar results have also been proposed by Tucker [8]. Brzutowski however proved that there exist winning strategies for restricted versions of TETRIS, in which for instance only one type of tetrominoes appear.

We are not aware of any previous scientific analysis of LUMINES. However, even though the game was released recently, many documents available on the Internet discuss tips and strategies. References and links to previous analyses of the game can be found in [11]. For instance, it is known that the game can be played continuously by compartmentalizing the board into different sections corresponding to each type of block [9,10].

Other types of results involve the application of complexity theory to puzzles and games. A survey of algorithmic and complexity results for two-player games and puzzles was published recently by Demaine [5]. He also provided a complexity analysis of TETRIS [2,6], showing that for a given sequence of blocks it is NP-hard to find a corresponding sequence of moves that minimizes the final height of the stack or maximizes the number of cleared rows. Further complexity and decidability results for TETRIS are given in the recent paper by Hoogeboom and Kosters [7]. TETRIS also served as a model for packing problems (see, e.g., [1]).

The known results on strategies for TETRIS are to be compared with our result for LUMINES in Section 2. There we propose an optimal winning strategy for LUMINES, that uses as few steps, rows, and columns as possible for each section of the board. We analyze the interferences which can occur between sections, and also the impact of the speed of the sweep-line on our strategy. We try to clear the board completely as often as possible.

Section 3 is dedicated question of whether any given terrain, made of previously fallen blocks, can be cleared, provided it does not contain any obviously indestructible structure. Although at first sight a positive answer seems within reach, we show that one must take care of traps.

2   Playing Forever

Unless mentioned otherwise, we assume that the sweepline moves quickly across the grid, with respect to the rate at which blocks fall. In effect, we assume that all marked blocks are deleted simultaneously when the sweepline appears.

Lemma 1. *On the original* LUMINES *grid, starting from the empty terrain, it is possible to continue playing indefinitely as long as the sweep-line appears after every block.*

Proof. We provide a simple repetitive strategy that can be used to play forever, and in which every type of block is placed in predefined columns.

First we examine the situation where the sweep-line appears at least four times after every block.

We allocate columns 1 and 2 for monochromatic and H-blocks, and maintain that the terrain will be identical in these two columns. Clearly in such conditions a monochromatic block is deleted as soon as it becomes part of the landscape. H-blocks are always rotated so that they form monochromatic rows, and so that the bottom row matches the top row in the terrain (which, by construction, is monochromatic). The two matching rows will be deleted immediately.

Since the top row of columns 1 and 2, as well as the bottom row of the falling monochromatic or H-block will always be "instantly" deleted, the height of the terrain in these two columns will be at most 2, in the case that an H-block falls on an empty terrain. Of course, the height can temporarily be 4, just before deletion.

Columns 4-6 are allocated for L-blocks that contain three white cells. We provide a sequence of operations so that every four of these blocks combine to yield an empty terrain. To avoid possible interference between monochromatic blocks and L-blocks, we always keep column 3 clear. Columns 8-10 are reserved for the alternate type of L-block, and column 7 is kept clear. Our sequence of operations for L-blocks is shown in Figure 3. This sequence produces a maximum height of 6, and when the last L-block has been added the triplet of columns is cleared in three sweeps.

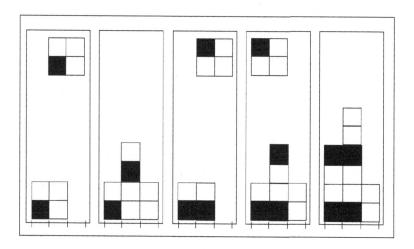

Fig. 3. Sequence of operations for L-blocks. Between the second and third frames, four white cells are deleted and part of the terrain collapses. After the last frame, three sweeps will clear the terrain.

Column 11 is kept clear and we allocate columns 12-14 for X-blocks. We demonstrate how to clear the terrain within these three columns after every four such blocks, in Figure 4. The maximum height in these columns is 8. The three columns are cleared in four sweeps after the last X-block falls.

A closer examination of these separate groups shows that we can place monochromatic and H-blocks in columns 1-2, one type of L-blocks in columns 3-5, X-blocks in columns 6-8, and the alternate L-blocks in columns 9-11. No interference occurs.

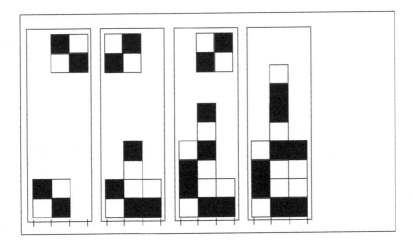

Fig. 4. Sequence of operations for X-blocks. No cells are deleted until after all four blocks have formed a terrain. Then four sweeps clear the terrain.

For a sweepline that appears only once after each block, we do not modify our method of dealing with each block type. We continue dropping blocks in the same patterns, but now within each triplet of columns the terrain does not follow one simple cycle. Depending on the sequence of blocks and sweep-line appearances, a triplet of columns may go through several other cycles. Each cycle has been verified and stay within a height of 8. For example, in Figure 3, if only one sweep-line appears after the last illustrated step, then just four white cells will be deleted instead of all three columns. Adding an L-block in the first two columns will just stack four cells onto the terrain, but then the four existing black cells will be deleted. In this way, proceeding with our normal pattern of dropped blocks, the sequence of deletions is not affected.

However, the new patterns that occur do cause interference between separate block-type regions. So if we wish to handle a slower sweep-line rate, we do need buffers and 14 columns overall. We have not yet examined whether this interference can be handled otherwise. Finally, we just mention a similar strategy which allows indefinite play within 16 columns and 4 rows. However, the cycle of each block type does not include the empty state.   □

The fact that one can play LUMINES forever is in sharp contrast with the opposite result known for TETRIS. This can be interpreted as LUMINES being "easier" than TETRIS. Yet it seems, based on recent reviews, that the game is no less addictive.

3   Clearing Terrains

A terrain is defined to be a *stable* configuration of cells, in the sense that no cells would collapse if a sweep-line were to scan through. In other words, no four cells of the same color forming a square exist in a terrain. We say that a terrain is *flat* if all columns have the same height. We make the observation that some terrains cannot be cleared, on a board of finite width. It suffices to construct a flat terrain where the top row consists of alternating colors. No cell in this top row can ever be part of a monochromatic square, so the row will never be deleted. A terrain with no alternating row is called a *legal terrain*.

The goal of this section is to determine whether all legal terrains can be cleared and how to clear them. In the algorithms that follow we assume that enough sweeps occur to delete the terrain. The following Lemma is useful for clearing terrains.

Lemma 2. *Let (A, B, C) be three consecutive columns with heights $(0, 0, k)$ respectively, where $k \geq 2$. The height of all three columns can be reduced to zero using an appropriate sequence of blocks, whether adjacent columns exist or not.*

Proof. To simplify the main proof assume that no adjacent columns exist, so that no interference occurs with the blocks that we will use within (A, B, C). This will be dealt with at the end of the proof.

Our general strategy is to add two new cells to B so that they duplicate those opposite to them in C. At the same time we add two white cells to A. The idea is to create duplicate columns in B and C, which can be easily eliminated using monochromatic or H-blocks. Following this, the white cells in A can be eliminated using white blocks.

Some deletions may occur as we add blocks in (A, B), and in fact it is clear that the rows in (B, C) will end up having alternating colors. When deletions occur, B and C both lose the same number of cells. Any deletion that happens in A also happens in (B, C), so A will always be at least as high as the other two columns.

We have no method of eliminating a single cell or single row within three columns, and suspect that it is not possible. Therefore it is important to avoid constructing these patterns during our procedure. There are several end-configurations which can lead to the two forbidden patterns if care is not taken. These are not included here due to space constraints. We just mention that we avoid being left with a single row by maintaining A higher than the other two columns.

Interference due to a column adjacent to A can cause its height to decrease. However we can always re-fill it in this case. Interference with C is not really an issue.   □

It is easy to take care of the two exceptional height sets $(0,0,1)$ and $(1,1,1)$ if a fourth column is used. It suffices to drop a block so that two cells land in C and the other two in its adjacent column.

Lemma 3. *Given four consecutive columns, including one column C of height zero, the height of a column adjacent to C can also be reduced to zero, with an appropriate sequence of blocks.*

Proof. Examine the case where C is the first column in the set (C, D, E, F):

It is easy to drop blocks in (E, F) so that the height of E reaches far above D, and the excess cells are all white. If E is initially higher than D, then we drop blocks in (D, E) to achieve the same effect. Thereafter we can drop blocks in (C, D), so that cells in C match the original ones of D, while the new cells in D are white and instantly eliminated along with part of E. Thus D does not grow in height as we add cells to C. We can even add one cell to D with the aid of (E, F), to ensure that there is no parity difference between C and D. Thus the alternating rows in (C, D) can be eliminated with monochromatic blocks.

If C is the second column, in the set (B, C, D, E), we do the following. Simply drop blocks in (C, D) so that the cells in C match those in B and then eliminate the alternating rows in (B, C) as mentioned above. We omit the details of dealing with a parity difference.   □

Theorem 1. *Any terrain of width at least four and containing a column of zero height can be cleared with an appropriate sequence of blocks.*

Proof. We use Lemma 3 to create two adjacent empty columns, and then repeat the uses of Lemma 2 to clear the remaining columns.   □

Corollary 1. *If a terrain has minimum height h, with an appropriate sequence of blocks the entire terrain can be flattened to height h, or the minimum height can be reduced.*

Proof. If the height cannot immediately be reduced, then use the column C which has height h to apply the technique of Lemma 3.   □

The preceding result almost leads one to conclude that all legal terrains can be cleared with an appropriate sequence of blocks. However, in Figure 5 we illustrate a legal terrain that cannot be cleared.

The top row is indestructible, unless some cell is dropped into the "well". If it is white, it will complete an alternating row instantly. If it is black, it will trigger a series of deletions which result in the bottom row becoming alternating. This does not contradict Corollary 1, since the terrain can be reduced for a few steps, but when the minimum height is one it can only be flattened, not reduced further. Corollary 1 does not specify that once the terrain is reduced to height h, it will also be legal.

We have an example of an indestructible terrain with a well of width two, shown in Figure 6. Again the top row is indestructible, and either of the two positions at the bottom awaits a cell which will trigger disaster or help to complete an alternating row at that level. For lack of space, an illustration of the triggered deletions is omitted. We just mention the interesting construction in which two separate triggers form different cascading deletions which ultimately affect the exact same position at lower height.

Let a *feasible* terrain be one which has at least one column of height 0. We have already proved that feasible terrains can be cleared with an appropriate sequence of blocks. Let a *fast* sweep-line be one which appears at least four times after every block falls.

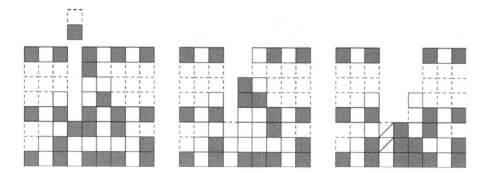

Fig. 5. Left: an indestructible terrain and a cell that triggers a series of deletions. Blocks with dashed sides have irrelevant color; Middle: after the first deletion; Right: after the second deletion. The two blocks with a diagonal may or may not be present. The final deletion which causes the lowest row to become alternating is not shown.

Theorem 2. *Any feasible terrain with at least 15 columns and arbitrary height can be cleared, assuming that all block types appear eventually after any given instant, and that the sweep-line is fast. If the sweep-line is not fast, then 3 more columns are needed.*

Proof. Since we allow arbitrary height, we can use two consecutive columns that are furthest from the column of height 0, in order to place blocks that are not useful to us at the moment that they appear. This allows us to assume that we have at our disposal exactly the sequence of blocks that we need. Since the two storage columns are at one extreme of the grid, there are 13 consecutive columns in which to apply the strategy of Theorem 1.

After clearing 13 columns, we can begin using the first 11 as a storage for useless blocks, according to Lemma 1. This allows us to clear the two initial storage columns, with the aid of the two columns nearest to them.

Of course, we will have to wait until a moment when each of our block types has "self-eliminated" within its allocated column space, for the terrain actually to be cleared. If blocks fall randomly, this will happen with probability one. There is simply no deterministic method of achieving this goal, due to adversarial arguments. □

To conclude, we claim that any feasible terrain with height h can be cleared without ever reaching a height greater than $2h$ (at least if the width of the grid is greater than 30), and we conjecture that this bound is tight. Terrains requiring $\frac{3h}{2}$ exist.

As we have seen, in LUMINES not all terrains can be cleared. In comparison, *any* TETRIS terrain can be cleared if an appropriate sequence of tetrominoes is given. In this sense, LUMINES is harder.

Future work may involve giving a complete characterization of clearable legal terrains.

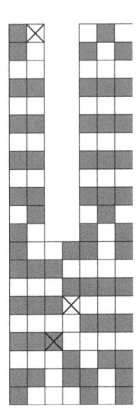

Fig. 6. An indestructible terrain. The black marked cell ultimately takes the place of the white cell below it in the lowest row. Depending on which side is triggered, one of the two marked white cells will also fall into the lowest row. The left trigger involves 9 deletions, and the right one involves 12.

Acknowledgments

This project was initiated during the algorithmic meetings at the ULB Computer Science department. The authors thank all participants, in particular Guy Louchard and Barry Balof, for helpful discussions.

References

1. Azar, Y., Epstein, L.: On Two-Dimensional Packing. Journal of Algorithms 25(2), 290–310 (1997)
2. Breukelaar, R., Demaine, E.D., Hohenberger, S., Hoogeboom, H.J., Kosters, W.A., Liben-Nowell, D.: TETRIS is Hard, Even to Approximate. International Journal of Computational Geometry and Applications 14, 41–68 (2004)
3. Brzutowski, J.: Can You Win at TETRIS? Master's thesis, The University of British Columbia (1992)

4. Burgiel, H.: How to Lose at TETRIS. Mathematical Gazette 81, 194–200 (1997)
5. Demaine, E.D.: Playing Games with Algorithms: Algorithmic Combinatorial Game Theory. In: Sgall, J., Pultr, A., Kolman, P. (eds.) MFCS 2001. LNCS, vol. 2136, pp. 18–32. Springer, Heidelberg (2001)
6. Demaine, E.D., Hohenberger, S., Liben-Nowell, D.: TETRIS is Hard, Even to Approximate. In: Warnow, T.J., Zhu, B. (eds.) COCOON 2003. LNCS, vol. 2697, pp. 351–363. Springer, Heidelberg (2003)
7. Hoogeboom, H.J., Kosters, W.A.: The Theory of TETRIS. Nieuwsbrief van de Nederlandse Vereniging voor Theoretische Informatica 9, 14–21 (2005)
8. Tucker, R.: TETRIS. Eureka Magazine 51, 34–35 (1992)
9. Whitelaw, C.: LUMINES: Killing the Fun (2005), http://caseyporn.com/blog/archives/000818.html
10. Whitelaw, C.: LUMINES: Killing the Fun, Part Two (2005), http://caseyporn.com/blog/archives/000819.html
11. Wikipedia contributors: LUMINES. Wikipedia, The Free Encyclopedia (2006)

Computing Proper Equilibria
of Zero-Sum Games

Peter Bro Miltersen and Troels Bjerre Sørensen

Department of Computer Science – Daimi,
University of Aarhus, Denmark
{bromille,trold}@daimi.au.dk

Abstract. We show that a *proper* equilibrium of a matrix game can be found in polynomial time by solving a linear (in the number of pure strategies of the two players) number of linear programs of roughly the same dimensions as the standard linear programs describing the Nash equilibria of the game.

1 Introduction

It has been known for more than fifty years that Nash equilibria of *matrix games* (i.e., two-player zero-sum games in normal form) coincide with pairs of maximin and minimax mixed strategies and can be found efficiently using linear programming. However, as is also well-established in game theory, the notion of a Nash equilibrium is too permissive to prescribe consistently any sensible behavior. For instance, consider the classical example of *penny matching*: Alice has to guess whether Bob hides a penny heads or tails up. If she guesses correctly, she gets the penny. The payoff matrix for this game, with Alice being the row player trying to maximize payoff and Bob being the column player trying to minimize it, is as follows.

	hide penny heads up	hide penny tails up
guess "heads up"	1	0
guess "tails up"	0	1

It is well known and easy to see that the unique pair of maximin, minimax strategies and therefore the unique Nash equilibrium in *penny matching* is for both players to mix up their two pure strategies uniformly, i.e., Alice guesses "heads up" with probability exactly $\frac{1}{2}$ and Bob hides the penny heads up with probability exactly $\frac{1}{2}$. Thus, the value of the game is $\frac{1}{2}$. There is not much more to say about this positively valued game, except that Bob clearly does not want to play it at all. Let us consider a modified version, *parsimonious penny matching*, where we give Bob the option of *teasing* Alice, by only pretending to hide a penny, but never really putting a penny at risk. This game is described by the following payoff matrix.

H.J. van den Herik et al. (Eds.): CG 2006, LNCS 4630, pp. 200–211, 2007.

	hide penny heads up	hide penny tails up	tease
guess "heads up"	1	0	0
guess "tails up"	0	1	0

It is clear that the value of *parsimonious penny matching* is 0 and that Bob's unique minimax strategy is to "tease" with probability 1. It is more interesting to consider the situation for Alice. *Any* mix of her two pure strategies is a maximin strategy (for instance, guessing "heads up" with probability 1 is a maximin strategy). The reason is that to be a maximin strategy, it is sufficient to guarantee that Alice achieves the value of the game. This value is 0 and Alice will achieve this no matter what she does. Thus, any strategy profile in which Bob chooses "tease" with probability 1 is a Nash equilibrium. But only one of these prescribes sensible (in an intuitive sense) behavior for Alice. The one where she uniformly mixes her two pure strategies as she did in the unmodified penny matching game. Indeed, it seems that Alice ought to hope that Bob (non-sensibly) chooses to hide a penny after all. Just in case he does, she should opportunistically try to get as much as she can out of such a situation and play as she would in the unmodified game.

To formalize considerations such as the above (and similar considerations for much more intricate games, including general-sum ones), *refinements* of the Nash equilibrium concept have been considered. A particularly appealing one is Myerson's notion of a *proper* equilibrium [15]. An equilibrium is said to be proper if it is a limit point of a sequence of ϵ-proper completely mixed strategy profiles for $\epsilon \to 0+$. Here, a strategy profile (i.e., a strategy for each player) is said to be completely mixed if it prescribes strictly positive probability to every pure strategy. It is said to be ϵ-proper if the following property is satisfied. If pure strategy x_i is a better reply than pure strategy x_j against the mixed strategy the profile prescribes to the other player, we have $p(x_j) \le \epsilon p(x_i)$ where $p(x_k)$ is the probability prescribed to pure strategy x_k. We refer to Myerson's paper for an intuitive justification of this refinement, but note for now that the unique proper equilibrium for *parsimonious penny matching* is the "sensible" strategy profile where Bob chooses "tease" with probability 1 and Alice uniformly mixes her two pure strategies.

In his seminal monograph on equilibrium refinements, van Damme [5], based on earlier work by Dresher [6], outlined a procedure for computing a proper equilibrium of a given matrix game. As he and Dresher describe the procedure, it is inherently exponential time. The main technical result of the present paper is a modification of the procedure so that it becomes a polynomial time algorithm. Thus, our main result is the following.

Theorem 1. *A proper equilibrium for a matrix game can be found in polynomial time in the size of the given payoff matrix.*

In addition, the algorithm we describe is also practical. We show that a proper equilibrium in a matrix game may be found by solving a linear (in the number of pure strategies of the two players) number of linear programs of roughly the same dimensions as the usual linear program describing the maximin/minimax strategies. Additional motivation for our result comes from Fiestras-Janeiro *et al.* [7] who define a notion of properness for solutions of general linear programs (not necessarily describing optimal strategies for matrix games) by a reduction to the matrix-game case and argue that this notion is relevant as a solution concept for general linear programs. They restate the exponential procedure of Dresher [4] and van Damme [16] as a way of finding the proper solutions. By the algorithm we present, we also get a polynomial time algorithm for finding a proper solution in the sense of their paper for general linear programs.

In the rest of the paper, we present our efficient algorithm for finding proper equilibria of matrix games. In Section 2, we present the original Dresher procedure which was shown by van Damme to find a proper equilibrium for a matrix game. Even though the philosophical motivation of Myerson's notion of a proper equilibrium [13] is beyond the scope of this paper, the reader unfamiliar with Myerson's notion should be able to intuitively see why the equilibrium Dresher's procedure finds is "sensible". In Section 3, we present our efficient modification and give an example of its execution. In Section 4, we conclude with a discussion on the relevance of our algorithm for AI applications and in particular, we ask if it can be extended to solving games in extensive form (i.e., game trees).

2   Background

We review some notation from van Damme [5]. A matrix game is a 3-tuple $\Gamma = (\Phi_1, \Phi_2, R)$, where Φ_i is a finite nonempty set and R is a mapping $R : \Phi_1 \times \Phi_2 \to \mathbb{R}$. The set Φ_i is the set of *pure strategies* of Player i and R is the payoff function for Player 1. We assume that the elements of Φ_i are numbered and, consequently, we will speak about the kth pure strategy of Player i. A mixed strategy s_i of Player i is a probability distribution on Φ_i. We denote the probability which s_i assigns to pure strategy k of Player i by s_i^k and we write S_i for the set of all mixed strategies of this player. If $s_i \in S_i$, then $C(s_i)$ denotes the *carrier* of s_i, i.e., $C(s_i) := \{k \in \Phi_i; s_i^k > 0.\}$ We denote the set of pure strategies which are in the carrier of some equilibrium strategy of Player i in Γ by $C_i(\Gamma)$. Note that $\Phi_i \backslash C_i(\Gamma)$ are those pure strategies for Player i that have probability 0 of being played in *all* equilibria. Thus, we can think of the strategies in $\Phi_i \backslash C_i(\Gamma)$ as the *superfluous* strategies. An alternative characterization is given below.

The payoff function R extends to mixed strategies by letting it denote the expected payoff when the mixed strategies are played. All equilibria of a matrix game Γ yield the same payoff to Player 1, and we denote this value $v(\Gamma)$ and call it the value of the game. We define $O_1(\Gamma) = \{s_1 \in S_1; R(s_1, l) \geqq v(\Gamma) \forall l \in \Phi_2\}$, and $O_2(\Gamma) = \{s_2 \in S_2; R(k, s_2) \leqq v(\Gamma) \forall k \in \Phi_1\}$. The set $O_i(\Gamma)$ is a convex polyhedron, the elements of which are called the *optimal strategies* of Player i in Γ. $O_1(\Gamma)$ and $v(\Gamma)$ can be determined by solving the linear programming problem (1):

$$\text{maximize}_{v, s_1 \in S_1} \ v \quad \text{s.t.} \ R(s_1, l) \geqq v \quad \text{for all } l \in \Phi_2 . \tag{1}$$

It was shown by Bohnenblust, Karlin and Shapley [3], and by Gale and Sherman [8] that $\Phi_2\backslash C_2(\Gamma)$ consists of exactly those pure strategies k so that for some mixed strategy $s_1 \in O_1$ we have that $R(s_1, k) > v(\Gamma)$. Thus, we can also think of $\Phi_2\backslash C_2(\Gamma)$ as the *exploitable* pure strategies for Player 2. The strategies for Player 2 against which it is possible for Player 1 to play optimally, yet get more than his "fair share" of the game. Both characterizations of $\Phi_2\backslash C_2(\Gamma)$ will be useful below.

We have now introduced the relevant notation to understand van Damme's reformulation [5, page 59] of the original procedure due to Dresher[4]

For a matrix game $\Gamma = (\Phi_1, \Phi_2, R)$ *Dresher's procedure* for selecting a particular optimal strategy of player 1 is described as follows.

(i) Set $t := 0$, write $\Phi_1^t := \Phi_1$, $\Phi_2^t := \Phi_2$ and $\Gamma^t := (\Phi_1^t, \Phi_2^t, R)$. Compute $O_1(\Gamma^t)$, i.e., the set of optimal strategies of player 1 in the game Γ^t.

(ii) If all elements of $O_1(\Gamma^t)$ are equivalent in Γ^t, then go to (v), otherwise go to (iii).

(iii) Assume that player 2 makes a mistake in Γ^t, i.e., that he assigns a positive probability only to the pure strategies which yield player 1 a payoff greater than $v(\Gamma^t)$. Hence, restrict player 2's pure strategies set to $\Phi_2^t\backslash C_2(\Gamma^t)$.

(iv) Determine the optimal strategies of player 1 which maximize the minimum gain resulting from mistakes of player 2. Hence, compute the optimal strategies of player 1 in the game $\Gamma^{t+1} := (\Phi_1^{t+1}, \Phi_2^{t+1}, R)$, where $\Phi_1^{t+1} := ext\ O_1(\Gamma^t)$ is the (finite) set of extreme optimal strategies of player 1 in Γ^t and $\Phi_2^{t+1} := \Phi_2^t\backslash C_2(\Gamma^t)$. Replace t by $t+1$ and repeat step (ii).

(v) The set of *Dresher-optimal* (or shortly *D-optimal strategies*) of player 1 in Γ is the set $D_1(\Gamma) := O_1(\Gamma^t)$.

It was shown by van Damme that if the above procedure is used to find a D-optimal strategy for Player 1 and the analogous procedure is used to find a D-optimal strategy for Player 2, then the strategy profile resulting from combining them is a proper equilibrium in the sense of Myerson.

We now discuss how to interpret the algorithm and analyze the implications for its complexity. First, strictly speaking, $O_1(\Gamma^t)$ is a set of mixed strategies for Player 1 in the game Γ^t, not the game Γ, so we need to understand how to interpret line (v). However, as is clear from the procedure, each pure strategy for Player 1 in Γ^i corresponds to a mixed strategy for Player 1 in Γ^{i-1}. Thus, each mixed strategy in Γ^i also corresponds to a mixed strategy in Γ^{i-1} and by iterating this interpretation, each mixed strategy in Γ^t can also be interpreted as a mixed strategy in Γ. In the following section, it will be convenient to have some notation for this interpretation. If s is a mixed strategy for Player 1 in Γ^i for some i, we let $\hat{s}$ be the corresponding mixed strategy in Γ and also extend this notation to sets of strategies. Second, it is not quite clear what is meant by "Compute $O_1(\Gamma^t)$" in line (i), i.e., what representation is intended at this point in the procedure for this infinite object. However, in line (iv) we are going to need ext $O_1(\Gamma^t)$, i.e., the set of all corners of $O_1(\Gamma^t)$, so we can assume that this is the finite representation we use. Indeed, van Damme is very explicit that the set of corners of the polytope $O_1(\Gamma^t)$ will be explicitly computed. He refers in the text following the procedure to an algorithm by Balinski [1] for performing such an enumeration and he notes that there are a finite number of extreme points. Also in Dresher's original formulation is it very clear that an enumeration is to be performed, and Dresher even carries out such an enumeration for a small example. *This explicit enumeration is the main source of inefficiency of the algorithm.* Indeed, it is well known and easy to see that in the worst case, the number of extremal points of a polytope defined by a linear program is exponential in the size of the linear program. Thus, the Dresher procedure as stated is an exponential procedure in the worst case sense. Also, in practice, enumerating all extremal optimal solutions to a linear program (even when this set is small) is a much more elaborate process than just finding an optimal solution. Finally, it is not explicitly stated by van Damme how to compute $\Phi_2^{t+1} := \Phi_2^t \backslash C_2(\Gamma^t)$ in line (iv) of the algorithm. In the original version by Dresher, it is done by letting $C_2(\Gamma^t)$ be the subset of Φ_2^t which yield the value of the game against every optimal strategy of Player 1, i.e., by using the characterization of $\Phi_2^t \backslash C_2(\Gamma^t)$ as the exploitable strategies for Player 2. As we have an explicit representation of ext $O_1(\Gamma^t)$, and it is enough to check for optimality against this finite set, this is one possibility. A second way to do compute $C_2(\Gamma^t)$, which is not very practical but at least polynomial, is to check each of $k \in \Phi_2^t$ for membership of $C_2(\Gamma^t)$. This could be done by solving $|\Phi_2^t|$ linear programs of the following form:

$$\max_{x,p} \; p$$

$$\text{s.t.} \quad {A'}^\top x \geq e_k p$$

$$x^\top \mathbf{1}_m = 1$$

$$x \geq \mathbf{0}_m$$

where e_k is the kth standard basis vector, and $\mathbf{0}_i$ and $\mathbf{1}_i$ are constant column vectors of height i, filled with 0s and 1s respectively, and A' is the $m \times |\Phi_2^t|$ payoff matrix of the game Γ^t with the *value* of the game Γ^t subtracted from each entry (i.e., the game matrix is "normalized" so that it has value zero). An optimal solution to the linear program with a positive value of p corresponds to an optimal strategy for Player 1 obtaining payoff strictly larger than v against the k'th pure strategy of Player 2, i.e, we have that k is exploitable by the characterization of Gale and Sherman mentioned above. This is the case if and only if k is not in $C_2(\Gamma^t)$. Alternatively, we could write a linear program whose set of feasible solutions is the optimal mixed strategies for Player 2 and with the objective function to maximize being the probability of choosing k. This formulation directly expresses whether k is superfluous or not. In both cases, we should solve $|\Phi_2^t|$ linear programs in the t'th iteration of the procedure, leading to a worst case *quadratic* number of programs being solved in total during the execution of Dresher's procedure.

3   Algorithm

To improve on the efficiency of Dresher's procedure, we have to change the way $O_1(\Gamma^t)$ is represented, since we cannot afford to enumerate the extreme points of this polyhedron. Since $O_1(\Gamma^t)$ is the set of optimal solutions to a linear program, it can be represented as a set of linear constraints. Our approach is to include the linear constraints of $O_1(\Gamma^{t-1})$ in the linear program used to obtain $O_1(\Gamma^t)$, or actually $\widehat{O_1(\Gamma^t)}$, i.e., the corresponding set of mixed strategies in the original game.

Lemma 1. *For all t, the set $\widehat{O_1(\Gamma^t)}$ is the set of x^*-parts and the value of the game Γ^t is the z^*-part of optimal solutions (x^*, z^*) to the LP:*

$$P_t: \qquad \max_{x \in \mathbb{R}^m,\, z \in \mathbb{R}} \quad z$$

$$\text{s.t.} \quad A_i'^{\top} x \geq 0_{n_i'}, \quad \forall i : 0 \leq i < t$$

$$A_t^{\top} x \geq 1_{n_t} z$$

$$x^{\top} 1_m = 1$$

$$x \geq 0_m$$

where m is $|\Phi_1|$, n_i is $|\Phi_2^i|$ and n_i' is $|C_2(\Gamma^i)|$, A_i' is the $m \times n_i'$ payoff matrix of the game $\Upsilon_i' = (\Phi_1, C_2(\Gamma^i), R)$ with the value of Γ^i (computed in a previous round) subtracted from each entry, and A_t is the $m \times n_t$ payoff matrix of the game $\Upsilon_t = (\Phi_1, \Phi_2^t, R)$.

The above lemma gives us an alternative way of computing $O_1(\Gamma^t)$. We next present an alternative way of computing $C_2(\Gamma^t)$.

Lemma 2. *Player 2's superfluous strategies in Γ^t, i.e., $\Phi_2^t \backslash C_2(\Gamma^t)$, are those k such that $p_k = 1$ in any (and all) optimal solutions to the LP:*

$$Q_t: \qquad \max_{x \in \mathbb{R}^m,\, p \in \mathbb{R}^{n_t}} \quad p^{\top} 1_{n_t}$$

$$\text{s.t.} \quad A_i'^{\top} x \geq 0_{n_i'}, \quad \forall i : 0 \leq i < t$$

$$A_t''^{\top} x \geq p$$

$$p \leq 1_{n_t}$$

$$p \geq 0_{n_t}$$

$$x \geq 0_m$$

with the same definitions as in Lemma 1 and with A_t'' being A_t with the value of Γ^t (found when solving P_t), subtracted from each entry.

Due to the space constraints, we omit the proofs of the lemmas. We are now ready to state our modification of Dresher's procedure.

Modified Dresher Procedure

  (i) Set $t := 0$, and let $\Phi_2^0 := \Phi_2$.

  (ii) Find an optimal solution to P_t.

  (iii) Find an optimal solution to Q_t. Let Φ_2^{t+1} be those $k \in \Phi_2^t$ where $p_k = 1$ in the optimal solution found.

  (iv) If $\Phi_2^{t+1} = \emptyset$ then go to (v) else replace t by $t+1$ and go to (ii).

  (v) The set of D-optimal strategies of Player 1 in Γ is the set of optimal solutions to P_t. Output any one of these optimal solutions.

Lemma 1 and 2 give us that the optimal solutions to P_t for the terminal value of t are indeed the D-optimal strategies for Player 1. By computing a D-optimal strategy for Player 1 and afterwards a D-optimal strategy for Player 2 by applying the procedure a second time, we have computed a proper equilibrium.

That the above given procedure runs in polynomial time, can be seen by observing that $|\Phi_2^t|$ decreases by at least 1 in each iteration. This means that we solve at most $|\Phi_2|$ linear programs of the form of P_t and just as many of the form of Q_t. The number of variables in P_t is $|\Phi_1| + 1$, and the number of constraints is $|\sum_{i=0}^{t-1} C_2(\Gamma^i)| + |\Phi_2^t| + 1 = |\sum_{i=0}^{t-1} C_2(\Gamma^i)| + |\Phi_2 \setminus \bigcup_{i=0}^{t-1} C_2(\Gamma^i)| + 1 = |\sum_{i=0}^{t-1} C_2(\Gamma^i)| - |\sum_{i=0}^{t-1} C_2(\Gamma^i)| + |\Phi_2| + 1 = |\Phi_2| + 1$. This is independent of t, and it is also the same number of constraints used to find just a Nash equilibrium in the standard way, i.e., the number of constraints in the linear program (1). The number of variables in Q_t is $|\Phi_1| + |\Phi_2^t|$, which is less than $|\Phi_1| + |\Phi_2|$ for all t. The number of constraints is the same as in P_t, not counting simple bounds on variables. We thus solve at most a linear number of linear programs of sizes comparable to the size of the linear program (1). Since linear programs are polynomial time solvable, the entire procedure is polynomial time. Also, from a more practical point of view, We notice that an optimal solution to P_i is a feasible solution to P_{i+1}, allowing us to "warm start" an LP-solver on P_{i+1}. We notice as well that the x-part of an optimal solution to P_i is a feasible solution to Q_i when the remaining variables are set to 0, again allowing for a "warm start".

3.1 Example

As an example of an execution of the algorithm, we will now find the proper strategy for Alice in the game of *parsimonious penny matching* from the introduction. The first linear program we need to solve, P_0 is the usual linear program for finding the Nash equilibria of the game.

$$P_0 : \quad \max_{x,z} z$$

$$\text{s.t.} \quad 1x_1 + 0x_2 \geq z$$

$$0x_1 + 1x_2 \geq z$$

$$0x_1 + 0x_2 \geq z$$

$$x_1 + x_2 = 1$$

$$x_1, x_2 \geq 0$$

Solving this, we find that the value of the game is $z^* = 0$. The next step is to decide which of Bob's strategies are superfluous. This is done by solving Q_0. Since z^* was 0, A_0'' is equal to A.

$$Q_0: \quad \max_{x,p} \; p_1 + p_2 + p_3$$

$$\text{s.t.} \quad 1x_1 + 0x_2 \geq p_1$$
$$0x_1 + 1x_2 \geq p_2$$
$$0x_1 + 0x_2 \geq p_3$$
$$p_1, p_2, p_3 \leq 1$$
$$p_1, p_2, p_3 \geq 0$$
$$x_1, x_2 \geq 0$$

Solving this, we find an optimal solution $x^* = [1,1]^\top, p^* = [1,1,0]^\top$, and therefore we may conclude that Bob's two first strategies are superfluous, i.e., that he would not willingly hide a penny. In the next iteration, Alice refines her strategy, trying to gain as much as possible from a mistake of Bob, while maintaining optimality in case no such mistake is made. Thus, we solve P_1:

$$P_1: \quad \max_{x,z} \; z$$

$$\text{s.t.} \quad 0x_1 + 0x_2 \geq 0$$
$$1x_1 + 0x_2 \geq z$$
$$0x_1 + 1x_2 \geq z$$
$$x_1 + x_2 = 1$$
$$x_1, x_2 \geq 0$$

The unique solution is $x^* = [\frac{1}{2}, \frac{1}{2}]^\top, z^* = \frac{1}{2}$. Thus, Alice can expect to gain half a penny if Bob makes the mistake of not teasing. We then check whether we can refine the strategy even further by solving Q_1:

$$Q_1: \quad \max_{x,p} \; p_1 + p_2$$

$$\text{s.t.} \quad 0x_1 + 0x_2 \geq 0$$
$$\tfrac{1}{2}x_1 - \tfrac{1}{2}x_2 \geq p_1$$
$$-\tfrac{1}{2}x_1 + \tfrac{1}{2}x_2 \geq p_2$$
$$p_1, p_2 \leq 1$$
$$p_1, p_2 \geq 0$$
$$x_1, x_2 \geq 0$$

The optimal solution has $p^* = [0,0]^\top$, and thus there are no further mistakes that can be exploited.

4   Discussion

Our main result deals with finding proper equilibria in zero-sum *normal form* games, i.e., games given by a payoff matrix. However, in many realistic situations where it is desired to compute prescriptive strategies for games with hidden information, in particular, the kinds of strategic games considered by the AI community, the game is given in *extensive form*. That is, the game is given as a *game tree* with a partition of the nodes into *information sets*, each information set describing a set of nodes mutually indistinguishable for the player to move. One may analyze an extensive form game by converting it into normal form and then analyzing the resulting matrix game. However, the conversion from extensive to normal form incurs an exponential blowup in the size of the representation. Koller, Megiddo, and von Stengel [10] showed how to use *sequence form* representation to compute efficiently minimax strategies for two-player extensive-form zero-sum games with imperfect information but perfect recall. The minimax strategies can be found from the sequence form by solving a linear program of size linear in the size of the game tree, avoiding the conversion to normal form altogether.

The Koller-Megiddo-von Stengel algorithm has been used by the AI community for solving many games, in particular variants of poker, some of them very large [16,2,9]. However, as was first pointed out by Koller and Pfeffer [11], the equilibria computed by the Koller-Megiddo-von Stengel procedure may in general be "non-sensible" in a similar sense as discussed above for matrix games. Alex Selby [16], computing a strategy for a variant of Hold'Em poker found similar problems. In a recent paper [14], we suggested that the notion of equilibrium refinements from game theory would be a natural vehicle for sorting out the insensible equilibria from the sensible ones, also for the application of computing prescriptive strategies for extensive-form zero-sum games, to be used by game-playing software. We showed how to modify the Koller-Megiddo-von Stengel algorithm so that a *quasi-perfect* equilibrium (an equilibrium refinement due to van Damme [4]) is computed, and we showed how computing such an equilibrium would eliminate the insensible behavior in the computed strategy alluded to in Selby's poker example and in many other examples as well.

An equilibrium for a zero-sum extensive form game is said to be *normal-form proper* if the corresponding equilibrium for the corresponding matrix game is proper. It was shown by van Damme that normal-form properness is a further refinement of quasi-perfection. Here, we show an example of an equilibrium for a fairly natural extensive-form game we call *Penny matching on Christmas morning*. The equilibrium arguably prescribes insensible play. However, it is quasi-perfect, and in fact, the algorithm of [14] gives the insensible equilibrium as output. However, the equilibrium is not normal-form proper, thus suggesting that this further refinement is also relevant for prescribing proper play in extensive-form zero-sum games. The game of Penny matching on Christmas morning is as follows. Recall from the introduction that in the standard penny matching game, Bob (Player 2) hides a penny and Alice (Player 1) has to guess if it is heads or tails up. If she guesses correctly, she gets the penny. If played on Christmas morning, we add a gift option. After Player 2 has hidden his penny but before Player 1 guesses, Player 2 may choose publicly to give Player 1 a gift of one penny, in addition to the one Player 1 will get if she guesses correctly. The

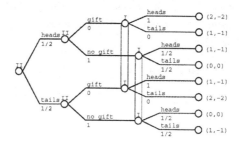

Fig. 1. *Penny matching on Christmas morning* - "bad" equilibrium

extensive form of this game as well as the pair of maximin/minimax behavioral strategies computed by the game-theory software tool textscGambit [13] using the Koller-Megiddo-von Stengel algorithm is given in Figure 1. We see that if Player 1 does not receive a gift, the strategy computed suggests that she randomizes her guess and guesses heads with probability $\frac{1}{2}$ and tails with probability $\frac{1}{2}$. This is indeed the strategy we expect to see. In contrast, if Player 1 *does* receive a gift, the strategy computed suggests that she guesses heads with probability 1. This does not seem sensible. Indeed, if she had randomized her guess, as in the "no-gift" scenario, her conditional expected payoff, conditioned by the fact that she receives the gift, would be guaranteed to be at least a penny and a half. Moreover, with the strategy suggested, this conditional expected payoff is only a penny in the case where the strategy of Player 2 happens to be the pure strategy of hiding the penny tails up and giving the gift. Thus, it seems that the unique sensible equilibrium for the game is the one where Player 1 randomizes her guess uniformly, even after having received a gift.

The "bad" equilibrium is quasi-perfect and a possible output of the algorithm for computing quasi-perfect equilibria of [14]. However, it is not normal-form proper and in fact the unique normal-form proper equilibrium for the game is the "good" equilibrium where player 1 randomizes her guess uniformly, even after having received a gift. This can be seen by converting the game to normal form and applying either the original Dresher's procedure or the version from this paper. We are not aware of any other equilibrium refinement notion that handles this and similar examples "correctly". It thus seems quite motivated to study methods for computing a normal-form proper equilibrium for a given extensive-form zero-sum game. We may do this by converting the game into normal form (incurring an exponential blowup in the size of the representation) and running Dresher's procedure. If the original version of Dresher's procedure were used, we would have a doubly-exponential time procedure. If the version of Dresher's procedure suggested in this paper is used, we have a singly-exponential time procedure. Ideally, we would like some way of combining the Koller-Megiddo-von Stengel algorithm with Dresher's procedure and obtain a polynomial time procedure, but we do not see an obvious way of doing this. We thus leave the following as a major open problem.

Open Problem 1. *Can a normal-form proper equilibrium of an extensive-form two-player zero-sum game with perfect recall be found in time polynomial in the size of the given extensive form?*

It is interesting to note that insisting on normal-form properness provides an intriguing and non-trivial solution to the problem of choosing between different minimax strategies even in *perfect information* games, a problem recently studied by Lorenz [12] using an approach very different from the equilibrium refinement approach. As an example, consider the game given in Figure 2 (payoffs are paid by Player 2 to Player 1).

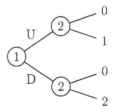

Fig. 2. *Up or Down?*

The value of the game for Player 1 is 0 and he[1] is guaranteed to obtain this value no matter what he does. However, if he chooses U and his opponent makes a mistake, he will receive a payoff of 1. In contrast, if he chooses D and his opponent makes a mistake, he will receive a payoff of 2. In the unique normal-form proper equilibrium for this game, Player I chooses U with probability 2/3 and D with probability 1/3 as can be seen by converting the program to normal form and applying Dresher's procedure. An intuitive justification for this strategy is as follows. Player 1 should imagine being up against a Player 2 that cannot avoid sometimes making mistakes, as otherwise the choice of Player 1 is irrelevant. Moreover, Player 1 should assume that Player 2 is still a rational player who can make an effort to avoid making mistakes, and in particular train himself to avoid making mistakes in certain (but not all) situations. Thus, Player 1's strategy should not be pure. In particular, if he chooses D with probability 1 (as is surely tempting), Player 2 may respond by concentrating his efforts to avoid making mistakes in his bottom node. Then, Player 1 will not get his "fair share" out of Player 2's mistakes. In conclusion, computing normal-form proper equilibria for zero-sum extensive-form games seems very interesting, even in the special case of perfect information games. Doing this special case efficiently might be easier than solving the general open problem above. It would also be interesting to compare this approach of selecting between different minimax solutions for such games with the very different approach by Lorenz.

References

1. Balinski, M.: An Algorithm for Finding All Vertices of Convex Polyhedral Sets. Journal of the Society for Industrial and Applied Mathematics 9(1), 72–88 (1961)
2. Billings, D., Burch, N., Davidson, A., Holte, R., Schaeffer, J., Schauenberg, T., Szafron, D.: Approximating Game-Theoretic Optimal Strategies for Full-Scale Poker. In: International Joint Conference on Artificial Intelligence (2003)

---

[1] For brevity and readability we use 'he' and 'his' whenever 'he' or 'she' and 'his' or 'her' are meant.

3. Bohnenblust, H.F., Karlin, S., Shapley, L.S.: Solutions of Discrete, Two-Person Games. Annals of Mathematical Studies, 37–49 (1950)
4. van Damme, E.: A Relation Between Perfect Equilibria in Extensive Form Games and Proper Equilibria in Normal Form Games. International Journal of Game Theory 13, 1–13 (1984)
5. van Damme, E.: Stability and Perfection of Nash Equlibria, 2nd edn. Springer, Germany (1991)
6. Dresher, M.: The Mathematics of Games of Strategy: Theory and Applications. Prentice-Hall, Englewood Cliffs (1961)
7. Fiestra-Janeiro, M.G., Garcia-Jurado, I., Puerto, J.: The Concept of Proper Solution in Linear Programming. Journal of Optimization Theory and Applications 106(3), 511–525 (2000)
8. Gale, D., Sherman, S.: Solutions of Finite Two-Person Games. Annals of Mathematical Studies, 37–49 (1950)
9. Gilpin, A., Sandholm, T.: Finding Equilibria in Large Sequential Games of Incomplete Information. Technical Report CMU-CS-05-158, Carnegie Mellon University (2005)
10. Koller, D., Megiddo, N., von Stengel, B.: Fast Algorithms for Finding Randomized Strategies in Game Trees. In: Proceedings of the 26th Annual ACM Symposium on the Theory of Computing, pp. 750–759. ACM Press, New York (1994)
11. Koller, D., Pfeffer, A.: Representations and Solutions for Game-Theoretic problems. Artificial Intelligence 94(1–2), 167–215 (1997)
12. Lorenz, U.: Beyond Optimal Play in Two-Person-Zerosum Games. In: Albers, S., Radzik, T. (eds.) ESA 2004. LNCS, vol. 3221, pp. 749–759. Springer, Heidelberg (2004)
13. McKelvey, R.D., McLennan, A.M., Turocy, Th.L.: Gambit: Software Tools for Game Theory, Version 0.97.0.7 (2004), http://econweb.tamu.edu/gambit
14. Miltersen, P.B., Sørensen, T.B.: Computing Sequential Equilibria for Two-Player Games. In: Proceedings of the Seventeenth Annual ACM-SIAM Symposium on Discrete Algorithms, Miami, Florida. ACM-SIAM (January 2006)
15. Myerson, R.B.: Refinements of the Nash Equilibrium Concept. International Journal of Game Theory 15, 133–154 (1978)
16. Selby, A.: Optimal Heads-Up Preflop Holdem (1999), http://www.archduke.demon.co.uk/simplex/index.html

Comparative Study of Approximate Strategies for Playing Sum Games Based on Subgame Types

Cherif R.S. Andraos, Manal M. Zaky, and Salma A. Ghoneim

Computer and Systems Engineering Department,
Faculty of Engineering, Ain Shams University, Cairo, Egypt
cherif.andraos@ieee.org, manalmourad@yahoo.com, salma_ghoneim@hotmail.com

Abstract. Combinatorial games of the form $\{\{A|B\}|\{C|D\}\}$ can be classified as either left excitable, right excitable, or equitable [2]. Several approximate strategies for playing sums of games of this form have been proposed in the literature [2,3,4]. In this work we propose a new approach for evaluating the different strategies based on the types of the subgames participating in a sum game. While previous comparisons [3,4] were only able to rank the strategies according to their average performance in a large number of randomly generated games, our evaluation is able to pinpoint the strengths and weaknesses of each strategy. We show that none of the strategies can be considered the best in an absolute sense. Therefore we recommend the development of type-based approximate strategies with enhanced performance.

1 Introduction

In [3], several strategies for playing sum games based on combinatorial game theory (cgt) were suggested, namely *BMove, MaxMove, Sente*, and *SenteQ*. Each of these strategies was compared to the others and to *HotStrat, ThermoStrat* [2] and *MiniMax* by allowing them to play against each other a large number of games where each game consisted of a sum of five randomly generated subgames of the form $\{\{A|B\}|\{C|D\}\}$ with D = rnd(50), C = D + rnd(50), B = C + rnd(50) and A = B + rnd(50), where rnd(N) is a function generating a random integer uniformly distributed in the interval [0,N]. The score of each strategy was calculated as the sum of its scores in the different tournaments.

HotStrat was shown to give the best performance [3] according to the evaluation method suggested by the author. Its score even surpassed that of optimal MiniMax search. This is due to the fact that the score used for comparing the strategies had been obtained by playing against other strategies, most of which are not optimal. This is not consistent with the assumption of a perfect opponent used in most game-playing algorithms. From our point of view a more objective test should be used to provide more reliable results.

In [4] another strategy, *HotStrat+*, was proposed for a more generalized game model. As the name implies, this strategy is an enhanced version of HotStrat.

H.J. van den Herik et al. (Eds.): CG 2006, LNCS 4630, pp. 212–219, 2007.

The authors used a more objective criterion for evaluating the new strategy based on the total number of points lost against a perfect opponent in a large number of randomly generated games of varying complexity. It was shown that HotStrat+ performs slightly but consistently better than HotStrat.

We have the following two reservations against the above results.

1. Combinatorial games of the form {{A|B}| {C|D}} can be classified [2] as left excitable (sente for left), right excitable (sente for right) or equitable (double sente or gote according to ambient temperature [5]):

   - Left excitable (type 1) if S > 4B,
   - Equitable (type 2) if 4B ≥ S ≥ 4C,
   - Right excitable (type 3) if S < 4C,

   where S = A+B+C+D. Skilful human players take into account the types of subgames in selecting the best move in a sum game. For example, in a Go endgame, double sente subgames are played immediately while one-sided sente allows a player to control the course of the game so as to execute a plan of several moves to his[1] advantage. HotStrat+ takes sente games into account in selecting the next move. SenteStrat and SenteQ give special consideration also to double sente games. Thermostrat implicitly takes subgame types into consideration in the process of decision making because of the more complete information available to it, while HotStrat, BStrat, and MaxMove are independent of subgame types. Consequently, the performance of these strategies must depend upon the types of the subgames participating in the sum.

2. In a given sum game, the number of subgames belonging to each type varies. The game model used in the comparison process should be able to produce sum games consisting of the most probable subgames types in real games. We computed the probability density function for sum games of all different type combinations of five subgames by generating a large number of sum games. We found that the game model proposed in [3] yields a sum game with subgames that are all equitable in 43% of the cases, four equitable and one excitable in about 40%, three equitable and two excitable in about 11% with the remaining 5% to cover all the remaining cases. In view of these results, it is obvious that the model used cannot objectively evaluate the performance of the approximate strategies.

No one, as far as we know, has ever considered the effect of subgame types on the performance of different strategies. In this work we investigate how the behavior of the proposed strategies varies when playing sums of games with imposed type patterns.

2   Subgame Types Effect

Experiments were conducted to investigate the effect of the types of subgames upon the performance of HotStrat+, HotStrat, ThermoStrat, MaxMove, Sente,

---

[1] For brevity and readability, we use 'he' and 'his' wherever 'he or she' and 'his' or 'her' are meant.

SenteQ, and BMove. All the experiments were conducted on sums composed of 4 subgames. The game generator was modified to produce subgames with specified type patterns as follows. Given a list of the required types of the subgames, for each type a random game is generated according to the model described above and has its type checked. If the type matches the desired value, the game is accepted; otherwise the process is repeated until the condition is satisfied.

We also used an alternate method for strategy evaluation. It consists of comparing the move selected by each strategy with the move(s) selected by Mini-Max search for a large number of random sums. The performance measure for the strategy is then calculated as the percentage of sums in which it selects the same move (or one of the moves) selected by the optimal strategy. We call this the percentage coincidence. Besides being an objective measure for comparing the strategies, this approach is consistent with the assumption of a perfect opponent used by all game-playing programs. Both this approach and the approach used in [4] are complementary in the sense that the former provides a measure of the number of errors of a certain strategy when compared to the optimal, while the latter provides a measure of the error magnitude. Table 1(a) lists the results for the strategies discussed in [3] using the proposed evaluation method. To arrive at these results we generated 1000 random sums (without forcing any type patterns) and computed the percentage coincidence for each strategy. The results obtained in [3] are shown in Table 1(b) for easy reference.

Table 1. Strategies Performance

(a) Percentage Coincidence		(b) Results from [3]	
Strategy	% of Optimal	Strategy	% of Optimal
BMove	58.70	BMove	88.39
SenteQ	64.70	MaxMove	95.40
MaxMove	66.80	SenteQ	97.63
Sente	72.30	Sente	97.71
MaxThreat	87.70	MaxThreat	99.17
ThermoStrat	89.70	ThermoStrat	99.60
HotStrat	91.50	Optimal	100.00
HotStrat+	91.60	HotStrat	100.33
Optimal	100.00		

3   Experimental Results

3.1   Objective

Below we study the performance of HotStrat+, HotStrat, ThermoStrat, Max-Move, Sente, SenteQ, and BMove relative to MiniMax for all possible subgames type patterns and players.

3.2 Experimental Setup

The test was performed using sums of 4 random subgames. Since the type of each subgame can take one of the three values 1, 2 or 3, there are exactly 15 possible type combinations. A number of 10,000 sum games of each type combination were generated and the moves selected by the different strategies were compared with the optimal move(s) for the case when it was left's turn to play. The whole process was then repeated for the case when it was right's turn to play. The percentage coincidence of the selected move by each strategy with the one(s) selected by MiniMax was recorded for each type combination, for each player.

3.3 Observations

Table 2 summarizes the obtained results when it is left's turn to start. In these results each type is represented by a number from 1 to 3 where 1 denotes left excitable, 2 equitable, and 3 right excitable games. For example, the pattern "1,1,2,3" represents a sum containing two left excitable games, one equitable, and one right excitable. The same notation is used in the remaining part of this document whenever we want to describe the subgame pattern of a certain sum. Almost equal results were obtained for right and are shown in Table 3.

Table 2. Percentage coincidence of different strategies for 10,000 sums of each type combination when left starts

Pattern	BMove	Sente	SenteQ	Max-Move	Hot-Strat	Hot-Strat+	Thermo-Strat	Optimal
1,1,1,1	92.77	92.92	94.76	89.70	**100**	**100**	99.29	100
2,2,3,3	54.93	83.29	62.96	79.91	**91.74**	**91.74**	89.36	100
1,2,2,3	60.39	82.52	64.74	77.96	**91.72**	**92.14**	90.60	100
1,1,1,2	80.48	84.27	72.90	79.68	**91.01**	**94.25**	89.20	100
1,1,2,2	71.27	81.76	66.66	77.22	**90.67**	**91.60**	90.34	100
1,1,2,3	61.79	84.33	68.90	81.08	**90.36**	**92.12**	88.39	100
2,3,3,3	47.96	87.95	59.91	**91.14**	90.28	90.28	**92.65**	100
1,2,3,3	52.26	85.78	64.17	85.05	**90.13**	**90.64**	88.69	100
2,2,2,3	57.03	76.88	65.43	70.70	**90.13**	**90.13**	88.80	100
1,2,2,2	63.88	76.19	65.76	69.37	**89.94**	**90.06**	89.06	100
2,2,2,2	57.41	70.72	65.93	63.66	**89.25**	**89.25**	88.69	100
1,1,1,3	57.97	77.96	74.40	72.89	**86.89**	77.20	**88.40**	100
1,1,3,3	46.66	73.65	64.30	67.48	**80.23**	65.52	**82.33**	100
1,3,3,3	45.03	**74.08**	55.62	72.90	72.54	61.94	**80.55**	100
3,3,3,3	48.60	79.21	48.76	**93.72**	60.59	60.59	**93.72**	100
Average	59.90	80.77	66.35	78.16	**87.03**	85.16	**89.34**	100
Std. dev.	13.15	**5.99**	10.04	8.94	9.45	12.53	**4.32**	0

Results were sorted in descending order of HotStrat performance. For each type combination, the best and second best strategies are marked with bold.

Table 3. Percentage coincidence of different strategies for 10,000 sums of each type combination when right starts

Pattern	BMove	Sente	SenteQ	Max-Move	Hot-Strat	Hot-Strat+	Thermo-Strat	Optimal
3,3,3,3	63.69	93.09	94.92	90.21	**100**	**100**	99.47	100
2,3,3,3	39.19	85.02	72.90	80.15	**91.36**	**94.55**	89.62	100
1,1,2,2	12.67	82.41	63.46	78.35	**91.32**	**91.32**	88.27	100
1,2,2,3	17.75	82.69	65.14	77.88	**90.98**	**91.72**	90.45	100
2,2,3,3	30.26	82.17	66.25	78.25	90.63	**91.66**	**91.01**	100
1,2,2,2	14.33	76.47	64.61	70.06	**90.47**	**90.47**	88.91	100
2,2,2,3	23.16	76.66	66.48	70.88	**90.44**	**90.54**	89.73	100
1,2,3,3	22.70	84.55	67.44	82.13	**90.37**	**92.29**	88.76	100
1,1,1,2	14.23	87.31	59.75	**91.23**	89.71	89.71	**92.41**	100
1,1,2,3	16.53	85.32	63.75	84.65	**89.42**	**90.04**	89.09	100
2,2,2,2	16.67	71.02	66.44	63.43	**89.13**	**89.13**	88.59	100
1,3,3,3	30.13	77.27	74.24	71.73	**87.80**	78.98	**86.65**	100
1,1,3,3	24.64	73.89	63.76	66.95	**81.09**	67.80	**82.04**	100
1,1,1,3	27.18	**74.24**	54.97	73.36	71.92	63.01	**79.33**	100
1,1,1,1	30.52	78.91	49.24	**93.54**	61.55	61.55	**93.54**	100
Average	25.58	80.73	66.22	78.19	**87.08**	85.52	**89.19**	100
Std. dev.	13.00	**5.93**	10.05	9.02	9.26	11.91	**4.60**	0

Figures 1 and 2 illustrate the results of Tables 2 and 3 respectively. To make visualization easier, the points in the figures are connected by line segments. Below we formulate eight observations.

1. As expected, the subgame types as well as the player affect the performance of the different strategies.
2. The high performance achieved by all the strategies for the combination "1,1,1,1" when it is left's turn to move and the symmetric case of "3,3,3,3" for right should not be understood as an improvement in their performance. The reason behind the high scores is that when all subgames are of type 1 (left excitable) and it is left's turn to move, left can select any subgame since the best choice for right will most probably be to respond in the same subgame (since it is left excitable) leaving the B value of that game with left free to choose the next subgame. This will be repeated for all subgames ending in a final value which is the sum of the Bs of all subgames regardless of which game was selected first by left. In fact, in these specific cases MiniMax returns a set of optimal moves instead of a single one. This causes the probability that a move selected by any strategy will coincide with optimal play to be very high, resulting in a high coincidence number for all strategies.
3. ThermoStrat achieves results that are consistently very close to best. It also has the highest average performance. Note however that this is not a true average since the probability of occurrence of each type combination is not the same. A more realistic measure of average performance would be a weighted average of the values listed.

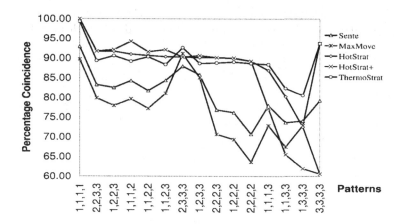

Fig. 1. Percentage coincidence of different strategies for 10,000 sums of each type combination when left starts

4. ThermoStrat's reliability is evident from the standard deviation of its results. It has the smallest standard deviation among all strategies. However, the higher standard deviation associated with the results of HotStrat and HotStrat+ show that they are less reliable than ThermoStrat. This was expected since ThermoStrat implicitly makes use of the complete information about the types of the subgames in the process of move selection.

5. HotStrat and HotStrat+ are sensitive to subgame type patterns. As seen from Tables 2 and 3 their performance varies for from 60.59% to 100%. They achieve the lowest performance for patterns that consist of only type-1 and type-3 subgames or type 3 only.

6. Tables 2 and 3 show that contrary to the results given in [4] the performance of HotStrat+ is not always better than HotStrat. We notice the following.
   – Their results are identical for patterns that do not contain type 1 subgames when it is left's turn to play and for those that do not contain type 3 subgames when it is right's turn.
   – HotStrat performance is much better than HotStrat+ for patterns that contain only type-1 and type-3 subgames where the difference varies between 8.8% & 14.7%.
   – For the rest of the patterns HotStrat+ performs slightly better than HotStrat.
   – Notice that there is no conflict between these results and the result in [4]. As mentioned above, the game generator used in both [3] and [4] is almost incapable of producing the patterns in which HotStrat's performance is better than that of HotStrat+, namely "1,3,3,3", "1,1,3,3", and "1,1,1,3". The total probability of producing these patterns has been computed and found to be equal to 0.05%.

– It is interesting to compare the average percentage coincidence for both HotStrat and HotStrat+ using our results and the result in [4], using the following formula:

$$(Ca)_s = \sum_i P_i.(Ci)_s \qquad (1)$$

where $i \in [1,15]$ is the pattern number, $(Ci)_s$ is the percentage coincidence for strategy s and pattern i as given in Table 2. P_i is the probability of the occurrence of pattern i and $(Ca)_s$ is the average percentage coincidence for Strategy s. Using (1) we get $(Ca)_{HotStrat} = 89.8\%$ and $(Ca)_{HotStrat+} = 89.9\%$, which is consistent with the results in [4] that shows that on the average HotStrat+ is slightly better than HotStrat.

7. An unexpected and very interesting result is that MaxMove, which is a very simple strategy, achieves best performance in the two oppositely symmetric cases: the case where it is left's turn to move in a sum game containing only right excitable (type 3) subgames (93.72%) and the case where it is right's turn to move in a sum of only left excitable (type 1) subgames (93.54%). Its results for these cases are equal to ThermoStrat's. We have developed a mathematical justification of this behavior [1].

8. For the same two cases in 7, HotStrat and HotStrat+ show a considerable degradation in performance.

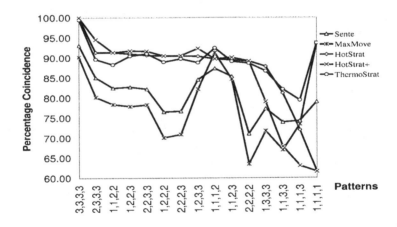

Fig. 2. Percentage coincidence of different strategies for 10,000 sums of each type combination when right starts

4   Conclusions and Future Work

We have shown that the game generator proposed in [3] and generalized in [4] is unable to produce sum games with games of different type patterns with equal probability. In fact, it is almost unable to produce certain patterns, resulting in a biased experimental setup and unreliable results. We proposed a modified version of the game generator that produces subgames with imposed type patterns. The results from our model and those reported in [3,4] agree if we compare the average performance computed as a probabilistic weighted average of the coincidence to previous ones. It would be very interesting to investigate the behavior of the strategies considered using our type-based game generator with different parameters such as the number of games or the interval used for the generation of random games. It would also be very interesting to test these strategies on values extracted from real games or on generators capable of producing realistic sum games with higher probability to sum games that are most likely to occur in real games. More experiments could be executed to compute the error magnitude of each strategy on typed subgames instead of calculating the percentage coincidence only.

We were able to prove for the first time that the performance of approximate strategies for playing sums of combinatorial games is highly dependent on the types of subgames. This dependence is minimumal in the case of ThermoStrat because of its *awareness* of the game types. This was proved experimentally by the low standard deviation in the percentage coincidence for all patterns in case of ThermoStrat. We have also shown the weakness of HotStrat and HotStrat+ in dealing with patterns that contain reverse sente subgames alone or when combined with only sente games. Our experiments were able to determine the patterns where HotStrat+ is better than HotStrat and vice versa. A very significant result is that the MaxMove strategy whose average performance is very low was shown to give the same result as ThermoStrat for the pattern with only reverse sente games. This work could be the basis for the development of type-based approximate strategies with enhanced performance.

References

1. Andraos, C.R.S.: Development of an Integrated Game Strategy Based on Classical AI Game Analysis Techniques and Combinatorial Game Theory. M.Sc. Thesis, Ain Shams University, Cairo, Egypt (2006)
2. Berlekamp, E.R., Conway, J.H., Guy, R.K.: Winning Ways. Academic Press, London, 1982. Revised version published by A.K. Peters, Natlick, MA, USA (2001–2004)
3. Cazenave, T.: Comparative evaluation of strategies based on the values of direct threats. Board Games in Academia V, Barcelona (2002),
   http://www.ai.univ-paris8.fr/~cazenave/ts.ps
4. Müller, M., Li, Z.: Locally Informed Global Search for Sums of Combinatorial Games. In: van den Herik, H.J., Björnsson, Y., Netanyahu, N.S. (eds.) CG 2004. LNCS, vol. 3846, pp. 273–284. Springer, Heidelberg (2006)
5. Spight, W.: Go Thermography - the 4/21/98 Jiang-Rui Endgame. In: Nowakowski, R. (ed.) More Games of No Chance, pp. 89–105. Cambridge University Press, Cambridge (2002)

On the Symbolic Computation of the Hardest Configurations of the Rush Hour Game[*]

Sébastien Collette[1,**], Jean-François Raskin[1], and Frédéric Servais[2]

[1] Département d'Informatique,
[2] Department of Computer & Decision Engineering, CoDE,
Université Libre de Bruxelles, Brussels, Belgium
{sebastian.collette,jraskin,frederic.servais}@ulb.ac.be

Abstract. Rush Hour is a sliding blocks game where blocks represent cars stuck in a traffic jam on a 6×6 board. The goal of the game is to allow one of the cars (the target car) to exit this traffic jam by moving the other cars out of its way. In this paper, we study the problem of finding difficult initial configurations for this game. An initial configuration is difficult if the number of car moves necessary to exit the target car is high. To solve the problem, we model the game in propositional logic and we apply symbolic model-checking techniques to study the huge graph of configurations that underlies the game. On the positive side, we show that this huge graph (containing $3.6 \cdot 10^{10}$ vertices) can be completely analyzed using symbolic model-checking techniques with reasonable computing resources. We have classified every possible initial configuration of the game according to the length of its shortest solution. On the negative side, we prove a general theorem that shows some limits of symbolic model-checking methods for board games. The result explains why some natural modeling of board games leads to the explosion of the size of symbolic data-structures.

1 Introduction

Rush Hour is a commercial sliding blocks puzzle. Pieces representing cars and trucks are placed on a 6×6 square board. The game starts by placing the vehicles according to an *initial* configuration as shown in Fig.1(a). The goal of the game is to get the red car to the board exit square, as shown in Fig.1(b), by moving the other vehicles out of its way. Cars and trucks take up two and three board squares, respectively. In an initial configuration, each car is positioned either vertically or horizontally and cannot steer from that direction during the game, i.e., each vehicle stays on its initial row or column, respectively. As a consequence, the target car must be facing the exit from the beginning. The commercial game provides 40 cards describing initial board configurations, with various number of cars. They are ranked in four levels: beginner, intermediate, advanced, and expert.

[*] Supported by the FRFC project "Centre Fédéré en Vérification" funded by the Belgian National Science Fundation (FNRS) under grant nr 2.4530.02.
[**] Aspirant du F.N.R.S.

H.J. van den Herik et al. (Eds.): CG 2006, LNCS 4630, pp. 220–233, 2007.
© Springer-Verlag Berlin Heidelberg 2007

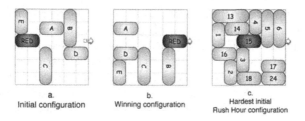

<table>
a.
Initial configuration | b.
Winning configuration | c.
Hardest initial
Rush Hour configuration
</table>

Fig. 1. RUSH HOUR

The motivation of this paper is to find the *hardest initial board configurations* of the game. We consider that a configuration is hard if the minimal number of moves necessary to exit the red car is high. This problem is challenging for two reasons. First, the number of possible initial configurations of the game is huge: around $3.6 \cdot 10^{10}$ configurations for 6×6 board. Second, finding the minimal length of a solution to a single initial configuration is already a hard problem. The RUSH HOUR algorithmic complexity was first studied by G. W. Flake and E. B. Baum in [6]. Their work inspired a more general sliding blocks complexity proof technique by R. A. Hearn and E. D. Demaine in [7]. These works show that the problem of deciding if an initial configuration is solvable for a generalized version of RUSH HOUR with arbitrary board size, is PSPACE-COMPLETE. This implies that there is no polynomial time algorithm to find a solution (unless P=PSPACE) and the length of the shortest solution of hard initial configurations grows exponentially with the board size n (provided that P $\neq$ NP and NP$\neq$ PSPACE).

In this paper, we present an elegant solution to compute the hardest configurations of the game RUSH HOUR. The solution relies on the propositional modeling of the (huge) graph of configurations that underlies the game and on the implicit exploration of this graph using symbolic model-checking techniques.

Symbolic model-checking techniques have been developed since the early 1990's by the *computer aided verification* research community and have shown successful in verifying logical properties of complex hardware circuits. Symbolic model-checking techniques are useful to explore very large graphs (with 10^{20} states for example, see [4]) and to compute properties of the paths in those graphs. The graphs are represented implicitly using symbolic data-structure such as Binary Decision Diagrams (BDDs) [3].

The three contributions of this paper are the following. First, we show that symbolic model-checking techniques can be used successfully to analyze the entire configuration space of 6×6 RUSH HOUR with reasonable computational resources. Symbolic model-checking techniques allow computing the hardest configurations of the game. Second, we show that, unfortunately, the most natural way of modeling the game into propositional logic leads to the construction of symbolic data-structures of which the size explodes. We prove that this phenomenon is not limited to the symbolic analysis of RUSH HOUR but will occur for any board game. Indeed, we show that the symbolic representation of simple constraints like "A position of the board can only hold one piece" requires BDDs of which size is exponential in the problem size. To avoid this phenomenon, we propose a dual modeling of RUSH HOUR which leads to more manageable symbolic data-structures. This modeling can be straigthforwardly translated into the input language of NuSMV [5], a state of the art symbolic model-checking tool. This second modeling allows to apply successfully symbolic model checking methods and to classify the entire set of configurations of the game according to the length of its minimal solution. This shows that the choice of modeling is crucial for the successful application of symbolic model checking techniques. The application of these techniques to other games, like chess or checkers, should be studied. Third, we show that the techniques proposed here cannot only be used to compute hard initial configurations but are also useful to analyze interesting structural properties of the game.

The rest of this paper is organized as follows. In Section 2, we formalize the problem, present a general breadth-first search algorithm and an estimation of the computing resources (time and memory) a classic explicit implementation would require. In Section 3, we recall the notion of Binary Decision Diagram and present a symbolic implementation of the algorithm. In Section 4, we propose a first modeling of the game into propositional logic, report on the explosion of the BDD for this modeling, and we develop a theoretical argument which explains why our first modeling leads to explosion in BDD size. In Section 5, we come up with a dual modeling of the game that takes into account the theoretical result of the previous section. In Section 6, we report on the success of the second modeling and present some interesting results on the hardest initial configurations of the game.

2   Formalization of the Problem

In this section, we show how the RUSH HOUR hardest configurations problem can be solved by a simple backward graph exploration equivalent to a (one-player) retrograde analysis. The possible configurations of the cars on the board define the vertices of the graph, and valid moves between configurations define the edges of the graph. After presenting this conceptually simple solution, we evaluate the cost of traversing the graph of configurations with an explicit algorithm that operates at the level of vertices of the graph (treats each vertex individually).

2.1   The Hardest Configurations

Let $G = (V, E)$ be a finite directed graph: V is the set of vertices and $E \subseteq V \times V$ is the set of edges. Let $v \in V$ and $U \subseteq V$. A path from v to U is a sequence of vertices $\rho = v_1 v_2 \ldots v_n$ such that $v_1 = v$, $v_n \in U$ and $\forall i \cdot 1 \leq i < n \cdot (v_i, v_{i+1}) \in E$. The length of the path $\rho = v_1 v_2 \ldots v_n$, noted $|\rho|$, is $n - 1$. The set of paths from v to U in G is noted $\mathsf{Path}_G(v, U)$. The distance from v to U is equal to $\mathsf{Min}\{|\rho| \mid \rho \in \mathsf{Path}_G(v, U)\}$ if $\mathsf{Path}_G(v, U)$ is non empty and equal to $+\infty$ otherwise, this value is noted $\mathsf{Dist}_G(v, U)$.

Let us now consider a graph, $G_{RH} = (V_{RH}, E_{RH})$, of which the vertices represent the valid configurations of RUSH HOUR (board configurations without collision) and the edges the valid transitions between those configurations. Let us denote the set of winning configurations as $\mathcal{W}$. A configuration $v \in V_{RH}$ is of index $n \in \mathbb{N}$ if $\text{Dist}_{G_{RH}}(v, \mathcal{W}) = n$. Our goal is to compute the set of configurations of the largest index n.

A classical breadth-first search algorithm can be applied. Starting from the set of the winning configurations, we successively compute the set of configurations of index n until reaching an empty set. The last non-empty set contains the hardest configurations of the game. In the sequel we name this algorithm *retrograde analysis* in analogy with the two-player game analysis method.

We are now equipped to compute in G_{RH} all the configurations that can reach a winning configurations and their index. The set of configurations of index k is noted C_k and it is computed inductively as:

$$C_0 = \mathcal{W}; R = \mathcal{W};$$
$$\text{for each } v \in (C_{i-1}) \text{ do} \qquad \text{for } i > 0$$
$$\quad \text{for each } w \text{ with } (w, v) \in E_{RH} \text{ and } w \notin R \text{ do}$$
$$\quad\quad \text{add } v \text{ into } C_i \text{ and into } R$$

Note that if $C_i = \emptyset$ then $C_j = \emptyset$ for all $j \geq i$. So our algorithm will start from C_0 and compute the sets C_i's until we get an empty set.

2.2   Explicit Implementation

Solving this problem with classical retrograde analysis is theoretically feasible. The challenge is to deal with the huge state space: $3.6 \cdot 10^{10}$ valid configurations.

The breath first search algorithm requires in this case a mapping between each configuration and a bit telling wether or not it has been visited before. A clever indexing scheme is not enough to make the map fit into the computer's limited memory, since it requires at least 4.5 GB ($3.6 \cdot 10^{10}$ bits). However, partitioning the problem to make it fit is straightforward. Since the vertical cars cannot leave their respective column and horizontal cars cannot steer from their respective line, the number of cars and trucks for each line is an invariant of transitions. Fixing these numbers defines a partition of our problem. We can solve each of these partitions independently of each other.

These partitions do fit easily into memory. Indeed, for a line with 1 car there are 5 possible positions, with 1 truck there are 4 possible positions, for 2 cars there are 6 possible positions, for a line with 1 car followed by a truck there are 4 positions. Let consider one of these partitions and let n_i be the number of possible positions for the i-th line and m_i the number of possible positions for the i-th column. We have $1 \leq n_i, m_i \leq 6$, thus there are at most $6^{12} = 2 \cdot 10^9$ possible configurations in a partition. We need only one bit telling if the configuration has been visited or not, so we need about $2 \cdot 10^9$ bits of memory which is about 270 MB. A closer look at the possible configurations of a partition would show a smaller memory requirement. For example, we do not need to include in the map the winning configurations nor configurations with more than 17 cars (since the board has 36 squares).

This takes care of the first obstacle: fitting the problem into the limited physical memory of a standard computer. However this is only half the solution. We broke down the problem into $6^{12} = 2 \cdot 10^9$ subproblems since there are 6 possible configurations of a line or column (no car, 1 car, 2 cars, 1 truck, 1 car followed by a truck, 1 truck followed by a car) and there are 6 lines and 6 columns. We now must solve these subproblems.

For each of these partitions we must generate the set of winning configurations. Then, for each of these winning configurations, we compute the configurations we can reach in one step, update the map accordingly and keep track of the newly reached configurations. Once all winning configurations have been treated, we apply the same operations to the set of newly reached configurations iteratively, until reaching an empty set. The tricky part that we do not tackle here is generating in an efficient manner the set of winning configurations of a partition.

This method has to treat the $3.6 \cdot 10^{10}$ configurations. According to preliminary experiments, this can be achieved in about 20 hours.

This section showed that we can use a classical retrograde analysis algorithm, but it needs a significant amount of computing resources. This also gives us a rough idea of expected performance to compare our results with.

3 Symbolic Implementation

In this section, we turn the explicit algorithm of previous section, whose basic operations treat vertices individually, into a symbolic algorithm whose basic operations treat set of vertices instead. To implement such an algorithm, we need a data structure to manipulates set of configurations efficiently. We present such a data structure and then we give our symbolic algorithm.

3.1 Symbolic Data Structure

Reduced Ordered Binary Decision Diagrams (ROBDD), introduced by R. Bryant [3] in 1986, are data-structures that canonically represent boolean functions as direct acyclic graphs. Equivalently they are a canonical representation of sets of valuations that can be exponentially smaller than the sets it represents. This data-structure has found tremendous success in verification of the logic of hardware circuits. So ROBDDs are natural candidates to represent symbolically the sets that we have to handle in the algorithm of next section.

As an illustration, the set of valid configurations containing $3.6 \cdot 10^{10}$ configurations is represented with a BDD containing 10 Million nodes in our second modeling (see Section 5). This is 3600 configurations per node. Furthermore, as we will see, the operations are performed on the compressed representation without decompression.

ROBDDs are essentially binary decision trees where the sequence of variables associated with the nodes of any path follows a given global order and where common subtrees have been shared across the tree. They provide computation of operations with interesting complexity.

More precisely, a *binary decision diagram*, or BDD, is a rooted acyclic graph with two terminal nodes of out-degree zero labeled 0 or 1 and a set of variable nodes of out-degree two. This is illustrated in Figure 2 where the dotted lines represent the low branches, i.e., variable is 0, while the solid lines represent the high branches. A BDD is ordered, OBDD, if for any path the sequence of variables associated with the nodes of this path follows a given global order. A BDD node is not unique if another of its nodes has the same variable name and low and high successors. Moreover, we say that a BDD node is a redundant test if it has identical low and high successors. Finally, an OBDD is said to be reduced, ROBDD, if all its nodes are unique and are not redundant tests.

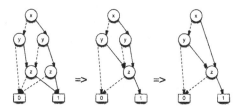

Fig. 2. BDD for the formula $z \wedge (x \vee y)$. Removal of non-unique node followed by removal of redundant test. The rightmost BDD is the canonical ROBDD for the ordering $x <$ $y < z$, both leftmost are OBDDs.

Low complexities of important operations is what makes ROBDDs attractive for verification methods and our problem. The central property of ROBDDs is that, given a variable ordering, they canonically represent Boolean functions. As a consequence, tautology, satisfiability, and equivalence are done in constant time. Let A and B be two BDDs, $|A|$ and $|B|$ are their respective size, i.e., their number of nodes. Reduction algorithms run in $O(|A|)$. Exhibiting a value that satisfies the function can be done in $O(n)$, where n is the number of variables. The SAT-count algorithm must output the number of assignments that satisfy the function, it has a running time of $O(|A|)$. Union and intersection (conjunction and disjunction) algorithms have running time of $O(|A| \cdot |B|)$. The complement implemented with a tree traversal has a running time of $O(|A|)$. Universal and existential quantification are done in $O(|A|^2)$. Finally the Pre operator, extensively used in verification techniques, consists of n existential quantification and thus has a running time of $O(|A|^{2n})$, diverse methods have been developed to make it as efficient as possible [9].

The complexity of the operations described above depends on the size of the BDD which may be exponentially smaller than the set it represents, but it may also vary between a linear and an exponential range depending on the ordering of the variables. It is therefore crucial to find a good ordering. However, finding the optimal ordering or even improving it has been proved to be a NP-COMPLETE problem. Thus efficient heuristics have been studied to tackle this problem.

While BDD have been introduced for Boolean formulas, this structure can easily be extended to finite integer domains through a Boolean encoding of the bounded integer variables. For efficiency reasons, the Boolean variables that encode an integer variable will be gathered in the variables ordering of the BDD. In the following, we will use BDD over a finite integer domain, since it is the structure used in NuSMVand other verification tools, and when considering x we will directly refer to that variable in the BDD and not to the binary variables that encode it.

3.2   Symbolic Algorithm

Let $G_{RH} = (V_{RH}, E_{RH})$ be the graph of the game defined above, and let $X = \{x_1, \ldots, x_k\}$ be a set of bounded integer variables representing the system (e.g., the position and direction of each car). To each vertex of G_{RH} corresponds a valuation of these variables. To a valuation may correspond a vertex of G_{RH}, provided the valuation defines a valid configuration.

Given a propositional formula ϕ, we note $[\![\phi]\!]$ the set of valuations that satisfy ϕ. For example, if $\phi \equiv x_1 \Rightarrow x_2$, then $[\![\phi]\!]$ is the set of valuations that maps the pair (x_1, x_2) to a pair in $\{(0,0), (0,1), (1,1)\}$.

A propositional formula ϕ over the variables $x_1, \ldots, x_k$ defines (via the set $[\![\phi]\!]$) a set of vertices or, equivalently, a set of configurations of the game. For any set of configurations, $U \subseteq V_{RH}$, considered as a set of valuations, there is a propositional formula ϕ_U such that $U = [\![\phi_U]\!]$. We note ϕ_W the proposition defining the winning configurations. In the same way if $X' = \{x'_1, \ldots, x'_k\}$ is a set of variables representing the game configuration after one transition, there is a propositional formula, ϕ_E over $\{x_1, \ldots, x_k, x'_1, \ldots, x'_k\}$ such that $E_{RH} = [\![\phi_E]\!]$.

Given a set U of vertices in G_{RH}, we define the set of one-step predecessors of U as

$$\mathsf{Pre}(U) = \{v \in V_{RH} | \exists u \in U : (v, u) \in E_{RH}\} \tag{1}$$

If U is defined by propositional formula ϕ over X', i.e. $U = [\![\phi(X')]\!]$, then $\mathsf{Pre}(U)$ is represented by the following propositional formula[1] over X:

$$\exists X' : \phi_E(X, X') \wedge \phi(X') \tag{2}$$

So we have:

$$\mathsf{Pre}([\![\phi(X')]\!]) = [\![\exists X' : \phi_E(X, X') \wedge \phi(X')]\!] \tag{3}$$

We can now symbolically apply the following algorithm:

$$\begin{aligned} C_0 &= [\![\phi_W]\!] \\ C_i &= \mathsf{Pre}(C_{i-1}) \setminus \bigcup_{0 \leq j \leq i-1} C_j \quad \text{for } i > 0 \end{aligned} \tag{4}$$

All these sets can be represented by BDDs and all these operations can be directly applied on these BDDs.

---

[1] The existential quantification is a shorthand for the disjunction over all variables over all their finite set of possible values.

4 First Propositional Model

We present in this section a first modeling of RUSH HOUR in propositional logic: we define ϕ_W and ϕ_E. We show that this first solution is not satisfactory and we give a mathematical argument that explains the phenomenon. This mathematical argument is general and has applications in the study of the symbolic analysis of other board games.

4.1 Formalization

Let n and m be two fixed parameters of the specification, n being the size of the board and m the number of cars. For the sake of readability, we make here the hypothesis that all vehicles have a length of 2, the modeling for vehicles of length 2 and 3 can be obtained from this one in straigthforward manner. Let the pair of variables (x_i, y_i) denote the cartesian coordinates of the upper-left square occupied by the i-th car, $(1, 1)$ being the lower-left corner of the board. Let the variable h_i indicate the orientation of the vehicle, i.e., h_i is 1 if the i-th car is horizontal and 0 if it is vertical. The target car uses index 1. We note $X = \{x_1, y_1, h_1, ..., x_m, y_m, h_m\}$ the set of the system variables and $X' = \{x_1', y_1', h_1', ..., x_m', y_m', h_m'\}$ the set of variables describing the configuration after one transition, this will be useful to specify the evolution of the system.

The set of all possible configurations is $S = (\{1, ..., n\}^2 \times \{0, 1\})^m$. We specify 3 relations on S: the invariant of the system $Invar \subseteq S$, which denotes the legal configurations, the transition relation between configurations $Trans \subseteq S \times S$, which does not check collision, and $Win \subseteq S$ the set of configurations with the target car on the exit square. We have: $\phi_W(X) = Win(X) \wedge Invar(X)$ and $\phi_E(X, X') = Trans(X, X') \wedge Invar(X')$

To specify those relations we use propositional formulas over finite integer domains.

Invariant. Proposition (5) states that cars are fully on the board, (6) states that cars do not overlap. The *Invar* relation is the conjunction of (5) and (6). $[\![Invar]\!]$ is the set of all valid states.

$$\bigwedge_{1 \leq i \leq m} (h_i < x_i + h_i \leq n) \wedge ((1 - h_i) < y_i \leq n) \quad (5)$$

$$\bigwedge_{\substack{1 \leq i, j \leq m \\ i \neq j}} (x_i > x_j + h_j) \vee (x_j > x_i + h_i) \vee (y_i > y_j + (1 - h_j)) \vee (y_j > y_i + (1 - h_i)) \quad (6)$$

In the same manner, we have propositional formulas for the set of winning configurations and for the transition relation. We omit them here, but we will give another complete formalization of RUSH HOUR in Section 5.

Having formalized the RUSH-HOUR rules in such a way that for any couple (m, n) we obtain a Boolean propositional specification that describes the game, we can apply our algorithm.

Table 1. Number of nodes in the *Invar* BDD relatively to board sizes and number of cars

# Cars	4×4	5×5	6×6
2	123	233	368
3	1,237	3,918	9,490
4	7,334	44,209	172,583
5	24,227	321,114	2,153,132
6	44,209	1,520,760	—
7	50,081	—	—
8	30,762	—	—
9	1	—	—

4.2 Results of the First Implementation

We ran our specification for board sizes n ranging from 4 to 6. For each board size the number of cars m ranged from 2 to the maximum our system memory could handle. As the number of cars increases the *Invar* BDD size explodes. This can be observed in Table 1. We only report on the *Invar* BDD, since the *Win* and *Trans* BDD sizes do not explode and are thus not relevant here.

The explosion in the memory consumption limited us to the exploration of boards of sizes 5 and 6, with a number of cars smaller than 6 and 5 respectively. More complex systems cannot be handled with this approach.

It is a fundamental BDD property that its size increases with the number of inter-variable dependencies. This is the reason why the *Invar* BDD size explodes. As mentioned above, the purpose of this BDD is to validate car positions against collisions. This puts in interdependency all car positions, since each car position must be checked against all others.

These first negative experiments motivate the next section where we show that board games are intrinsically difficult to model with ROBDDs.

4.3 Limitation of ROBDD-Based Methods for Board Games

In this Subsection, we abstract the collision problem to linear boards filled with tokens. We exhibit a lower bound on the size of the ROBDD that detects a collision on this linear board. We obtain a two-dimension result as a direct corollary.

An assignment of a set of tokens modeled by variables in $X = \{x_1, \ldots, x_m\}$ on the board of size n is a function $v : \{x_1, \ldots, x_m\} \to \{1, \ldots, n\}$. There is no collision iff this function is injective. This is formalized by the following propositional formula over a finite integer domain:

$$\phi_{coll} = \bigwedge_{1 \leq i,j \leq m} i \neq j \to (x_i \neq x_j) \tag{7}$$

Lower-bound results for ROBDDs are generally based on the concept of a fooling set. Fooling sets were introduced by Sedgewick for VLSI and then applied by Bryant to ROBDD [2]. We adapt this notion here for finite integer domains.

Let $X = \{x_1, \ldots, x_m\}$ be the set of variables of which the domain of values is $\{1, \ldots, n\}$.

Definition 1. *An input assignment is a function* $v : X \rightarrow \{1, \ldots, n\}$. *Given* (L, R) *a partition of* X. *We call a **left (right) input assignment** any function* $l : L \rightarrow \{1, \ldots, n\}$ $(r : R \rightarrow \{1, \ldots, n\})$. *We denote by* $l \cdot r$ *the input assignment defined by* l *and* r *on* X. *We say that an OBDD is compatible with a partition* (L, R) *iff all variables of* L *precede all variables of* R *in this OBDD variable ordering.*

Before defining the notion of fooling set, we need an additional notion. Let $V = \{v \mid v : X \rightarrow \{1, \ldots, n\}\}$ be the set of valuations for variables in X. A function $f : V \rightarrow \{0, 1\}$ partitions the valuations as making the function true or false. We compactly note the type of such a function by $f : [X \rightarrow \{1, \ldots, n\}] \rightarrow \{0, 1\}$.

Definition 2. *Let* (L, R) *be a partition of* X *and* f *be a function such that* $f : [X \rightarrow \{1, \ldots, n\}] \rightarrow \{0, 1\}$. *A **fooling set** F for f over L is a set of left assignments such that: for any* $l, l' \in F, l \neq l'$, *there exists a right assignment* r *with* $f(l \cdot r) \neq f(l' \cdot r)$. *Such a right assignment is said to distinguish between* l *and* l'.

Lemma 1. *Given a partition* (L, R) *over* X, *a function* $f : [X \rightarrow \{1, \ldots, n\}] \rightarrow \{0, 1\}$ *and a fooling set* F *for* f *over* L, *then any OBDD compatible with* (L, R) *has more than* $\#F$ *nodes.*

Proof. For any two distinct left assignments, l and l', of F there exists, by definition of F, a right assignment r that distinguish them, i.e., such that $f(l \cdot r) = 0$ and $f(l' \cdot r) = 1$. It follows that l and l' must lead to two different "intermediate" nodes in the OBDD and thus that there is at least as many nodes as there are elements in F.

We now prove a lower bound on the size of any OBDD that detects the collision of tokens on a linear board.

Theorem 1. *Let* $f : [X \rightarrow \{1, \ldots, n\}] \rightarrow \{0, 1\}$ *such that* $f(v) = 1$ *iff* $v \models \phi_{coll}$, *where* ϕ_{coll} *is defined in (7). Let* A *be an ROBDD over the set of variables* $X = \{x_1, x_2, \ldots, x_m\}$ *representing* f. A *has at least* C_n^{m-1} *nodes.*

Proof. Let $\mathcal{N}$ be the set of subsets of $m - 1$ values from $\{1, \ldots, n\}$ and let $X_1 = \{x_1, \ldots, x_{m-1}\}$ and $X_2 = \{x_m\}$. To each $N \in \mathcal{N}$ we associate one bijection $f_N : X_1 \rightarrow N$. We note $\mathcal{F}$ to be this set of functions. $\mathcal{F}$ is a fooling set for f over X_1: let $l_1, l_2 \in \mathcal{F}$ with $l_1 \neq l_2$. By construction of $\mathcal{F}$, we know that $codom(l_1) \neq codom(l_2)$, and let n_1 be a value such that $n_1 \in codom(l_1)$ and $n_1 \notin codom(l_2)$. Let $r : X_2 \rightarrow \{1, \ldots, n\}$ be an injective function such that $codom(r) \cap codom(l_1) = n_1$ and $codom(r) \cap codom(l_2) = \phi$. We have $f(l_1 \cdot r) = 0$ and $f(l_2 \cdot r) = 1$. Applying lemma 1 finishes the proof since $\mathcal{N}$ has C_n^{m-1} elements.

Table 2. Lower bounds on the size of the ROBDDs detecting pieces collisions for chess, draughts and american checkers board

# of Pieces	Chess	Draughts	American checkers	RUSH HOUR Invar BDD size (observed)
2	64	49	32	368
3	2,016	1,176	496	9,490
4	41,664	18,424	4,960	172,583
5	635,380	211,876	35,960	2,100,000
6	7,600,000	1.900,000	201,376	—
7	75,000,000	14,000,000	906.192	—
8	620,000,000	86,000,000	3,400,000	—

Since a two-dimension board is equivalent to a linear board with n^2 squares, we have the following corollary.

Corollary 1. *Let A be a BDD over the position variables $X = \{x_1, y_1, x_2, y_2, ..., x_m, y_m\}$ for a two-dimension board of size n with m tokens. If, for every $1 \leq i \leq m$ the variables $\{x_i, y_i\}$ are gathered in the BDD variable ordering, then A has at least $C_{n^2}^{m-1}$ nodes.*

Note that this result is fundamentally connected to the chosen encoding of the problem. The token positions are encoded in a cartesian-like board coordinates style. Applying this result to board games we obtain a lower bound on any ROBDD representing the collision of pieces on a chess board or on draughts board with the afore mentioned encoding. The lower bounds in Table 2 suggests that this technique is not suitable to explore chess and draughts with more than 5 to 6 pieces (additional complexity will be brought in with the complex rules of these games).

A dual encoding is to use a boolean variable for every square of the board that indicates if the square is occupied or not. For board games with more complex tokens, like american checkers or chess, an integer, that indicates which kind of pieces occupied the board if one, is required.

Using this dual encoding, Baldamus et al., explored the possibility to solve American checkers with ROBDD[1]. They observed an explosion in the size of ROBDD preventing them to solve American checkers for boards with size greater than 4 × 4. American checkers is the simplest game considered here. However, the number of legal positions is estimated to be 10^{18}.

5   Dual Propositional Model

In the light of the previous section, we propose now a dual encoding of the RUSH HOUR board which limits the explosion of the size of the symbolic structure. Because vehicles take more than one square, we work on a line and column level, instead of on a square level as in [1].

We have shown that interdependencies between the variables lead to huge ROBDDs. Here, we try to limit these interdependencies using a specific property of RUSH HOUR: two horizontal cars on different lines can never collide. Similarly, vertical cars cannot collide with other vertical cars that are not on the same column. This is the basic idea behind our second model. Again, for the sake of readability, the model below is limited to vehicle of size 2. Our actual implementation is general, it takes into account vehicles of size 2 and 3.

Let n be the size of the board. On each column and each line we have at most $k = \lfloor n/2 \rfloor$ cars. Let $h_{i,j} = (o^h_{i,j}, p^h_{i,j})$, $1 \leq i \leq n$ and $1 \leq j \leq k$ represents the j-th horizontal car of the i-th row, such that this car is on the board if $o^h_{i,j} = 1$ and out of the board if $o^h_{i,j} = 0$. If on the board, its leftmost square is on the $p^h_{i,j}$ square (from the left) of the i-th row. Similarly, let $v_{i,j} = (o^v_{i,j}, p^v_{i,j})$, $1 \leq i \leq n$ and $1 \leq j \leq k$ represents the j-th vertical car of the i-th column, such that this car is on the board if $o^v_{i,j} = 1$ and out of the board if $o^v_{i,j} = 0$ and its upper square is on the $p^v_{i,j}$ square (starting from the bottom) of the i-th column. We have $0 \leq p^v_{i,j}, p^v_{i,j} < n$. Let $(o'^h_{i,j}, p'^h_{i,j})$ and $(o'^v_{i,j}, p'^v_{i,j})$, for $1 \leq i \leq n$ and $1 \leq j \leq k$, describe the configuration after a transition.

Invariant. Proposition (8) states that cars on the same line, column, do not overlap and (9) that any horizontal car does not collide with any vertical car. The *Invar* relation is the conjunction of (8) and (9).

$$\bigwedge_{\substack{d \in \{h,v\} \\ 1 \leq i \leq n, 1 \leq j,j' \leq k}} (o^d_{i,j} = 1 \wedge o^d_{i,j'} = 1 \wedge j < j') \rightarrow p^d_{i,j} < p^d_{i,j'} - 1 \quad (8)$$

$$\bigwedge_{\substack{1 \leq i,i' \leq n \\ 1 \leq j,j' \leq k}} \left(\begin{array}{c} (o^h_{i,j} = 1 \wedge o^v_{i',j'} = 1) \\ \rightarrow \\ ((p^h_{i,j} \leq i' - 2) \vee (p^h_{i,j} > i') \vee (p^v_{i',j'} \leq i - 2) \vee (p^v_{i',j'} > i)) \end{array} \right) \quad (9)$$

Transition. Proposition (10) states that only one car is moving during one transition. Proposition (11) states that vehicles on the board stay on the board, and vehicles out of the board stay out of the board. Finally, proposition (12) states that cars move one square at a time. The transition relation is the conjunction of these propositions.

$$\bigwedge_{\substack{d,d' \in \{h,v\} \\ 1 \leq i,i' \leq n, 1 \leq j,j' \leq k}} p^d_{i,j} \neq p'^d_{i,j} \rightarrow p^{d'}_{i',j'} = p'^{d'}_{i',j'} \quad (10)$$

$$\bigwedge_{\substack{d \in \{h,v\} \\ 1 \leq i \leq n, 1 \leq j \leq k}} o^d_{i,j} = o'^d_{i,j} \quad (11)$$

$$\bigwedge_{\substack{d \in \{h,v\} \\ 1 \leq i \leq n, 1 \leq j \leq k}} |p^d_{i,j} - p'^d_{i,j}| \leq 1 \quad (12)$$

Table 3. Number of configurations in the furthest frontiers

	Index							
	93	92	91	90	89	88	87	86	85
# Configurations	1	6	14	26	47	80	123	172	223
# Dominating Configurations	1	0	0	2	2	2	1	3	6

Winning Configuration. One of the cars of the exit square row is on the board and positioned on the exit square. Let e be the number of the exit square row, we have:

$$\bigvee_{1 \leq j \leq k} [(p_{e,j}^d = n - 1) \wedge (o_{e,j}^d = 1)] \tag{13}$$

6   Results: RUSH HOUR Hardest Configuration

Contrary to the first encoding, the second encoding gives rise to manageable symbolic data-structures. Using NuSMV, we were able to analyze the entire configuration space of RUSH HOUR. Here are some representative results of our analysis.

Hardest Configuration. The hardest configuration of RUSH HOUR is given in Figure 1c and it requires 93 steps to reach a winning configuration. From that initial configuration 24132 configurations can be reached. This gives a good idea of the difficulty of this configuration: please give it a try.

Besides finding the hardest configuration, our analysis has classified every solvable configuration according to the length of its minimal solution. Table 3 presents the number of configurations for the greatest indexes. We learn that there are 10^{10} winning configurations and $2.98 \cdot 10^{10}$ solvable configurations, thus, about 7 billions of valid configurations have no solution, while about 19 billion non-winning configurations have one. The vast majority of the latter are very easy (shortest solution is very short).

Our symbolic solution can also be used to isolate, what we think are, the most interesting configurations of the game. We say that one configuration *dominates* another if the latter is reachable from the former and the index of the former is greater than the index of the latter. We present in Table 3 for each index the number of configurations that are not dominated.

Performance. The results were obtained on an Intel(R) Xeon(TM) CPU 3.06GHz and our symbolic algorithm used up to 1.5GB. It took about 10 hours to complete the total exploration of RUSH HOUR. Since there are about $3.6 \cdot 10^{10}$ valid configurations this is about 10^6 configurations each second. While the running time is comparable to the explicit implementation shown in Subsection 2.2, we stress that the symbolic method is more generic (standard data-structure, generic algorithms), while the explicit implementation had to be conceived for

RUSH HOUR. Moreover, at the end of the computation, we keep the whole structure in memory. This allows us to perform various kinds of queries easily. For instance, we can retrieve all hard configurations without trucks, with exactly 3 cars, etc.

These results should encourage to study more deeply the application of symbolic methods to other games such as chess or checkers.

References

1. Baldamus, M., Schneider, K., Wenz, M., Ziller, R.: Can American Checkers be Solved by Means of Symbolic Model Checking? In: Workshop on Formal Methods Elsewhere, Technical Report 00-11, pp. 3–17. University of Kent at Cantebury, UK (2000)
2. Bryant, R.E.: On the Complexity of VLSI Implementations and Graph Representations of Boolean Functions with Application to Integer Multiplication. IEEE Transactions on Computers 40(2), 205–213 (1991)
3. Bryant, R.E.: Graph-based Algorithms for Boolean Function Manipulation. IEEE Transactions on Computers 35(8), 677–691 (1986)
4. Burch, J.R., Clarke, E.M., McMillan, K.L., Dill, D.L.: Symbolic Model Checking: 10^{20} States and Beyond. In: Proceedings of the Fifth Annual IEEE Symposium on Logic in Computer Science, pp. 1–33. IEEE Computer Society Press, Los Alamitos (1990)
5. Cimatti, A., Clarke, E., Giunchiglia, E., Giunchiglia, F., Pistore, M., Roveri, M., Sebastiani, R.: NuSMV 2: An OpenSource Tool for Symbolic Model Checking. In: Proceedings of the International Conference on Computer-Aided Verification (2002)
6. Flake, G.W., Baum, E.B.: RUSH HOUR is PSPACE-complete, or Why You Should Generously Tip Parking Lot Attendants. Theoretical Computer Science 270(1-2), 895–911 (2002)
7. Hearn, R.A., Demaine, E.D.: PSPACE-Completeness of Sliding-Block Puzzles and Other Problems through the Nondeterministic Constraint Logic Model of Computation. Theoretical Computer Science 343(1–2), 72–96 (2005)
8. Holzmann, G.J.: SPIN Model Checker, The: Primer and Reference Manual. Addison-Wesley, Boston (2004)
9. Meinel, C., Theobald, T.: Algorithms and Data Structures in VLSI Design: OBDD - Foundations and Applications. Springer, Heidelberg (1998)

Cheat-Proof Serverless Network Games

Shunsaku Kato[1,*], Shuichi Miyazaki[2], Yusuke Nishimura[1], and Yasuo Okabe[2]

[1] Graduate School of Informatics,
[2] Academic Center for Computing and Media Studies,
Kyoto University, Kyoto, Japan
{shunsaku,yusuke}@net.ist.i.kyoto-u.ac.jp
{shuichi,okabe}@media.kyoto-u.ac.jp

Abstract. We consider playing online games on peer-to-peer networks, without assuming servers that control the execution of a game. In such an environment, players may cheat the opponent by, for example, illegally replacing the cards in their hands. The aim of this paper is to examine a possibility of excluding such cheatings. We show that by employing cryptographic techniques, we can exclude some types of cheating at some level. Finally, based on our discussion, we implement the cheat-proof network "Gunjin-Shogi", which is a variant of Japanese Chess.

1 Introduction

Because of recent progress of technology in computer networks, the number of Internet users has rapidly increased, which made it popular to play games on the Internet. However, it is difficult to prove that players act honestly since we cannot see the opponents' actions physically. So, there may arise a risk of being cheated. In fact, it is reported that in some survey, the rate of online game players who have never cheated nor encountered other players' cheating is only 10% [6,10]. One natural and simple solution to resolve this problem is to instal a server that controls the whole execution of the game. Actually, there are a plenty of client-server type network games in which servers play the role of a judge. However, there are several inconveniences in this server-client model. First, we have to believe that the servers work perfectly. Whatever the case, there still remains room for being cheated by the server. This pessimistic assumption is somehow realistic when we recall recently prevailing crimes on the Internet. Second, using servers may restrict the number of players because of the capability of servers. For example, there is a Japanese Chess site that allows up to about 3,000 users to login[1] simultaneously. This motivates us to consider cheat-proof *serverless* network games.

There is a fair amount of research work on cheating in online games [1,2,7,9, 10,11,12]. Yan and Choi [10] consider cheatings actually happening or possible to happen in online games. They classify cheatings into 11 categories, which includes disconnecting PCs from the Internet when players are to lose, cracking

---

[*] Currently at Nintendo Co., Ltd. Entertainment Analysis & Development Division.
[1] Internet Shogi Dojo (in Japanese) http://www.shogidojo.com/dojo/dojoindex.htm

H.J. van den Herik et al. (Eds.): CG 2006, LNCS 4630, pp. 234–243, 2007.

passwords of game sites, beating opponents by DoS(Denial of Service) attacks, and so on, and also propose methods to resolve them. Later, Yan [11] focuses on Contract Bridge and enumerate cheatings possible in online play. Furthermore, Yan and Randell [12] extend categories described in [10] into 15 categories, and also gives a systematic taxonomy of cheatings in online games. Baughman and Levine [1] deal with real-time online games, such as fighting games and shooting games. They focus on the way of mitigating a cheating caused by the difficulty of distinguishing packet delays by network congestion from one by cheating. From the theoretical point of view, several cryptographic protocols have been proposed to perform communication securely on peer-to-peer networks, which includes Poker [8, 14], and comparing the amount of money [13]. In 1987, Goldreich, Micali, and Wigderson [5] generalized above protocols and proposed a multi-party protocol that can perform any task one may reasonably consider. However, their protocol is too general and it includes computational overhead, so research on developing or sophisticating protocols for specific applications has followed, e.g., electronic election [4], and electronic bidding [3].

In this research, we consider applying such protocols to network games, inspired by the following idea. Note that, although cheating in an electronic election or electronic money is a serious problem, its effect in case of games is smaller in some cases, or only players feel uncomfortable. Hence, we may relax the condition for preventing cheatings in our case, by which we may be able to obtain an advantage such as simplifying protocols, or reducing computational complexity.

In this paper, we start by formalizing games. Here, we focus on static games, which means that each player makes a decision in turn, or, more precisely, each player's action is synchronous to discrete time steps, such as Poker, Chess, and Go. We define games by using sequential machines.

Based on this model, we will propose a method of executing games. We first consider an easier case where players act honestly by obeying protocols. Even in this case, however, execution is not trivial in peer-to-peer assumption; there are two major problems. First, states of a sequential machine must be maintained only by the players since there is no server. However, no player is allowed to know the states since a state contains information that must be kept secret from the players. Second, to proceed a game, the next state of the machine must be computed only by the players. Again, however, the current state, actions (inputs) of other players, and the next state must be hidden from the players. To resolve these problems, we exploit cryptographic tools. For the first problem, we use a secret sharing scheme, and for the second, we use a multi-party protocol.

Next, we impose a stronger assumption that players may cheat for their benefit. We first enumerate the cheatings possible in our game model. We consider cheatings only by exploiting game rules; hence, cheatings like DoS attacking are out of our scope. We then define three levels of protection, and examine which type of cheating can be protected at which level.

Finally, based on the above observation, we implement Gunjin-Shogi using Java 1.4. Gunjin-Shogi is a variant of Japanese Chess, but a player is not allowed to know the type of opponent's pieces, namely, each player has information secret to the opponent. Our program equips a verification phase at the end of the game, by which the players can detect a cheating if it is committed during the game.

2   Definition of the Game

In this section, we define games using sequential machines. A game is a nine-tuple $G = (n, \mathcal{I}, \mathcal{O}, S, s_0, F, R, \delta, \Lambda)$. Here, n is a positive integer, which represents the number of players. $\mathcal{I} = \{I_1, I_2, \cdots, I_n\}$ is a class of input sets, where each I_i is a set of player P_i's inputs. Similarly, $\mathcal{O} = \{O_1, O_2, \cdots, O_n\}$ is a class of output sets. $S = \{s_0, s_1, \cdots, s_\ell\}$ is a set of states. $s_0 \in S$ is the initial state, and $F \subseteq S$ is a set of final states. R is a set of random strings. $\delta : S \times I_1 \times I_2 \times \cdots \times I_n \times R \to S$ is a state transition function. Finally, $\Lambda = \{\lambda_1, \lambda_2, \cdots, \lambda_n\}$ is a set of output functions, where each $\lambda_i : S \to O_i$ is a function for player P_i.

Formally, a state is nothing more than an element in a set, which could be the set of natural numbers, or the set of strings on $\{0, 1\}$. Intuitively, a state is an encoding of all information at each time-step of the game. For example, in Chess, a state is an encoding of the player's turn and the position of pieces on the board, or in Poker, it consists of information on the player's turn, each player's hand, talon cards, and discarded cards.

The game proceeds as follows. The state of the machine is initially s_0. Assume that in the current step, the machine is in the state s. At this step, each player P_j $(1 \leq j \leq n)$ selects an input $i_j \in I_j$ according to his[2] strategy. Also, the machine selects a random string r from R. The machine computes its next state $s' = \delta(s, i_1, i_2, \cdots, i_n, r)$, and then, $\lambda_j(s')$ for each j, and informs it to player P_j. The game ends if the machine reaches a state in F.

The above formalization is general enough in the following sense.

- By employing output functions, we can deal with different types of secret information. Consider Poker, for example. The state s of the machine contains information on all players' hands and the talon cards. However, no player is allowed to see the other players' hands and talon cards. The output $\lambda_i(s)$ extracts partial information from the state corresponding to the information that player P_i is allowed to know.
- In games such as Cards and Chess, only one player acts at each step. To represent this type of games, we simply have to restrict a transition function as follows. If, in the state s, P_1 is the player to make an action, then $\delta(s, i_1, i_2, \cdots, i_n, r) = \delta(s, i_1, i'_2, \cdots, i'_n, r')$ for any $i_1, i_2, \cdots, i_n, r, i'_2, \cdots, i'_n$ and r'. This means that δ depends only on the second argument if the first argument is s.
- For games that have no randomness, we may simply restrict the transition function so that it does not depend on r as before. Namely, $\delta(s, i_1, i_2, \cdots, i_n, r_1) = \delta(s, i_1, i_2, \cdots, i_n, r_2)$ for any $s, i_1, i_2, \cdots, i_n, r_1$, and r_2.

---

[2] For brevity and readability, we use 'he' and 'his' wherever 'he or she' and 'his or her' are meant.

- For perfect information games, where there is no need to hide game information, such as Chess, we may set $\lambda_i(s) = s$ for every i. If states have to be secret but all players may share the same information, such as in the Nervous breakdown game, we may set $\lambda_i = \lambda_j$ for all i and j.

One may check that most of the popular static games, such as Chess, HIT & BLOW, Backgammon, etc., can be formalized in our model by appropriately choosing δ, λ, and so on.

3   Playing Games on Peer-to-Peer Environment

Now, we propose a protocol to simulate sequential machines only by players. In this section, we assume that each player acts honestly, obeying protocols. Even assuming this, it is non-trivial to execute games because some information must be kept secret from every player, while information itself must be maintained only by players since there is no server playing a role of a game manager. In the following, we show how to resolve such problems. Without loss of generality, we assume that states, inputs, and outputs are strings in $\{0,1\}^*$.

Maintaining states. To simulate computation of a sequential machine, it is necessary to maintain its states. However, for games in which players are not allowed to know the game states, there arises a difficulty; players must maintain game states without knowing the actual value. We will use a simple secret sharing algorithm. Recall that a state s is a string in $\{0,1\}^*$. Determine strings s_i $(1 \leq i \leq n)$ so that they satisfy $s = s_1 \oplus s_2 \oplus \cdots \oplus s_n$. Each player P_i keeps s_i secret. Here, each s_i has the same length with s, and XOR is a bitwise operation. This secret sharing method is convenient in using multi-party protocols for computing the next state, which we will see later.

Secrecy of inputs. At each step, players have to give an input to the machine. If there is a server that simulates the machine, each player has only to send an input to the server. However, in our case, all players must simulate the computation of a machine, which means that each player has to reveal a part of one's input to other players. To keep the input secret, we use the same technique as above. If an input of the player P_j is i_j, then P_j determines i_{j_k} $(1 \leq k \leq n)$ that satisfy $i_j = i_{j_1} \oplus i_{j_2} \oplus \cdots \oplus i_{j_n}$, and sends i_{j_k} to player P_k $(1 \leq k \leq j-1, j+1 \leq k \leq n)$. It is easy to see that as long as P_j keeps i_{j_j} secret, no player can obtain information on i_j.

Generating random strings. For a game that involves randomness, a string r must be chosen at random, and in some cases, it must be kept secret from all players. To perform it without servers, each player generates a random string r_i, and keeps it secret. The real random string is defined as $r = r_1 \oplus r_2 \oplus \cdots \oplus r_n$.

Computing δ. Recall that inputs for δ are s, $i_j(1 \leq j \leq n)$, and r, which are not known to players. However, players must compute the function δ by cooperating in order to proceed the game. Furthermore, the output of the function must be kept secret again. To compute δ, we use a multi-party protocol [5]. Let us briefly explain the multi-party protocol. Let $f(x_1, x_2, \cdots, x_n)$ be a function from $\{0, 1\}^n$ to $\{0, 1\}$. There are n players P_i $(1 \leq i \leq n)$, and each P_i possesses an input bit x_i. Using the multi-party protocol, every player can know the correct value of $f(x_1, x_2, \cdots, x_n)$, while the player P_i cannot know the value of any x_j $(j \neq i)$. To perform the multi-party protocol, each player P_i computes x_{i_j} $(1 \leq j \leq n)$ such that $x_i = x_{i_1} \oplus x_{i_2} \oplus \cdots \oplus x_{i_n}$, and sends x_{i_j} to P_j. So, P_i has x_{j_i} for $1 \leq j \leq n$ at the beginning of the protocol. Computation is performed step by step by following the protocol, and finally, each player obtains a bit y_i. If every player follows the protocol, it is guaranteed that $f(x_1, x_2, \cdots, x_n) = y_1 \oplus y_2 \oplus \cdots \oplus y_n$. At the final step, each player P_i broadcasts y_i to all other players, so that all players can compute the output by themselves.

Recall that all inputs for δ are shared by the secret sharing method described before. So, this protocol can be easily adopted to compute δ. In our case, however, each player does not broadcast the output y_i, in order to keep the next state secret again.

Computing λ. This can be done by a protocol similar to the case of computing δ. After the completion of the multi-party protocol for computing λ_i, each player P_j knows secret information o_j. In the case of δ, P_j keeps it secret, but this time, P_j sends o_j only to P_i, so that only P_i can compute the output of λ_i.

4   Excluding Cheating

In the previous section, we assumed that all players follow the protocol. In this section, we assume the existence of malicious players who try to get unfair advantage by cheating. Under this assumption, we examine which type of cheating can be excluded to what extent. We first enumerate the types of cheating possible in our game model.

Cheating 1: A player P_j gives an illegal input $x \notin I_j$.
Cheating 2: A player leaks his input to other players.
Cheating 3: A player eavesdrops other players' input.
Cheating 4: A player modifies other players' input.
Cheating 5: A player eavesdrops a random string.
Cheating 6: A player modifies a random string.
Cheating 7: A player leaks output (his secret information) to other players.
Cheating 8: A player eavesdrops outputs for other players.
Cheating 9: A player modifies outputs for other players.
Cheating 10: A player eavesdrops the state of the game.
Cheating 11: A player modifies the state of the game.
Cheating 12: Two or more players collude (e.g., share secret information).

Next, we define the levels of excluding cheatings. A cheating c is *eventually detectable* if the fact that cheating c is committed, and the player who committed it, can be detected after the game is over. A cheating c is *instantly detectable* if the fact that cheating c is committed, and the player who committed it, can be detected right after c is committed. Cheating c is called *preventable* if c cannot be committed, or even if c is committed, other players can continue the game without being affected by c.

We examine which cheating among 1 through 12 listed above can be excluded at which level by the game execution protocol proposed in Sec. 3. Before doing so, we clarify assumptions on networks. We assume that for each pair of players, there is a secure communication channel between them. This may easily be realized by using standard cryptographic protocols. Also, we assume that when a player sends information to other players, he attaches a digital signature.

Consider Cheating 1. After the game is over, we verify all computations performed during the game execution. If someone sends an illegal value to some player, he cannot deny this fact since the message is accompanied by his signature. So, this is eventually detectable. Cheating 2 cannot be excluded at any level since any player can send an input via a channel independent of the game system, e.g., e-mail, fax, and telephone. Cheatings 7 and 12 can be considered similarly. Note that these cheatings cannot be excluded even if we use servers. Cheating 3 is preventable by the assumption of secure communication channels. This is the same for Cheatings 5, 8, and 10.

There are two possibilities in Cheating 4. The first case is the following: when P_i sends a string x to P_j, P_k modifies it to a different string x'. This is preventable under the assumption on the network. The other case is that after P_j has received a string x from P_i, P_j modifies it to x', and uses it instead of x for future computation. This is eventually detectable by the same manner as described in the case of Cheating 1. Cheatings 6, 9, and 11 can be considered similarly.

There is one non-trivial case in Cheating 6. When creating a random string r, each player P_i is expected to generate a random string r_i. Yet, it can happen that P_i selects r_i not at random but with some intention. However, by the definition of r, if at least one player selects it at random, then r is also random. Hence, this is preventable.

5 Implementing Gunjin-Shogi

Based on the discussions so far, we implemented Gunjin-Shogi, which is a variant of Japanese Chess, but each player has secret information.

5.1 Rules of Gunjin-Shogi

Gunjin-Shogi is a two-player game. It needs a judge. A game board consists of 8×9 cells as shown in Fig. 1. Each player uses 31 pieces, each of which has a "type". There are 15 types and thus there are pieces that have the same type. On one face of a piece, its type is drawn, and the other face is null. For any pair of types, there is a strength relation between them, as shown in Fig. 2. A "$\bigcirc$" sign in row a and column b means that a is stronger than b. A "$\times$" in row a and column b means that a is weaker than b. "$\triangle$" means a draw, which occurs only when comparing the same type of pieces.

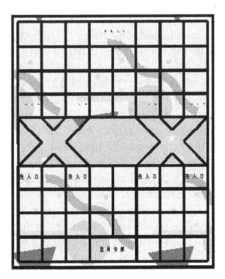

Fig. 1. A board of Gunjin-Shogi

	大将	中将	少将	大佐	中佐	少佐	大尉	中尉	少尉	騎兵	飛行機	タンク	工兵	スパイ	地雷
大将	△	○	○	○	○	○	○	○	○	○	○	○	○	×	×
中将	×	△	○	○	○	○	○	○	○	○	○	○	○	○	×
少将	×	×	△	○	○	○	○	○	○	○	○	○	○	○	×
大佐	×	×	×	△	○	○	○	○	○	○	×	×	○	○	×
中佐	×	×	×	×	△	○	○	○	○	○	×	×	○	○	×
少佐	×	×	×	×	×	△	○	○	○	○	×	×	○	○	×
大尉	×	×	×	×	×	×	△	○	○	○	×	×	○	○	×
中尉	×	×	×	×	×	×	×	△	○	○	×	×	○	○	×
少尉	×	×	×	×	×	×	×	×	△	○	×	×	○	○	×
騎兵	×	×	×	×	×	×	×	×	×	△	×	×	○	○	×
飛行機	×	×	×	○	○	○	○	○	○	○	△	○	○	○	○
タンク	×	×	×	○	○	○	○	○	○	○	×	△	○	○	○
工兵	×	×	×	×	×	×	×	×	×	×	×	×	△	○	○
スパイ	○	×	×	×	×	×	×	×	×	×	×	×	×	△	×
地雷	○	○	○	○	○	○	○	○	○	○	×	○	×	○	△

Fig. 2. A relation on strength of type of pieces

Initially, each player places 31 pieces on one's side. Pieces are placed with null face top, so that the opponent cannot see which piece is of which type. After the game has started, each player alternately moves one of his pieces. Positions in which each piece can be moved are determined by its type, but a piece cannot be moved to the cell which is already occupied by an own piece. If two pieces (each of each player) meet at the same cell, a "battle" begins. The judge looks at their types, and removes the weaker piece from the board. If both pieces are of the same type, both will be removed. A cell of size 2×1 is called a "base". When a player p proceeds a piece into the opponent's base, then p wins the game (only six types are allowed to proceed to the base). The challenge of the game is to guess which piece is of which type, by observing the movement of pieces and the result of battles.

5.2   Implementation

We implemented Gunjin-Shogi using Java 1.4. A game window is shown in Fig. 3. For the cheat-proof issue, we realized only eventual detection, namely, if a cheating is committed during the game execution, it can be detected at the verification phase after the game is over. To perform this, the program records all messages exchanged during the play. At the beginning of the verification phase, the players exchange their secret keys, and each player replays the whole game execution. Fig. 4 shows a window after the verification phase has been completed. The verification program informs a player that the opponent has cheated. In this

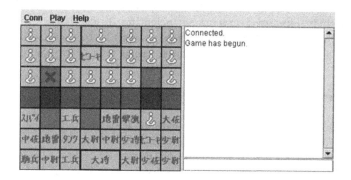

Fig. 3. A game window of Gunjin-Shogi

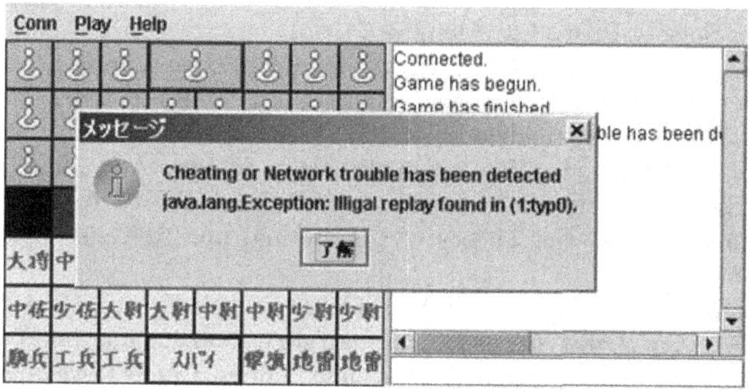

Fig. 4. An example of the verification phase

example, the opponent replaced weaker pieces by stronger ones, and placed many strong pieces against the game rule.

6   Conclusion

In this paper, we considered a possibility of playing cheat-proof serverless network games. We formally defined games, and proposed a general protocol by which we can execute games in a serverless environment. Then, we listed cheatings possible in our game model, and examined a possibility of excluding those cheatings. We finally developed a program of cheat-proof network Gunjin-Shogi.

As we have shown in Sec. 3, every game in our scope can be implemented by combining several basic protocols. Our possible next step is to develop a library of fundamental tools, so that games can be implemented easily by combining them. A second future direction is to employ deeper cryptographic tools, such as zero-knowledge proofs to guarantee higher security levels.

Acknowledgments

This research is supported in part by Scientific Research Grant, Ministry of Education, Japan, 17650016. The authors would like to thank the referees for their valuable comments, and the editors for their help in improving this paper.

References

1. Baughman, N.E., Levine, B.N.: Cheat-Proof Playout for Centralized and Distributed Online Games. In: Proc. of the 20th IEEE INFOCOM, pp. 104–113. IEEE Computer Society Press, Los Alamitos (2001)
2. Davis, S.B.: Why Cheating Matters: Cheating, Game Security, and the Future of Global On-line Gaming Business. In: Proc. of the Game Developer Conference 2001 (2001)
3. Franklin, M.K., Reiter, M.K.: The Design and Implementation of a Secure Auction Service. In: Proc. of the IEEE Symposium on Security and Privacy, pp. 2–14. IEEE Computer Society Press, Los Alamitos (1995)
4. Fujioka, A., Okamoto, T., Ohta, K.: A Practical Secret Voting Scheme for Large Scale Elections. In: Zheng, Y., Seberry, J. (eds.) AUSCRYPT 1992. LNCS, vol. 718, pp. 244–251. Springer, Heidelberg (1993)
5. Goldreich, P., Micali, S., Wigderson, W.: How to Play Any Mental Game, or a Completeness Theorem for Protocols with Honest Majority. In: Proc. of the 19th ACM Symposium on the Theory of Computing, pp. 218–229. ACM Press, New York (1987)
6. Greenhill, R.: Diablo, and Online Multiplayer Game's Future. Games Domain Review(1997), http://www.gamesdomain.com/gdreview/depart/jun97/diablo.html
7. Kirmse, A., Kirmse, C.: Security in Online Games. Game Developer 4(4), 20–28 (1997)
8. Shamir, A., Rivest, R.L., Adleman, L.M.: Mental Poker. In: Klarner, D.E. (ed.) Mathematical Gardner, pp. 37–41 (1981)

9. Smed, J., Kaukoranta, T., Hakonen, H.: Aspects of Networking in Multiplayer Computer Games. In: Proc. of the International Conference on Application and Development of Computer Games in the 21st Century, pp. 74–81 (2003)
10. Yan, J., Choi, H.J.: Security Issues in Online Games. The Electronic Library international journal for the application of technology in information environments 20(2), 125–133 (2002)
11. Yan, J.: Security Design in Online Games. In: ACSAC 2003, pp. 286–297. IEEE Computer Society, Los Alamitos (2003)
12. Yan, J., Randell, B.: A Systematic Classification of Cheating in Online Games. In: Proc. of the 4th Workshop on Network & System Support for Games (NetGames'05). ACM Press, New York (2005)
13. Yao, A.C.: Protocols for Secure Computation. In: Proc. of the 23rd IEEE Symposium on the Foundation of Computer Science, pp. 160–164. IEEE Computer Society Press, Los Alamitos (1982)
14. Zhao, W., Varadharajan, V., Mu, Y.: A Secure Mental Poker Protocol Over The Internet. In: Proc. of the Australasian Information Security Workshop, pp. 105–109 (2003)

Monte-Carlo Methods in Pool Strategy Game Trees

Will Leckie[1] and Michael Greenspan[2]

[1] Department of Electrical and Computer Engineering,
[2] Department of Electrical and Computer Engineering, School of Computing,
Queen's University, Kingston, Canada
will.leckie@ece.queensu.ca, michael.greenspan@queensu.ca

Abstract. An Eight Ball pool strategy algorithm with look-ahead is presented. The strategy uses a probabilistically evaluated game search tree to discover the best shot to attempt at each turn. Performance results of the strategy algorithm from a simulated tournament are presented. Players looking further ahead in the search tree performed better against their shallower-searching competitors, at the expense of larger execution time. The advantage of a deeper search tree was magnified for players with greater shooting precision.

1 Introduction

Pool and billiards are cue sports that have recently enjoyed a resurgence of interest worldwide. Pool was recognized as a demonstration sport by the International Olympic Committee at the 1998 Nagano Olympics, and it is estimated that in 2004 over 30 million people picked up a cue in the U.S. alone. Computational and robotic billiards have also enjoyed increased interest and research attention in recent years. There are currently a number of robotic pool systems under development worldwide, including DEEP GREEN [15], POOL SHARC [4], and others [5,7,14]. The 10th Computer Olympiad in Taipei, Taiwan, 2005, featured a computational Eight Ball tournament which pitted pool strategy algorithms against one another in matches hosted by a game server running a simulated game environment [11]. The computational Eight Ball tournament was again a feature of the 11th Computer Olympiad in Turin, Italy, 2006.

Pool is a game of physical technique and strategy. Placement of the cue ball following a shot is considered one of the key elements to successful play, and even moderately accomplished human players tend to plan several shots ahead. In this way pool bears a similarity to other games of strategy, such as chess and checkers. One significant difference is that chess and checkers are played on a board with discrete positions so that the number of possible board states is finite, although huge. In contrast, a pool table is a continuous domain with a truly infinite number of possible table states; no two pool games will ever be the same.

Another significant difference between chess or checkers and pool is that there is no element of chance or randomness in chess or checkers; the game state resulting from a given move is always certain. Human pool players are capable

H.J. van den Herik et al. (Eds.): CG 2006, LNCS 4630, pp. 244–255, 2007.

of learning advanced strategies and can hone their technique through hours of practice, but will never achieve perfectly repeatable shot making because of the inherent minute errors in human perception and muscle control when attempting to strike precisely the cue ball with the cue. Similarly, in a robotic pool system, a non-negligible amount of systemic noise imposes a physical limitation on the repeatability of the robot's shot making. Pixelation/quantization noise in image processing and positioning error in robot actuation can be minimized through careful calibration [15], but can never be eliminated completely. In the computational Eight Ball tournaments played at the Computer Olympiad, the simulated game environment adds Gaussian noise to the parameters that a pool strategy algorithm uses to specify its shots, in order to simulate the small but non-negligible amounts of uncertainty faced by human and robotic players. The addition of this noise not only makes the simulated tournament more realistic, but also makes the computational pool strategy problem more interesting and challenging. The strategy algorithm for a computational or robotic pool player must therefore consider the effects of additive noise in its decision-making in order to play competitively.

A pool strategy algorithm can be characterized by three main features; (1) method of shot selection, (2) learning ability, and (3) use of look-ahead. Shot selection is the task of choosing which ball to sink in which pocket given a current table state (i.e., the positions of all balls on the table). Shot selection methods fall into the two categories of *shot specification* and *shot discovery*. Shot specification involves identifying a desired final position for the cue ball following a shot, and employs numerical optimization methods to determine the shot parameters required to achieve this objective [9]. Shot discovery involves applying geometric rules to a current table state to enumerate potential shots. An evaluation function, such as a quantitative estimate of the difficulty of each shot, is then applied to select a shot from the options available. Evaluation of shot difficulty has been implemented using methods such as fuzzy logic [2,3,6,8], neural networks [5], and grey decision making [14]. Alian also explored the use of reinforcement learning [3] and genetic optimization [2] to train a fuzzy-based shot selection system.

Look-ahead is a key component of any two player turn-based game and is widely implemented using game-tree methods for games, such as chess, checkers and Backgammon. Look-ahead is also an important feature in a pool strategy program because a player continues shooting after successfully sinking a ball. Even moderately accomplished human players choose a shot by considering the positions of the other balls and potential future shots, usually to a depth of two or three shots ahead. Skilled human players strike the cue ball with the appropriate velocity and English (spin) in order not only to pocket one of their object balls, but also to leave the cue ball in a desired region on the table from which they will subsequently have an advantageous shot. Professional-calibre human Eight Ball players try to visualize the entire run of eight balls, at the start of a rack after the break. These players place great importance on identifying the *key ball*, the last ball of the player's color group to be pocketed, to plan their desired cue ball position for shooting the Eight ball [1]. Strong human pool players therefore use a combination of shot specification and look-ahead to select their shots.

A second strategic consideration in the game of Eight Ball and other pool game variations is a tactical play called the *safety*, which can be utilized by a player on any shot. When a player declares a safety shot, the turn will be passed to the opponent after the current shot, even if a ball was legally pocketed on the shot. The safety play can be used both as a defensive and an offensive tactic. From a defensive standpoint, a player declares a safety when there is no shot that results in an advantageous table state for the next shot. After successfully potting the ball, the turn is passed to the opponent, preventing the player from facing a difficult shot on a disadvantageous table state. From an offensive standpoint, a player declares a safety and makes a legal shot, possibly pocketing one of his own balls, but is careful to leave the cue ball in a location making the opponent's next shot especially difficult. If the opponent then makes an illegal shot, for example by failing to cause the cue ball to strike a ball of his own color group first, then the player is awarded *ball-in-hand* and is entitled to place the cue ball anywhere on the table and begin a turn, which is very advantageous. By looking ahead in the game and examining potential future table states resulting from a given shot, a pool strategy algorithm can therefore identify situations in which its own player will be left with a difficult shot warranting a defensive safety call, and can also examine the options available to an opponent to identify potential ball-in-hand opportunities resulting from an offensive safety.

Despite its important role in human play, none of the pool strategy algorithms reported in the literature contain the look-ahead feature [2,3,5,6,8,14]. Rather, all of these methods strive to identify the *easiest* shot given the current table state. However, the easiest shot on the table may not be (and often is not) the best shot to choose strategically. Pocketing the easiest shot may leave the cue ball in such a position that the player has no subsequent shots with acceptable difficulty, or worse yet no legal shots at all, which is advantageous to the opponent because of the potential for fouling and surrendering ball in hand to the opponent. Considering the additive noise in the simulated tournament environment and the systemic noise in robotic play, it is also important to note that the easiest shot determined from purely geometric considerations may not be the most likely shot to succeed.

In this paper, we present a pool strategy algorithm that uses shot discovery and look-ahead to choose the best shot given the current table state. The algorithm accounts for the additive noise using Monte-Carlo methods when simulating shot outcomes, and expands a game tree to look ahead and choose the best shot option on each turn. *-Expectimax game trees have been used for the game of Backgammon [12], in which the element of chance involved in the roll of the dice necessitates statistical sampling in the game tree. We have adapted the *-Expectimax approach to obtain a suitably efficient and compact tree representation and search algorithm for pool strategy.

This paper continues in Section 2 with a brief review of *-Expectimax and details of its adaptation for use in our pool strategy algorithm. Section 3 presents test results demonstrating the performance improvements and execution times for various degrees of look-ahead. The paper concludes in Section 4 with a description of future work.

2   Pool Strategy Based on *-Expectimax Game Trees

2.1   Modeling Shot Outcomes

To generate a game tree for pool, the outcome of each shot must be determined, which in turn requires a simulation of the physics of a shot. There are five *shot parameters* that describe the impact between the cue and the cue ball at the start of a shot. These five parameters uniquely specify the resulting trajectory of the cue ball and therefore the outcome of the shot, once the cue ball and object balls collide. The parameter a is the horizontal offset of the cue tip from the center of the cue ball, b is the vertical offset of the cue tip from the center of the cue ball, θ is the elevation of the cue stick above the horizontal, ϕ is the aiming direction of the cue stick in the table plane, and v is the speed of the cue stick immediately before impact with the cue ball. In addition to linear velocity, the collision with the cue also imparts an angular velocity, or *spin* to the cue ball whenever a, b, or θ are nonzero. By varying the spin (the values of a, b, and θ), the aiming direction ϕ, and the speed of the cue strike v, a player controls the trajectory of the cue ball and the final position of the ball after the shot.

The POOLFIZ library has been developed to simulate the outcome of a pool shot given the five shot parameters and the current table state (*i.e.*, the positions of all the balls on the table) [13].

2.2   Additive Random Noise Model

In the simulated pool environment for the computational Eight Ball tournaments at the Computer Olympiad, a noise model is used to simulate the uncertainty involved in executing a shot. When a pool strategy algorithm specifies its chosen shot in the form of the five shot parameters, the game server then adds random noise to each of the five parameters, simulates the shot using POOLFIZ, and analyzes the resulting table state to referee the match. The noise model is a zero-mean Gaussian distribution with a standard deviation specific to each of the five parameters. The standard deviations for each of the shot parameters were empirically determined and are published to all tournament participants in advance, allowing the details of the noise model to be coded directly into a competitor's pool strategy algorithm.

2.3   *-Expectimax Modified for Pool Strategy

To search a game tree and determine the best move from the current game state, each node in the tree is assigned a numerical utility value. The utility of a game state is a measure of how advantageous that game state is for the player, or how disadvantageous it is for the opponent. In a chess program, for example, the utility might be calculated based on the positional advantage, the material advantage, and the safety of the King. The game tree is a minimax tree and the Alpha-Beta Search algorithm is commonly used to search the tree to a certain depth and identify the best move [16].

Games such as chess and checkers are completely deterministic; the outcome of a given move is certain. The minimax approach has been adapted to the game of Backgammon using the *-Expectimax game tree [12]. Since Backgammon involves randomness in the rolling of the dice, the *-Expectimax game tree inserts a layer of *chance nodes* between each layer of Min and Max nodes in the tree. Statistical sampling is performed at the Chance nodes to examine possible outcomes of a given move based on the sampled outcomes of dice rolls. In *-Expectimax trees, the utility of a Chance node in the tree is the sum of the utilities of each of the child nodes, weighted by the probability of each particular outcome occurring.

The *-Expectimax game tree, with a few modifications, is well suited for use in pool strategy because of the element of uncertainty imposed by the noise model, and because the table is a continuous domain. Due to the additive random noise, a set of shot parameters $\{a, b, \theta, \phi, v\}$ could result in a range of possible final table states, and the weighted-sum approach of *-Expectimax is very useful for analyzing and "averaging" these possible final table states numerically.

The game tree structure for pool is illustrated in Fig.1. A given table state is represented by a *State node*, denoted by St_i. Each table state St_i has a set of N_i possible shots, with each shot represented by a child *Shot node*, denoted by Sh_j^i, $j = 1 \ldots N_i$. A pool shot is the arc of the tree between a State node St_i and a Shot node Sh_j^i and is represented by its set of five shot parameters: $\{a, b, \theta, \phi, v\}_j^i$.

A player's turn begins at the root State node St_0. The player has a set of N_0 shots available, each of which arcs to a child Shot node $\{Sh_j^0\}_{j=1}^{N_0}$. A child State node of a Shot node results in a player either continuing the turn and shooting again after successfully pocketing an object ball, losing the turn after failing to pocket an object ball, fouling and surrendering ball-in-hand, or winning the game by pocketing the Eight ball.

One difference between *-Expectimax for Backgammon and its adaptation for pool follows from the difference in turn ordering between the two games. In Backgammon players strictly alternate turns at each dice role, so the *-Expectimax tree contains alternating layers of Min and Max nodes, with the Min nodes representing the opponent's turn. In contrast, in pool the player's turn continues and they keep shooting so long as an object ball is legally pocketed at each shot.

To accommodate this difference, we have modified the *-Expectimax tree and the search method accordingly. The tree is recursively expanded by pre-order traversal as long as a player's shot is legal and successful. The traversal is terminated either when the specified search depth is reached, or at a leaf node. A leaf is reached either when a player loses his turn (by failing to pocket an object ball or by fouling), or when a player pockets the Eight ball to win the

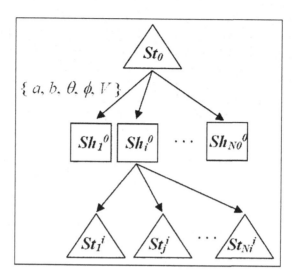

Fig. 1. Example game tree for pool

game. No forward modeling of an opponent's possible shots is performed, and the advantage given to an opponent at the loss of a player's turn is estimated by calculating the utility of that leaf State node for the opponent.

The Main Difference. The main difference of the scheme described above and *-Expectimax for turn-based games is that, in the case of pool, the opponent's State nodes are not expanded and explored further. In Backgammon, enumerating all of a player's possible moves is trivial because the game space is discrete. In pool, however, an opponent's shot is specified by five continuous parameters, which makes enumerating all of an opponent's possible shots impossible. Even if the strategy algorithm used its own "common sense" rules to explore a few of an opponent's expected shots, there is no guarantee that the opponent will select any of those shots because the opponent's shot selection algorithm is a different program. For example, a shot-discovery-style shot-selection algorithm would almost certainly fail to predict the shot chosen by an opponent's shot specification-based selection algorithm. Therefore, no time is spent modeling an opponent's potential shots by expanding a leaf State node, because it is fairly likely to be an inaccurate projection and therefore a waste of time. This adaptation of *-Expectimax also simplifies the search algorithm and results in a more compact tree and a faster shot-selection time, because far fewer nodes are added to the tree.

Let the utility measure of a State node St_i be denoted by $V(St_i)$. The state utility $V(St_i)$ is a static measure of how advantageous a given table state is, and could be calculated many different ways. One approach is to estimate the difficulty of a shot based on purely geometric considerations, and then sum the inverses of the shot-difficulty measures for all shots available to the player to obtain a static utility measure for the table state based on the number and difficulty of shots available. Our pool strategy uses a more complicated function to calculate $V(St_i)$ based on positional (cue ball location and number of available shots) and material (number of balls sunk) advantage of the table state St_i.

Since there is some uncertainty in executing the shot parameters due to the additive random noise, the utility of a Shot node should account for this uncertainty and the range of possible table states that could result, which motivates the use of the weighted sum method from *-Expectimax. The utility $U(Sh_i)$ of Shot node Sh_i could also be calculated many different ways. In our strategy, $U(Sh_i)$ depends on the utility of the resulting table states following the shot, since this affects the number and quality of shots subsequently available.

The utility of a given shot Sh_j is estimated at each Shot node by performing a Monte-Carlo simulation of N_δ shots, each of which is randomly perturbed from the nominal shot parameters: $\{a+\delta a, b+\delta b, \theta+\delta\theta, \phi+\delta\phi, v+\delta v\}$.

The Monte-Carlo simulation is performed by repeated calls to the POOLFIZ library. To obtain a reasonably accurate estimation of the utility of the shot, N_δ should be as large as possible given the constraints on the total time allowed to evaluate the tree.

The utility $U(Sh_j)$ of the Shot node is calculated as:

$$U(Sh_j) = \sum_{k=1}^{N_\delta} P_k V(St_k^j) \tag{1}$$

where P_k is the joint probability of occurrence of the noisy parameter set $\{a+\delta a, b+\delta b, \theta+\delta\theta, \phi+\delta\phi, v+\delta v\}$, and $V(St_k^j)$ is the utility of the kth child table state St_k^j of Sh_j. In the additive noise model the probability density function for the noise added to each shot parameter is a continuous Gaussian distribution, so the joint probability distribution of the noisy shot parameters is itself a Gaussian distribution. The probability of occurrence of a given set of noisy shot parameters can therefore be estimated by taking the product of the values of each of the probability densities for a given noisy parameter set.

A Second Difference. A second difference between *-Expectimax for Backgammon and our adaptation for pool strategy lies in the statistical sampling. In *-Expectimax for Backgammon, the statistical sampling is of the roll of two dice so it is possible to enumerate all possible outcomes of the dice roll. The roll of each die follows a discrete uniform probability density function, and the joint probability distribution of the sum of both dice rolls is a discrete triangular function. All possible game states resulting from the statistical sampling are enumerated and added to the tree, and the utility measure of a Chance node accounts for the fact that different states have different probabilities of occurrence by taking the weighted sum of their utilities. In pool strategy, the joint probability distribution for a set of noisy shot parameters is a continuous Gaussian distribution. In the discrete triangular distribution for dice rolls in Backgammon, even the least likely occurrences have a non-negligible probability of occurrence (3%), whereas in the continuous Gaussian distribution for pool the least likely occurrences have a vanishingly small probability of occurrence and can safely be discounted from the tree search. For reasons of memory compactness and search efficiency, rather than adding a child State node St_k^j to the tree for all N_δ Monte-Carlo samples performed at a Shot node, the Shot node utility $U(Sh_j)$ is estimated by treating each of the N_δ resulting table states as *leaves* of the tree, and calculating their utilities $V(St_k^j)$ appropriately. Then, the shot is simulated once with no additive noise and the resulting table state St_k^j is added to the tree as the child State node St_k^j of the Shot node Sh_j. This single child State node is then recursively visited until the true leaves of the tree are encountered. Adding only the noiseless child table state is a necessary simplification to minimize the branching of the tree, but necessitates the approximation of calculating the Shot node utility from the static utility of the "leaf" table states resulting from the Monte-Carlo simulation.

Similarities and Differences. To summarize, there are several similarities and differences between *-Expectimax for Backgammon and its adapted form for pool. State nodes in the pool strategy are analogous to the Min or Max nodes of *-Expectimax, and Shot nodes are analogous to the Chance nodes. The tree search is terminated at the desired depth or at the leaves of the tree, which in the pool strategy occur when a player loses his turn or the game ends. No forward modeling of an opponent's shots is done to save time and memory, since the continuous five-dimensional shot parameter space makes it very difficult to predict an opponent's probable shot selection. Finally, results of the Monte-Carlo simulation are used to estimate the score of a Shot node, and only the State node resulting from the noiseless shot parameter set is added as a child of each Shot node.

This adapted *-Expectimax approach integrates the shot selection task with look-ahead in the game, so that the focus at the start of each turn is on choosing the *best* shot, not necessarily the *easiest* shot. Identification of the best shot results from an examination of potential shot outcomes and the resulting table states, and upon a loss of turn at the leaf of the tree, the advantage given to the opponent is estimated by calculating the utility of the game state for the opponent.

3   Experimental Results

The performance improvement of a strategy algorithm using look-ahead was quantified by playing a computational Eight Ball tournament. There were three competitors in the tournament, all with identical shot selection and strategy algorithms. The only difference between the competitors was the maximum search depth of the game tree used by each player. Player 1 used a search depth of 1, meaning that it looked one shot ahead in the game by examining the table states resulting from its potential shots. Similarly, Player 2 used a search depth of 2. Player 0 had a search depth of 0, and chose its shots based solely on the probability of success of the shot. A shot was deemed successful if it resulted in a continuation of the player's turn due to a correctly pocketed ball, or of it resulted in the game being won by pocketing the Eight ball. In the case of Player 0, the score returned from a Shot node was simply the probability of success of the shot, and the player selected the shot with the highest probability of success with no regard for the resulting table state.

Each of Players 1 and 2 played a 100-game match versus Player 0, to examine the improvement in play against a reference player as the search depth was increased. Games were scored as follows: the winning player was awarded awarded a total of 10 points, and the losing player was awarded 1 point for each ball of their color group that was pocketed at the end of the game (*i.e.*, a player pocketing all of their balls but losing at the Eight ball would score 7 points). The match score was simply the sum of all a player's points from the games in the match. Each player was allowed a total of 1200 seconds (20 minutes) per game for shot selection. The number of Monte-Carlo samples at a Shot node, N_δ, was set to 25 for all players.

Two tournaments were played in this format, with different noise models (values of the standard deviations σ) in each tournament. The Gaussian random number generator in the GNU Scientific Library [10] was used to generate additive noise. The two different noise models used reflect the technical skill (precision in making shots) of the players involved; tournament 1 had higher sigma values and modeled human players with less technical skill who missed more shots, while tournament 2 had lower sigma values and modeled human players with higher technical skill who missed relatively fewer shots. Since all players involved in each tournament used the same noise model, the results of a given tournament show the performance versus search depth. Comparing the results of the two tournaments illustrates how beneficial a given search depth is for a player of a certain technical skill level.

The situation is similar to comparing two human players by categorizing their play in two areas: technical skill (precision in making shots) and level of strategic play (how far ahead in the game the player looks). A human player with relatively low technical skill (or, a strategy algorithm in a computational tournament with relatively high σ values for the noise model) will not play well against any player, no matter how strategically they play (or, how deep the strategy searches in the game tree). Similarly, a human player with very high technical skill (or, an algorithm in a tournament with low σ values for the noise model) will probably not play as well as a player with equally high technical skill who has a greater strategic sense for the game (or, an algorithm that searches more deeply in the game tree). In analyzing a player's performance, it is important to understand which factor limits their overall competitiveness, technical skill or search depth.

The results of the tournaments are summarized in Tables 1 and 2. The results for each player shown are the games won, points scored, total number of shots attempted during the match, number and percentage of shots missed during the match, and the average time taken by the player to select its shot. A missed shot was defined as a shot that resulted in the loss of a player's turn. The number of points scored by each player in a match is noted because the point differential in a match is a good indication of the overall competitiveness of the players.

Table 1. Results of Tournament 1 (noise model: $\sigma_a=0.6$, $\sigma_b=0.6$, $\sigma_\theta=0.1$, $\sigma_\phi=0.15$, $\sigma_V=0.1$)

Player	Wins	Points	Shots taken	Shots missed (%)	Avg. time/shot (sec.)
0 vs.	37	678	778	154 (19.8%)	0.45
1	63	830	858	159 (18.5%)	3.27
0 vs.	44	709	752	122 (16.2%)	0.38
2	56	733	742	146 (19.7%)	27.8

Table 2. Results of Tournament 2 (noise model: $\sigma_a=0.1$, $\sigma_b=0.1$, $\sigma_\theta=0.02$, $\sigma_\phi=0.025$, $\sigma_V=0.02$)

Player	Wins	Points	Shots taken	Shots missed (%)	Avg. time/shot (sec.)
0 vs.	33	602	591	55 (9.3%)	0.52
1	67	821	749	52 (6.9%)	4.4
0 vs.	27	477	474	46 (9.7%)	0.55
2	73	829	717	32 (4.5%)	34.1

In tournament 1, which modeled a decent amateur player, Player 1 beat Player 0 and missed fewer shots than Player 0. Player 2 also beat Player 0, but not by as high a score. Interestingly, Player 2 missed a higher proportion of shots than Player 0, the opposite of the result from the match between Players 0 and 1. This could be due to a combination of Player 2 starting some games with a more difficult table state, and the performance of the Gaussian random number generator used for the noise model. It could also simply reflect the effect of the relatively high noise; with high enough uncertainty in predicting shot outcomes, a player choosing shots based on cue ball positioning and future game states may be hampered by the effects of the noise and play poorer than a player placing more importance on the present shot at hand.

In tournament 2, which modeled a skilled player, both Player 1 and Player 2 defeated Player 0 convincingly. Furthermore, Player 2 missed fewer shots than both Players 0 and 1, and scored more wins and a higher point differential against Player 0 than did Player 1. This illustrates the advantage of a deeper search for a more skilled player. Player 2 missed fewer shots than both Players 0 and 1 presumably because Player 2 was leaving itself easier shots in general by looking further ahead in the game and having good control of the cue ball positioning from shot to shot.

Overall, it is clear that players looking further ahead in the game performed better, but were limited by both the time taken up by their search algorithm, and more importantly, by the technical skill level of the player. Players in tournament 1, who had relatively weak shot-making skill, enjoyed limited benefit from searching further ahead in the game. Meanwhile, the advantage of a deeper search tree was magnified in tournament 2 by a more forgiving noise model, which modeled human players with greater technical skill.

4   Conclusion

There are several unique challenges in the study of computational and robotic billiards that differentiate these games from other games of strategy like chess and checkers. Uncertainty in shot outcomes and a five-dimensional continuous domain for shot-selection parameters motivate the use of statistical simulation methods to examine potential shot outcomes. Past work in the area of pool strategy has focused on the development of algorithms for selecting the easiest shot available to a player. We have presented a new strategy involving shot discovery by Monte-Carlo simulation that looks ahead in the game in order to select the most strategically advantageous shot from the current table state.

In a computational tournament, players looking further ahead in the game scored more wins with a higher point differential, illustrating the advantage of looking ahead in the game. The trade-off between search depth and the shot precision afforded by the noise model was also highlighted. Higher technical skill in the players motivates more strategic play through deeper tree searches. In contrast, the performance of very strategic players with deeper search trees was limited by the technical skill of the player, which in this work was simulated by the additive noise model. It is a very interesting result of this work, no less so because it applies to both humans and computers alike.

In future work, the Monte-Carlo shot simulation will be parallelized on a cluster of workstations in order to speed up the shot selection task. In addition, shot specification using genetic optimization will be explored and will be integrated with the look-ahead game tree presented here. The shot specification method is expected to be especially effective for executing offensive safety shots. Various optimizations of the tree formulation will be explored, including an adaptive Monte-Carlo sample size based on tree depth, and the development of a pruning method to reduce the time required to evaluate the tree. Finally, there is potential for learning in this strategy algorithm, either by reinforcement learning or a genetic optimization of the various parameters of the strategy algorithm, including the search depth, number of Monte-Carlo samples used, and scoring function for the State utility.

Acknowledgments

The authors would like to thank the Natural Sciences and Engineering Research Council of Canada (NSERC) for their support, and Jonathan Schaeffer for suggesting the application of *-Expectimax to the pool strategy problem.

References

1. Alciatore, D.: Personal communication (2006)
2. Alian, M.E., Lucas, C., Shouraki, S.B.: Evolving Game Strategies for Pool Player Robot. In: 4th WSEAS Intl. Conf. on Sim. Mod. and Opt. (2004)
3. Alian, M.E., Shouraki, S.B.: A Fuzzy Pool Player Robot with Learning Ability. WSEAS Trans. on Electronics 2(1), 422–425 (2004)

4. Alian, M.E., Shouraki, S.B., Manzuri, M.T.: Robotshark: A Gantry Pool Player Robot. In: ISR 2004: 35th Intl. Sym. Rob. (2004)
5. Cheng, B.R., Li, J.T., Yang, J.S.: Design of the Neural-Fuzzy Compensator for a Billiard Robot. In: IEEE Intl. Conf. Networking, Sensing & Control, pp. 909–913. IEEE Computer Society Press, Los Alamitos (2004)
6. Chua, S.C., Wong, E.K., Tan, A.W.C., Koo, V.C.: Decision Algorithm for Pool Using Fuzzy System. In: iCAiET 2002: Intl. Conf. AI in Eng. & Tech., pp. 370–375 (2002)
7. Chua, S.C., Wong, E.K., Koo, V.C.: Pool Balls Identification and Calibration for a Pool Robot. In: ROVISP 2003: Proc. Intl. Conf. Robotics, Vision, Information and Signal Processing, pp. 312–315 (2003)
8. Chua, S.C., Wong, E.K., Koo, V.C.: Performance Evaluation of Fuzzy-based Decision System for Pool. Applied Soft Computing 7(1), 411–424 (2005)
9. Dussault, J.-P., Landry, J.-F.: Optimization of a Billiard Player – Position Play. In: van den Herik, H.J., Hsu, S.-C., Hsu, T.-s., Donkers, H.H.L.M. (eds.) CG 2005. LNCS, vol. 4250, pp. 263–272. Springer, Heidelberg (2006)
10. GNU Scientific Library (2007), http://www.gnu.org/software/gsl/
11. Greenspan, M.: Pool at the 10th computer olympiad (2005), http://www.ece.queensu.ca/hpages/faculty/greenspan/papers/8ball.pdf
12. Hauk, T., Buro, M., Schaeffer, J.: *-Minimax Performance in Backgammon. In: van den Herik, H.J., Björnsson, Y., Netanyahu, N.S. (eds.) CG 2004. LNCS, vol. 3846, pp. 51–66. Springer, Heidelberg (2006)
13. Leckie, W., Greenspan, M.: An Event-Based Pool Physics Simulator. In: van den Herik, H.J., Hsu, S.-C., Hsu, T.-s., Donkers, H.H.L.M. (eds.) CG 2005. LNCS, vol. 4250, pp. 247–262. Springer, Heidelberg (2006)
14. Lin, Z.M., Yang, J.S., Yang, C.Y.: Grey Decision-Making for a Billiard Robot. In: IEEE Intl. Conf. Systems, Man and Cybernetics, pp. 5350–5355. IEEE Computer Society Press, Los Alamitos (2004)
15. Long, F., Herland, J., Tessier, M.-Ch., Naulls, D., Roth, A., Roth, G., Greenspan, M.: Robotic Pool: An Experiment in Automatic Potting. In: IROS 2004: IEEE/RSJ Intl. Conf. Intell. Rob. Sys., pp. 361–366 (2004)
16. Schaeffer, J., Plaat, A.: New Advances in Alpha-Beta Searching. In: ACM Conference on Computer Science, pp. 124–130 (1996)

Optimization of a Billiard Player – Tactical Play[*]

Jean-Pierre Dussault and Jean-François Landry

Département d'Informatique,
Université de Sherbrooke, Sherbrooke (Québec), Canada
{Jean-Pierre.Dussault,Jean-Francois.Landry}@USherbrooke.CA

Abstract. In this paper we explore the tactical aspects needed for the creation of an intelligent computer-pool player. The research results in three modifications to our previous model. An optimization procedure computes the shot parameters and repositions the cue ball on a given target. Moreover, we take a look at possible heuristics to generate a sound selection of targets repositioning. We thus obtain a greedy but rather good billiard player.

1 Introduction

Our work concerns the development and improvement of an automated billiard player modestly named PoolMaster. The billiard game presents mechanical continuous issues (how the balls roll, bounce, etc.), and planning issues (strategic play). Most classical games (chess, checker, go, etc.) deal only with a planning difficulty, monthly addressed by some form of tree search. Billiard requires planning, but also accuracy. Following the main line of our research, we concentrate on the physical accuracy of our player. We identify the point where planning will be unavoidable in order to improve the player, but as we show, the accuracy improvement to be presented yields a much stronger POOLMASTER than its predecessor, that was runner up in the first Computer Pool tournament [7].

In [6], we presented a key computation to develop an optimal billiard player: the skill to sink a ball while achieving good position for the next shot. This is a very basic skill. We then developed an optimization model to compute the actual shot parameters required to sink the ball, and to reach a precise target without addressing the non trivial aspect related to how we choose the target point.

In this paper, we do not discuss the basic optimization methodology developed in [6]. We focus on building some tactical behavior within a player exploiting the aforementioned optimization model. Of course, we will refine the optimization model now and then to represent new items of the positional play unveiled by tactical issues that are investigated.

To begin with, we delineate precisely the context in which the player evolves. We state the basic assumptions on which we rely to construct the player.

Next, we present a model (similar to [4]) to compute the actual difficulty of sinking a given ball, based on the maximum allowable error in the shots'

---

[*] This research was partially supported by NSERC grant OGP0005491.

H.J. van den Herik et al. (Eds.): CG 2006, LNCS 4630, pp. 256–270, 2007.

parameters still ensuring the ball is sunk; we consider direct shots as well as kick and bank shots. This basic model allows us to rank the remaining balls from the easiest one to the most difficult one (other approaches have also been developed in [2,3,5,10]).

By identifying the best spot we choose where to put the cue ball on. This will be used as a target, or as the spot on which we put the cue ball when having ball in hand. Actually, we develop an optimization heuristic to build an ordered list of a few good spots. We discuss variants of the evaluation function.

Finally, we develop a generalization of the basic optimization model [6] to take into account three related deficiencies of the original proposition. Sometimes, the cue ball scratches (gets sunk), other times, the object ball does not reach the pocket, and finally, in some situations, it is not required to pin the cue ball at a given target, but to hit some cluster (to break it) with a minimum speed. Those three situations call for incorporating terms in the optimization model that reflect non vanishing speeds of either the cue or the object ball.

Before concluding, we present some limited statistical observations confirming that POOLMASTER is indeed a quite strong billiard player.

2   Context of Computer Pool

In order to produce an automated billiard player, the physics of the game must be understood, modeled, and simulated. For physical modeling we refer to [1,7]. We take as a black box a *simulator* [9] which "executes" the proposed shots. If the game is played using the same simulator, the computed shots and their outcome will match. If the game is played on a real table, the behavior of the shots will certainly differ as no simulator can ever be perfect. The simulator's input consists of the following five parameters:

- a and b represent the horizontal and vertical offset of the cue tip from the center of the cue ball;
- θ the elevation of the cue stick;
- ϕ the aiming direction;
- V, the speed of the cue stick.

Our player specifies legal values for those five parameters. Given the values, the simulator provides information called the table state, indicating every ball position, or flagging pocketed ones.

The first computer pool tournament was held during the 10th computer Olympiad, a parallel event with ACG 11 in Taipei [7]. The simulator was used by the "computer referee", and acted as the table as well. In the tournament, some Gaussian noise is added (with [8]) to the prescribed shot parameters to represent the unavoidable inaccuracies of any player, human or robot; this also allows, albeit indirectly, to acknowledge the limitations of the simulator which neglects uneven cloth thickness, imperfect cue stick tip, and so many other real-life details.

Our actual implementation uses the simulator as a black box. Therefore, any improved version of the simulator is likely to improve the realism of our player.

We plan to develop our own simulation engine in order to allow some cooperation between the simulator and the optimizer. In particular, the simulator could be enriched to yield some additional information useful for sensitivity analysis. The optimizer ([11]), which handles non linear constrained least-squares models, relies on some derivative information. Since we now use the simulator as a black box, derivatives are obtained via finite differences.

3   Shot Difficulty

This section provides functions to compute the difficulty of a shot. We are given the cue ball's position, the object ball position, and the aimed pocket. It is based on the margins of the error concept described in [1,4], and the related technical proofs available from the book's WWW page. We also provide extensions to estimate the difficulty of bank and kick shots. We use the estimate of shot difficulty to obtain a priority order, since we consider not only the easiest shot to pocket, but also the repositioning. Therefore, works like [2,3,5,10] are not directly applicable for us. Below, we present a straightforward model to perform the ranking we require.

3.1   Direct Shots

The difficulty of pocketing a given object ball is proportional to the error margin the player has, compared to the perfect shot. The object ball need not reach the center of the pocket. The possible angle θ depends on the distance of the object ball to the pocket and the pocket width. Once this angle is known, we may compute a target window that the cue ball must reach so that the object ball's trajectory remains within the angle θ_p. The maximum angle θ_c allowing the cue ball to reach the target window is a measure of the shot difficulty. So we have

$$\cos\theta_c = \hat{\mathbf{c}}_1 \cdot \hat{\mathbf{c}}_2 \tag{1}$$

with

$$\mathbf{c}_1 = p_{c1} - p_c \tag{2}$$
$$\mathbf{c}_2 = p_{c2} - p_c \tag{3}$$

where p_{c1}, p_{c2} are the two positions where the cue will have to hit the object ball to pocket it at the two extremities of the pocket. As in [6], vectors like $\mathbf{c}_1$ and $\mathbf{c}_2$ are denoted in bold, their norm is in math italics c_1 and c_2 and their normalization inherits a hat: $\hat{\mathbf{c}}_1 = \frac{\mathbf{c}_1}{c_1}$. $\mathbf{c}_1$ and $\mathbf{c}_2$ are the two vectors going from the cue ball position (p_c) to the cue ball hit points, as seen in Fig. 2.

However, as can be seen in Fig. 3, this approach may give undesired results in situations where we have a very big cut angle. Further calculations would need to be added to take this into account and that is why we choose to use another method instead, to estimate this difficulty which will require fewer operations and still give a good approximation.

We can easily compute the cosine of the cut angle $\cos \alpha = \hat{\mathbf{v}}_{op} \cdot \hat{\mathbf{v}}_{co}$ where $\mathbf{v}_{op}$ represents the object ball-pocket vector, and $\mathbf{v}_{co}$ the cue ball-object ball vector. For a corner shot, we have our shot difficulty coefficient κ_c:

$$\kappa_c = \frac{\cos \alpha}{v_{op} v_{co}} \tag{4}$$

For a side shot, we still use a similar method but this time, we add a new parameter γ which will represent the angle between the object ball and the x-axis of the pocket. This may make sure that we account for the fact that a shot in a direct line with the pocket is much easier than one having an angle. The parameter γ is defined as: $\cos \gamma = \hat{\mathbf{v}}_{op}.\hat{\mathbf{x}}$ with $\hat{\mathbf{x}} = \mathbf{x} = [1, 0, 0]$.

We now arrive at the side shot difficulty with:

$$\kappa_s = \frac{\cos \alpha \ \cos \gamma}{v_{op} v_{co}} \tag{5}$$

Here we will explicitly notice the fact that a side shot often needs more precision than a corner shot.

It is also important to note that a cut-off is used for γ to make sure that in the case of a very small angle, which would result in an impossible shot, we ignore that shot. We remark that in ideal cases this should never happen since a good detection of possible shots is to foreseen in the future. In our case, with the help of empirical results, we defined the cut-off value at 0.25.

Our κ is an intuitive measure of the shot difficulty. We present some statistical observations in Subsect. 7.1.

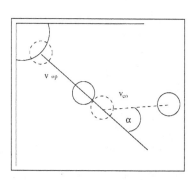

Fig. 1. Difficulty coefficient estimated with cut angle and distance

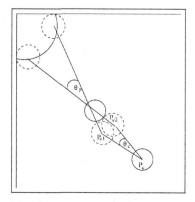

Fig. 2. Difficulty coefficient represented by α

3.2 Kick/Bank Shots

In some cases, often when there is a very big cut angle, it is possible that the easiest of the direct shots is still harder to perform than a bank or kick shot. To assess the difficulty of these indirect shots, we create a mirror image of the

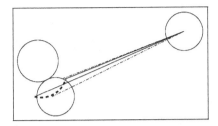

Fig. 3. Problems with the exact difficulty coefficient

current table on the corresponding side. In the case of a kick shot, we mirror the position of every ball on the table except the cue ball. For a bank shot we do the same except for the object ball. This enables us to calculate the difficulty of the shot as if it was a direct shot. We do so by taking the coordinates and aiming for the pockets of the mirror table. Of course, the coefficient calculated are only valid if we do a standard shot with no spin added. However, this still gives us a good idea of how hard the shot is. The mirror approach is incorporated in our optimization model and can be very helpful when there are no possible direct shots. Fig. 4 shows an example of a kick shot.

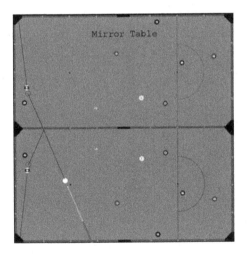

Fig. 4. Since in this case there is no direct shot possible (stripes), we will do a kick shot using the north rail by computing a mirror image of the table

We use the mirror table to estimate the shot's difficulty. When actually computing the shot, the optimization process benefits from an initial guess obtained by the mirror paradigm; moreover, the actual optimized shot will take into account the rebound properties of the rail. Since the rebound properties of the rail certainly affect the shot difficulty, we do not use it in our estimate.

3.3   Combinations

At times, it may be possible that even a hard shot like a combination is the best shot to perform. To determine correctly how hard this shot will be, we may use the same approach as for direct shots, but we need to add two more parameters since we now have two cut angles and three distances. The formula (5) now becomes (for a corner shot):

$$\kappa = \frac{\cos \alpha_1 \cos \alpha_2}{v_{o_2 p} v_{co_1} v_{o_1 o_2}} \tag{6}$$

where the notation $\mathbf{v}_{p_1 p_2}$ stands for the vector joining the positions p_1 and p_2, and $v_{p_1 p_2}$ is its norm; o_1 is the position of the first object ball, o_2 the second, c the cue ball, and p the pocket; α_1, α_2 are the two cut angles. Of course, for a side shot, we must scale κ by $\cos \gamma$.

4   Global Table Difficulty

Below, we present variants of a global table difficulty coefficient. We are given the balls' positions, and we provide a way to combine the individual object balls' difficulties into a global measure. This ability is quite important, especially at the beginning of the game if the table is still open as it will let the player choose the best balls to aim for (low or high) based on how each ball is positioned. It will also be quite favorable when the player has a ball-in-hand as we will see in Section 5.

Since we already have a coefficient of difficulty for any particular shot from Sect. 3, we can use this as part of our calculations. We consider two strategies and combine individual shot difficulty coefficients into a global table difficulty.

A naive approach would be to list and sum up all the shot difficulty coefficients, but as we were able to find out, this would seem an approach that is too general and will not lead to the appropriate positioning of one particular ball.

Instead, we suggest to use the average and the maximum of the shot difficulty coefficients, which will provide a more appropriate value if the positioning is closer to one ball. We can see the best cue ball positions as outlined by the contours in Fig. 6. The main difference between the sum formula and the average lies in the fact that the number of considered balls increases the sum, but not the average. Therefore, the average focuses slightly more on the best candidates. This can be seen in the Figs. 6 and 5. The same comparison can be made with the maximum formula, but to a higher degree since we now only reposition for one ball and forget the other possible ones, thus we have a more aggressive repositioning. Those represent the level sets of the function. In Fig. 5, the sum clearly shows a preferred position while in Fig. 6, the average measure shows two other rather good spots. If we look at the max in Fig. 7, its also very straightforward to see how we reposition for one ball only, which would definitely of use if we want to look a few shots in advance.

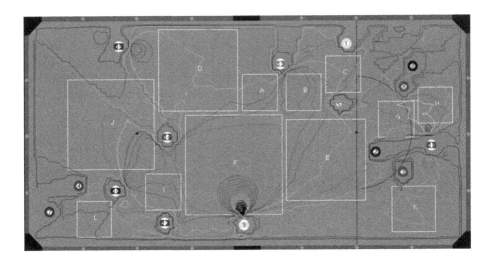

Fig. 5. Outline of the best regions for the cue ball positioning (aiming for the stripes) using the sum measure. In this case, it is obvious that the best area is close to the 9 ball as it will be an easy direct shot and also has all of the other stripes in direct view. Other good spot where fewer balls are in direct view are simply not competitive.

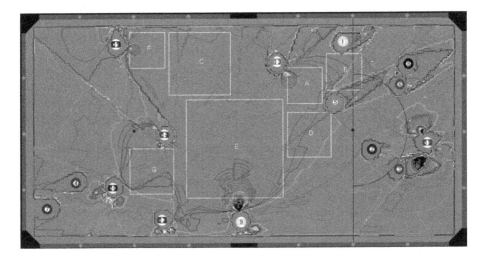

Fig. 6. Outline of the best regions for the cue ball positioning (aiming for the stripes) using the average measure. In this case, the number of relatively easy balls increases the measure. Apart from the spot close to the 9, two other rather good spot appear: close to the 11, where the average considers only the 11 since all other balls are hidden, and close to the 13. Observe that the spot close to the 11 ball is small, since positions where another ball becomes visible but more difficult because of the distance affect the average.

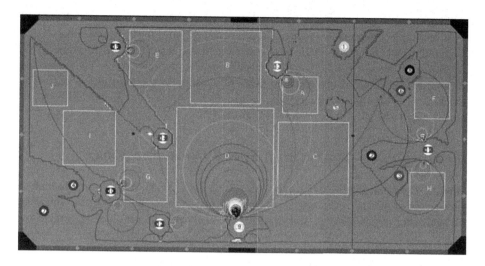

Fig. 7. Outline of the best regions for the cue ball positioning (aiming for the stripes) using the maximum measure. In this case, we immediately see each zone is focused on one ball for one pocket and nothing else. It is a much more aggressive play but one that a more talented player will probably use since he will rarely miss his shots.

In the event of the cue ball not having a clear path to the object ball, we still have to take it into account by adding a small value to our coefficient, since the path might clear out after the first shot is made. When that happens we will evaluate the shot difficulty as if the path was clear, but we add only half of that value to our computation.

5   Best and Worst Spots on the Table

In this section, we provide an heuristic optimization procedure to identify the best spots to aim for on the table. Moreover, we discuss the simplifying assumptions on which our optimization relies.

Spacially (seen on the table) the table difficulty is a non-linear complex function. We are interested in approximating good spots, corresponding to good local maxima. There might be several good spots. Clearly, we are not only interested in the global maximum of this function. Therefore, we propose to search (using an ascent method) local maxima starting from points on a grid. Several starting points usually lead to the same local maximum, and we are left with a short list of good spots. The relevant function reads

$$\max_{x,y} f(x,y). \tag{7}$$

with $f(x,y)$ representing our evaluation function.

We use this technique to find the best spot on the table: in case it is not possible to reach one of these spots, we can use the inverse of the function to find the worst spot and put the opponent in a tough position with a safe shot.

There are many other ways to explore a 2D function over a space described by a rectangle. We may evaluate f on a coarse grid, and perform a similar search on refined promising sub-grids (displayed by the square zones on Figures 6, 5, 7). However, even a 100×50 grid yields an accuracy of only $2cm$. We prefer the technique described above, since it takes advantage of the continuous nature of the function. Further enhancements will include heuristics to restrict the search and to facilitate attaining preferred regions.

6   Optimization with Ball Speed Considerations

Having solved the problem of finding a good target to reach with the cue ball after sinking an object ball, we may concentrate on assembling a much more complete objective function. We will use this section to refine our previous optimization model and solve some important issues to create a really solid player.

6.1   Sinking Object Ball with Minimum Speed

We start by addressing one of the most important matters, which consists in making sure that the object ball is actually sunk in a pocket during the optimization of the cue ball spin and velocity. Since our optimization method will try to do everything in its power to get the cue ball at the designed target, it is possible that in doing so it will reduce its velocity which in turn might cause the object ball not to be sunk. In our previous model, we had the objective function:

$$\min_{\mathbf{v}_0,\boldsymbol{\omega}_0,\mathbf{c},\bar{t},\tilde{t}} (\|\mathbf{p}(\bar{t}) - \mathbf{c}\|^2 - 4R^2)^2 + \|\bar{\mathbf{p}}(\bar{\tau}_f) - \mathbf{s}\|^2 + \|\tilde{\mathbf{p}}(\tilde{t}) - \mathbf{b}\|^2 \qquad (8)$$

which we will simplify (for the sake of this article) to :

$$\min_{\mathbf{v}_0,\boldsymbol{\omega}_0} \frac{1}{2}(\|\mathbf{p}_c(\tau_f) - \mathbf{c}\|^2 + \|\mathbf{p}_o(\tau_f) - \mathbf{p}\|^2). \qquad (9)$$

$\mathbf{p}_c$, $\mathbf{p}_o$ are the positions of the cue and object balls at their final resting time, $\mathbf{c}$ the cue ball target, and $\mathbf{p}$ the pocket aimed.

By adding a new component to this objective function to penalize the object ball speed, we are able to ensure it will reach the aimed pocket with the desired minimum speed. Let $\mathbf{v}_o(\tau_f)$ be the speed of the object ball at its final time τ_f and $\mathbf{v}_{min}$ the minimum speed. We can rewrite equation (9)into

$$\min_{\mathbf{v}_0,\boldsymbol{\omega}_0} \frac{1}{2}(\|\mathbf{p}_c(\tau_f) - \mathbf{c}\|^2 + \|\mathbf{p}_o(\tau_f) - \mathbf{p}\|^2 + \max(v_{min} - v_o(\tau_f), 0)^2). \qquad (10)$$

6.2   Cue Ball Scratching

A second problem is the cue ball scratching. More often than not, if the cue ball is pocketed after sinking the object ball, it will stay that way because the objective function will then hit a stationary area while minimising its position and will consider it as a local minimum. To help the function get out of this bad situation, we introduce a new term $v_c(\tau_f) - v_0$ which represents the velocity of the cue ball at its final resting time (v_0 representing a null speed). If the cue ball is immobilised on the table at its final time, then this component will not affect the rest of the function. However, if the cue ball is pocketed, its final velocity will not be null and the optimization will continue to find a better minimum and climb out of the pocket. We slightly complicate our previous equation (10) to

$$\min_{\mathbf{v}_0,\boldsymbol{\omega}_0} \frac{1}{2}(\|\mathbf{p}_c(\tau_f) - \mathbf{c}\|^2 + \|\mathbf{p}_o(\tau_f) - \mathbf{p}\|^2 + \max(v_{min} - v_o(\tau_f), 0)^2 + (v_c(\tau_f) - v_0)^2).$$

(11)

6.3 Breaking Clusters

A small trick will help us to solve a third problem, i.e., breaking clusters of balls. Sometimes in a game it may happen that there is simply no way to reposition the cue ball correctly. When that problem arises, it may be beneficial to try and break a cluster containing one or more of our balls to plan ahead for future shots. We have most of the necessary components available, and all we need to add is a time of impact τ_i to our minimization:

$$\min_{\mathbf{v}_0,\boldsymbol{\omega}_0,\tau_i} \frac{1}{2}(\|\mathbf{p}_c(\tau_i) - \mathbf{c}\|^2 + \|\mathbf{p}_o(\tau_f) - \mathbf{p}\|^2 + \max(v_{min} - v_o(\tau_f), 0)^2 + (v_c(\tau_i) - v_{cmin})^2).$$

(12)

We can now easily define a minimum speed at which the cue ball should hit the aimed cluster with v_{cmin}.

Actually, POOLMASTER does not wait until being without a good position shot to attempt breaking the clusters; we use a straightforward heuristic to trigger the choice of breaking the annoying cluster instead of aiming for a reposition. This strategic issue is one that will benefit from a planification approach, perhaps using a game tree.

7 Improved PoolMaster

Now it is the time to look at the actual benefits obtained by incorporating the features described above into our player. We will formulate observations to assess the gain in strength, and also will document some of our choices by comparisons. We present in Subsect. 7.1 empirical qualitative comparisons between the actual success rate and our κ difficulty coefficient. Thereafter, all our results will be assessed by mini-tournaments between competing variants. The results were obtained on a local server with a referee created to the exact image of the one used for the official tournaments (exact same rules and noise). As mentioned before, the server adds some additive noise to the shot parameters. The higher the noise level, the closer the different competitors. This makes sense, since a very high noise level makes the probability to miss a shot important, and so buries the expected reward of the refined player. We thus present two classes of results. Small noise to simulate a professional player, and higher noise to simulate a good amateur player.

Adding noise evaluates the robustness of the strategy used. We believe that no line of play will have a significant success under sufficiently high noise levels, but some form of stochastic optimization will certainly improve under small noise levels.

7.1 Shot Difficulty

As discussed in Sect. 3, the difficulty of a given shot depends on the error margin of the player when executing the shot. This is a complex function of all the shot parameters. Moreover, given a noise specification on the shot's parameters,we focus on the probability of success.

At the time to select a shot, we wish to rank the different options. In order to do so, let us illustrate that the measure we propose, $\frac{\cos \alpha}{p_{op} p_{co}}$ is qualitatively reasonable. In Figs. 8 and 9, we compare the actual success rate of several cases labeled $(d_1$—$d_2)$, with d_1 the distance between the object ball and the pocked, and d_2 from the object ball to the cue ball. We may conclude that our formula is a very good predictor for ranking purposes. As seen on the graphs, $\sqrt{\cos \alpha}$ would seems to fit slightly better to the observations, but we still prefer the plain cos, having an intuitive physical interpretation (reduction of a window with increasing angle). We remark that the fit cannot be perfect: for sufficiently small values of the noise, the success may well be 100% for all but the most severe cut angles. This does not mean that the difficulty is constant, but that the difficulty lies within the player's accuracy.

To fine tune defensive play, i.e., to decide when to attempt a shot, and when to use rather a defensive line, we need a refined estimate. A greedy approach is simulating the shot N times, recording the successes, and using this as an estimate.

7.2 Choice of the Global Table Coefficient

In Tables 1 and 2, we compare the three table coefficients introduced in Sect. 4, using both the optimisation method and a 2d grid zone scan method. Our simulated tournaments indicate that all target solution variants are more or less equivalent for the moment being. We also tested one match of Optim vs Zones to see which would be the real winner but the results weren't enough descriptives to do every matchup which would've taken 36 tournaments of 100 games. It is quite possible that the differences are so slight that they are covered by other flaws of the player or simply by the noise added to the shots.

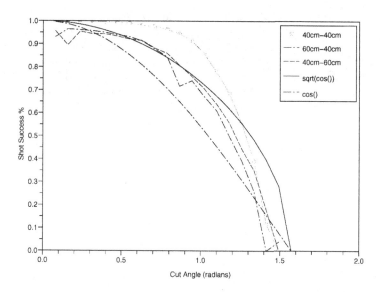

Fig. 8. Shot percentage for (40–40), (40–60) and (60–40). We compare the actual percentage of success for the tournament noise level. The qualitative fit with $\sqrt{\cos(\alpha)}$ is good. In abscissa, we sample a quarter of a circle.

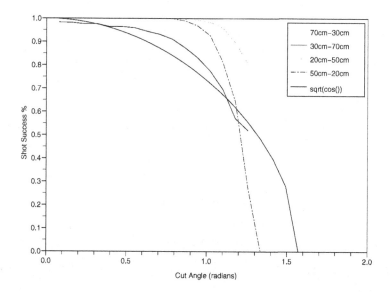

Fig. 9. Shot percentage for (20–50) and (30–70). Here, we consider easier shots involving situations where the distance of the object ball to the pocket differs significantly from the distance of the object ball to the cue ball. Again, the fit with $\sqrt{\cos(\alpha)}$ is still plausible, and the equivalence between (20—50) and (50—20), as well as (30—70) and (70—30) is observed.

Table 1. Optim vs 2d zone scan

Player	Small noise		Higher noise	
	Zones	Optim	Zones	Optim
Average vs Sum	50 vs 50	50 vs 50	46 vs 54	57 vs 43
Average vs Max	50 vs 50	45 vs 55	50 vs 50	44 vs 56
Max vs Sum	60 vs 40	46 vs 54	50 vs 50	51 vs 49

Table 2. Zones Max vs Optim Max (small noise)

Player	Wins
Zones Max	56
Optim Max	44

7.3   PoolMaster Old vs. PoolMaster New

PoolMaster Old vs. PoolMaster New. The original POOLMASTER used a straightforward global table coefficient, and only a back-top spin. No ball speed was incorporated in the optimization model. Therefore, it might happen that sometimes the object ball was going in the right direction, but did not reach the pocket.

PoolMaster—improved uses the refined global table coefficient, any possible spin, bank and kick shots, and is able to break clusters. Under the higher noise model, though, the differences are quite a deception. The noise destroys the positioning computed by both players; since the differences lie in the refined positioning capabilities, the results are very close. This is not a surprise: a refined player relies on precision, destroyed by too much noise. And this is also why a realistic virtual pool tournament would probably benefit from a much better/advanced noise model instead of just approximate values.

In Tables 3 and 4, by small noise level, we mean Gaussian noise with mean zero and standard deviation $a : 0.08, b : 0.08, \theta : 0.003, \phi : 0.0185, v : 0.0085$ while higher level is $a : 0.8, b : 0.8, \theta : 0.03, \phi : 0.185, v : 0.085$.

Table 3. POOLMASTER vs. POOLMASTER NEW

Player	Small noise		Higher noise	
	Wins	Points	Wins	Points
POOLMASTER	37	699	52	740
POOLMASTER NEW	63	800	48	767

Simple Player vs PoolMaster New. Here we compare POOLMASTER NEW to a simple greedy player, one which always chooses the easiest ball on the table, and does not deal with indirect or safety shots. The results are quite obvious, the simple player rarely manages to win a game since he never tries to correctly

Table 4. Simple Player vs. POOLMASTER NEW

Player	Small noise		Higher noise	
	Wins	Points	Wins	Points
Simple Player	3	321	12	443
POOLMASTER NEW	97	970	88	880

reposition for the next shot. We can also see he won a few more games playing on a higher noise table, which probably indicates the other player also missed a few more shots due to noise.

8   Conclusion

In this paper we pursued the development of an optimized billiard player. Building on the strength provided by accurate position play, as described in [6], we studied choices of target and aimed at repositioning. We use new optimization ideas to set up a short list of good targets based on estimating shot difficulties, and on a global evaluation function.

We further refined the positioning model in several items. First, we took into account non vanishing final velocities, allowing to include mini breaks of clusters of balls as a position target; in this case, the cue ball must hit the cluster with some velocity in order actually to break it. Also, we considered kick, bank, and combination shots.

The resulting player is rather strong, but we did not take into account any look ahead strategy yet. Without noise in the execution of the optimized shot, the player never misses but sometimes gets stuck with nothing to play. With kick, bank, combinations shot added to the possibilities, it is very rare that the player finds itself without play, but it happens, most often when he fails to break a cluster.

Of course, a realistic simulation will incorporate some random noise, whatever small, which may dictate, in some circumstances, to opt for defensive play instead of taking the risk of the easiest shot, still too difficult. The use of a look ahead strategy will allow to avoid to recourse to a defensive shot most of the time. This is the next step toward an optimal player, namely the planning part of the game, which will probably be addressed by stochastic dynamic programming.

Acknowledgments

We wish to warmly thank the referees, whose comments were so helpful in improving the original version of this paper.

References

1. Alciatore, D.G.: The Illustrated Principles of Pool and Billiards. Sterling Publishing (2004)
2. Alian, M.E., Lucas, C., Shouraki, S.B.: Evolving Game Strategies for Pool Player Robot. In: 4th WSEAS Intl. Conf. on Sim., Mod. and Opt. (2004)
3. Alian, M.E., Shouraki, S.B.: A Fuzzy Pool Player Robot with Learning Ability. WSEAS Trans. on Electronic 1, 422–425 (2004)
4. Chua, S., Wong, E., Tan, A.W., Koo, V.: Decision Algorithm for Pool using Fuzzy System. In: iCAiET 2002: Intl. Conf. AI in Eng. & Tech. (2002)
5. Chua, S.C., Wong, E.K., Koo, V.C.: Performance Evaluation of Fuzzy-Based Decision System for Pool. Applied Soft Computing 7(1), 411–424 (2005)
6. Dussault, J.-P., Landry, J.-F.: Optimization of a Billiard Player – Position Play. In: van den Herik, H.J., Hsu, S.-C., Hsu, T.-s., Donkers, H.H.L.M. (eds.) CG 2005. LNCS, vol. 4250, pp. 263–272. Springer, Heidelberg (2006)
7. Greenspan, M.: Pool at the 10th Computer Olympiad (2005), http://www.ece.queensu.ca/hpages/faculty/greenspan/papers/8ball.pdf
8. GSL - GNU Scientific Library (2007), http://www.gnu.org/software/gsl/
9. Leckie, W., Greenspan, M.: An Event-Based Pool Physics Simulator. In: van den Herik, H.J., Hsu, S.-C., Hsu, T.-s., Donkers, H.H.L.M. (eds.) CG 2005. LNCS, vol. 4250, pp. 247–262. Springer, Heidelberg (2006)
10. Lin, Z.M., Yang, J.S., Yang, C.Y.: Grey Decision-Making for a Billiard Robot. In: IEEE Intl. Conf. Systems, Man and Cybernetics. IEEE Computer Society Press, Los Alamitos (2004)
11. Nash, S.: A Truncated-Newton Optimization Package (2006), http://iris.gmu.edu/%7Esnash/nash/software/software.html

Gender and Cultural Differences (If Any!): South African School Children and Computer Games

Lizette de Wet and Theo McDonald

University of the Free State, Bloemfontein, South Africa
{Lizette,Theo}SCI@mail.uovs.ac.za

Abstract. When studying computer games several factors come into play. The issue of gender inequality has been a topic of many research projects in the past. The issue of culture is still in its infancy. Previous research regarding game-playing and gender issues seem to indicate that boys play more computer games than girls, that boys prefer more violent, action-oriented games in comparison to girls and that girls would prefer to play games with a feminine appeal. Intuitively it can be assumed that different cultures play different existing games at different frequencies. In this study grade ten school children (ages sixteen to seventeen) from one city in South Africa were questioned in order to establish if the same results hold true. The results indicate that there are no major differences in game playing between genders and cultures for this group. The conclusion is reached that especially with regard to gender the situation changed quite a bit over the past few years in comparison to research results found in the literature.

1 Introduction

Since Mr. Nelson Mandela became head of state in 1994 (which indicated the end of the "Apartheid" era), South Africans have been known as the "Rainbow Nation". One of the characteristics of this Rainbow Nation is eleven official languages. Therefore, the South African society can be described as truly multi-cultural. Another very important point to make, especially considering technology and the adaptation (or lack) thereof, is that South Africa is considered to be a third world country. This obviously results in different circumstances when compared to first world countries in general – also with regard to computer game playing.

Different factors come into play with computer games, namely the game genre, game content, age of the gamer, motivation, gender, culture, and gaming venue (private/public/online spaces) [3]. Gaming should therefore be investigated from a broad perspective.

The issue of gender inequality has been a topic of many a research project. This is not surprising when realizing that men and women's social, cultural, physical, and chemical make-up are quite different. In some cases different cultures have also been brought into the equation in game-playing research. Bearing South Africa's multi-culturalism in mind, very little, however, has been done to determine differences in the frequency and manner of computer game playing among different genders within

H.J. van den Herik et al. (Eds.): CG 2006, LNCS 4630, pp. 271–282, 2007.
© Springer-Verlag Berlin Heidelberg 2007

different cultural groups. In this research project a group of grade 10 school children (ages varying between 8 and 11) participated in order to attempt to address the issues of gender and culture in computer games. In particular, the study focused on the current differences in terms of gender and culture in the playing of computer games.

2 Background

One of the requirements for the utilization of gaming technology to its full potential is that it should be suited to cognitive and psychological needs of all the major user groups. This can only be done effectively if the views of *all* user groups are given adequate consideration [1]. These views would include both genders, as well as all the cultural groups (in this case in South Africa). However, in scrutinizing literature in this regard, this does not seem to be the case. Especially as far as gender inequality is concerned, much has been reported. In the following sub-sections the gender and the cultural issues in existing literature will be highlighted.

2.1 Gender Issues in Computer Games

According to Meunier [16] males in general tend to be more interested in computers than females and males use computers more than females at a younger age. While girls do play computer games, girls of all ages play less than boys. Hardcore gaming (over 15 hours a week) is largely a male preoccupation. Girls like and play different games than boys, and girls spend less on games [9,12]. Many psychologists state that girls are disadvantaged in the long run by playing far less games. They continue by saying that regardless of violence, games have been envisioned as potentially effective tools for learning and that gaming opens a door to computer literacy leading to potential technology careers [4].

Possible Reasons for gender inequality in computer games
When investigating the reasons for gender inequality in computer games, many possible reasons or explanations can be found in existing literature. According to the well-known study sponsored by the American Association of University Women [2], "...most computer games today are designed by men for men. They often have subject matter of interest to boys, or feature styles of interaction known to be comfortable to boys." Computer games are therefore rarely generated with females in mind. It also seems that girls shy away from computers because they are underrepresented in games. Females are either totally excluded from games, or depicted as passive weak individuals.

When considering game-playing styles Chu et al. [7] refer to differences in masculine and feminine styles in the following aspects: risk-taking, beating the game, non-linear flow, genre preferences, exercising reflexes, action content, role-playing games (social interaction, story, characters, adventure), creation, and destruction as themes and the platform. They also refer to biological differences in spatial skills between the sexes.

The literature also suggests that current marketing efforts can be held responsible for the observable segregation between genders in computer games. Marketing efforts are primarily done through print and TV. Approximately ninety-five percent of gaming magazine subscribers are male. This means that a girl might not even see a game ad that might potentially appeal to her. Most girls seem to find out about new or good games from friends, brothers, or male friends who are keen gamers. TV gaming ads generally also depict only boys as gamers – a situation that reinforces the stereotyping [6].

Different Game Preferences

According to Swanson [6] many young girls use their computers primarily for communication with friends and for the gathering of information. In gaming they follow the same patterns and are able to name character relationships and describe storylines more often than their male counterparts. Results of focus groups conducted with female school children indicated that they preferred qualities like racing, challenge, mystery, adventure, and winning cooperatively. They look for games that offer decision-making control and require strategy [17]. They have a greater desire to solve puzzles and use their creative skills through drawing and problem-solving.

Generally girls tend to be uncomfortable with violence in games [12]. However, in a study involving small single-gender focus groups [10], the girls reported that they preferred games with fantasy (cartoon) violence and an adventure theme, as well as abstract violence (e.g., crashing cars, hence their preference for racing as mentioned above). In contrast, the boys on the other hand, preferred the more concrete, realistic human violence, especially "first person shooters" which emphasized direct competition that often lead to death.

Martin [15] divides the primary game genres available to girls into two categories, namely traditional games and girls' games. In the traditional games females (if they are present) typically require rescue (the "princess" role), assist the male protagonist in his quest, or are themselves the reward upon completion of the mission. If one looks at the majority of the traditional games that are devoid of stereotypical representation of women, they are often games without concrete characters, such as abstract pattern games (e.g., TETRIS and BAKU BAKU), puzzle games (e.g., MYST), and exploration games (e.g., NIGHTS INTO DREAMS) [5]. In girls' games girls seem to be associated with the "Sugar and Spice" in the world of gaming. Games tend to focus on physical appearance or fashion style (e.g., BARBIE® REIGNS). Studies also show that girls receive strong messages emphasizing an unrealistic and unattainable standard of beauty from the images they see in the media.

Another trend regarding topics of computer games directed toward girls are tough, powerful, and independent woman characters as protagonists, like Lara Croft (in appearance an exaggerated Barbie-doll) in TOMB RAIDER and Joanna Dark in PERFECT DARK. These characters attempt to appeal to teenage girls but they fall short because of the violent nature of the games themselves. Another example, but this time as a non-violent character, is Ulala, heroine of SPACE. She is offered as a strong, independent role model with her strengths being her body and dance movements. Her body, however, also resembles Barbie's and is worlds apart from that of most teenage girls (or what is possible for them to achieve) [6]. The goal of girl's games is reported to offer girls an alternative to the technological environment traditionally believed to be a boys' domain. Unfortunately, the portrayal of females in these games does not seem to challenge traditional beliefs, but rather provide additional stereotyping. Categories relating to fashion, body types, food calories, friends, and love mirror the traditional stereotyping of teenage girl magazines.

Briefly comparing boys' preferences to girls', games perceived to be popular with boys include first person shooters and martial arts or other forms of fighting. These games generally end with one character (or a group of characters) either incapacitated or dead. Boys favor games with repetitive sequences, characters that have varying physical skills (as opposed to personalities), and fiercely competitive objectives [14].

A concept mentioned in research that "women prefer games that are less demanding than those that men prefer" and that "women prefer more whimsical, less aggressive" games than men, seems to be a misinterpretation of the true situation. Role playing games (a game genre that women seem to enjoy) are no less demanding than the more masculine fighting games. They are simply demanding in a different way [13].

2.2 Cultural Issues in Computer Games

In 1994 over 70% of digital games (especially video games) released in America were originally designed in Japan. The majority of characters in the video game universe are male. It is also obvious that the majority of human characters (excluding non-human characters such as robots, beasts, vegetables, and worms) are primarily white. It is therefore easy to leap to the conclusions that such video games were programmed by white males with politically incorrect values. This conclusion, however, quickly becomes invalid when one finds a string of Asian names in the credits [13].

The immediate answer to why Asian programs programmed primarily by Asian programmers would contain mainly Caucasian people is that they are trying to appeal to the Caucasian market. However, as almost 60% of the world gaming market is dominated by the Japanese gamer, it comes to light that Caucasian people, especially Caucasian Americans, are almost seen as fictitious characters in the popular Japanese culture. They apparently have the potential to help sell products in Japan [13]. In this context the gender inequalities in the gaming industry should be addressed from a clearer understanding of gender differences in e.g., Japan (and not from a Western point of view). To elaborate, the term feminism was imported into Japanese culture, but altered to fit more of traditional Japanese beliefs, thus loosing its original meaning. Examples of female portrayal that might seem blatantly chauvinistic to Westerners might be a true example of Japanese feminism to Japanese consumers [13]. Therefore it is obvious that Western-centric beliefs and morals cannot simply be applied to computer games across the board. It can also not simply be applied to South Africa's multi-cultural circumstances – hence part of the motivation for this paper.

When considering the literature regarding game-playing and the gender issues related to it, one cannot help to be left with the following assumptions: boys use computers more than girls; boys play more computer games than girls; computers are more accessible to boys than to girls; boys prefer more violent, action-oriented games in comparison to girls; girls play (or would prefer to play) more feminine games.

Intuitively the following pre-assumptions were made by the researchers regarding cultural issues: there is a difference in the frequency of playing games among different cultures; different cultures play different existing games; different cultures prefer different games / aspects of games.

This study attempts to determine if the above-mentioned assumptions are also valid for the South African youth (or at least for a section of them).

3 Methodology

Grade 10 pupils (ages varying between 8 and 11) from five schools in the Bloemfontein region (a city in the center of South Africa) were brought to the computer laboratories of the University of the Free State in order to be tested on various aspects of computer usage. Two schools from the traditionally white (Euro-centric culture) population and three from the traditionally black population (Afro-centric culture)[1] were selected for the study. All the participants were brought to the university in groups of 50 to 80 to be tested. In total 652 pupils were tested.

On arrival at the computer laboratory, the purpose of the research was explained to the participants. A questionnaire was administered to each to determine how they differed in their experiences in playing computer games. The basic demographic information obtained from the pupils included age, gender, and home language. Because of the sensitivity of the race-issue in South Africa, race was not included in the demographic information, but by using the home language it could be deduced. There were not enough cases in the different language groups to use them separately in the statistical analysis and therefore they were grouped. Those pupils that had Afrikaans and English as their home language were categorized as European. Those that had Sotho, Tswana and Xhosa as home language were categorized as African. There were a couple of languages that could not be classified in the above two groupings and they were omitted in the culture analysis.

The questionnaire started off with a question on the frequency of playing computer games. Those participants that played more than a few times per month (experienced

Table 1. Categories of different games (in order of popularity)

Category	Examples
Adventure	SACRED, HIDE-AND-SEEK, DINO CRISIS 2, CSI, DEUS EX, TOMB-RAIDER, SEPTERRA CORE, DIABLO, etc.
Sports	FIFA 2000-2005, EURO 2004, BASKETBALL, NFL 2004, SOCCER, NETBALL, SKATEBOARDING, RUGBY, CRICKET, WRESTLING, etc.
Family	WHINNY THE POOH, FINDING NEMO, LION KING, ALADDIN, POKÉMON, JIMMY NEUTRON, HARRY POTTER, etc.
Action	TEKKEN, DRAGON BALL Z & BUDOKAI, QUAKE 3, BATTLE REALMS, ROAD FIGHTER, HIT MAN 2, STREET FIGHTER X, etc.
Classics	PAC MAN, HANGMAN, HANGAROO, TIC-TAC-TOE, GOB MAN AND BLOCKS, TETRIS, MEMORY, MIND MAP, SNAKE, etc.
Educational	EDUMATH, ABC, etc.
Strategy	STAR KINGDOMS, CAESAR, THE EMPEROR, PHARAOH, ZOO TYCOON, AGE OF EMPIRES, ROME, CALL OF DUTY, etc.
Windows	HEARTS, SOLITAIRE, SPIDER SOLITAIRE, FREECELL, PINBALL, MINESWEEPER.
Racing	NEED FOR SPEED UNDERGROUND 1 & 2, NEED FOR SPEED PORCH 2000, F1 RACER, GRAND PRIX F1, MOTO GP, ETC.
Simulation	SIMCITY, BUILD YOUR OWN TOWN, FLIGHT SIMULATORS, LIGHTNING 3.

[1] For simplicity reasons there will be referred to "European" and "African" in this paper.

gamers) then had to complete a section on the types of games (educational, sport, action, etc.) that they played and how often (often, occasionally, or almost never) they played those games. They also had to indicate how different reasons (fun and relaxation, proficiency, stimulation, etc.) for playing computer games applied to them and had to state their opinions on the importance of several gaming characteristics (action, proficiency, strategy, etc.). There was also an open-ended section where they could list their favorite games and propose a new idea for a computer game. Those participants that never played computer games (inexperienced gamers) had to complete a section where they had to provide their reasons (no time, not interested, no access, etc.) for not playing.

One of the questions requested the participants to list their favorite computer games. A very long list was provided. To make the analysis more meaningful, the games were classified into eleven different categories as shown in Table 1.

Because of the categorical nature of the data, Chi-square analysis was done to determine if there were any differences between gender and culture and several variables related to the playing of computer games. Gender and culture were the independent variables, while the variables that corresponded to the different questions of the questionnaire were considered as the dependent variables. The null hypotheses in all cases were that the dependent variable was not related to the independent variable. The level of significance was 0.05 in all cases. Gender and culture were analyzed separately.

4   Results

4.1   Influence of Gender and Culture on the Frequency of Game Playing

Table 2 shows how often the different genders played computer games. A Chi-square analysis showed that ($\chi^2 = 7.07$, df = 4, p = .1322) gender does not account for how often pupils play computer games. Table 3 shows how often the different cultures play computer games. A Chi-square analysis showed that ($\chi^2 = 40.22$, df = 4, p = .00) culture can be associated with how often pupils play computer games.

Table 2. How often different genders played computer games

	Daily	Weekly	Weekends	Monthly	Never	Total
Male	25	54	22	169	51	321
Female	27	54	13	200	37	331
Total	52	108	35	369	88	652

Table 3. How often different cultures played computer games

	Daily	Weekly	Weekends	Monthly	Never	Total
European	21	41	19	231	32	344
African	29	63	16	123	55	286
Total	50	104	35	354	87	630

4.2 Influence of Gender and Culture on the Type of Games Played

Table 4 shows the results of the statistical analysis of the influence of gender and culture on the playing of different types of computer games. No significant statistical differences between gender and culture could be found for any of the types tested.

Table 4. The influence of gender and culture on the playing of different computer games

Dependent	Independent	N	χ^2	DF	p
Play educational games	Gender	232	0.78	2	.68
Play educational games	Culture	227	0.91	2	.63
Play sport games	Gender	237	2.14	2	.34
Play sport games	Culture	232	2.62	2	.27
Play action games	Gender	244	0.29	2	.87
Play action games	Culture	238	0.13	2	.94
Play strategic games	Gender	225	0.26	2	.88
Play strategic games	Culture	220	1.67	2	.43
Play adventure games	Gender	237	0.35	2	.84
Play adventure games	Culture	230	1.44	2	.49
Play puzzle games	Gender	239	1.80	2	.41
Play puzzle games	Culture	232	0.07	2	.97
Play simulator games	Gender	215	0.06	2	.97
Play simulator games	Culture	210	2.59	2	.27

4.3 Influence of Gender and Culture on the Reasons for Playing Games

Table 5 shows the results of the influence of gender and culture on the reasons for playing computer games. Both gender and culture were significant in the case of "playing for proficiency". No significant statistical differences between gender and culture could be found for any of the other reasons.

Table 5. The influence of gender and culture on the reasons for playing computer games

Dependent	Independent	N	χ^2	DF	p
Play for fun and relaxation	Gender	273	1.59	2	.45
Play for fun and relaxation	Culture	265	1.49	2	.48
Play for proficiency	Gender	222	14.27	2	.00
Play for proficiency	Culture	216	7.16	2	.03
Play for stimulation	Gender	210	1.02	2	.60
Play for stimulation	Culture	205	0.87	2	.65

4.4 Influence of Gender and Culture on the Reasons for not Playing Games

Table 6 shows the results of the influence of gender and culture on the reasons for not playing computer games. Only in one case, "the violence puts me off" was a significant association established for the independent variable culture. All the other reasons were insignificant for both gender and culture.

Table 6. The influence of gender and culture on the reasons for not playing computer games

Dependent	Independent	N	χ^2	DF	p
I do not have time	Gender	49	2.91	2	.23
I do not have time	Culture	48	4.18	2	.12
I am not interested	Gender	47	2.17	2	.34
I am not interested	Culture	46	5.22	2	.07
I do not have access	Gender	49	0.08	2	.96
I do not have access	Culture	48	0.47	2	.97
I do not like type of games	Gender	46	.11	2	.95
I do not like type of games	Culture	45	1.80	2	.41
The violence puts me off	Gender	44	2.25	2	.32
The violence puts me off	Culture	43	6.08	2	.05
My friends do not play games	Gender	44	.25	2	.88
My friends do not play games	Culture	43	0.99	2	.61
Would you like to play games	Gender	60	2.41	1	.12
Would you like to play games	Culture	59	2.43	1	.12

4.5 Influence of Gender and Culture on Different Characteristics of Games

Table 7 shows the results of the influence of gender and culture on the importance of different characteristics of computer games. The "Strategy" characteristic was significant for gender. No significant statistical differences between gender and culture could be found for any of the other characteristics.

Table 7. The influence of gender and culture on the importance of different characteristics of computer games

Dependent	Independent	N	χ^2	DF	p
Action	Gender	322	3.62	2	.16
Action	Culture	313	0.99	2	.61
Proficiency	Gender	300	3.73	2	.15
Proficiency	Culture	291	1.72	2	.42
Strategy	Gender	313	6.41	2	.04
Strategy	Culture	303	.60	2	.74
Realism	Gender	308	.85	2	.65
Realism	Culture	299	5.75	2	.06
Color	Gender	311	.19	2	.91
Color	Culture	303	2.09	2	.35
Movement	Gender	311	2.84	2	.24
Movement	Culture	302	1.59	2	.45
Sound	Gender	314	3.32	2	.19
Sound	Culture	305	1.23	2	.54
Score	Gender	309	.13	2	.93
Score	Culture	302	.94	2	.63
Fun	Gender	297	1.32	2	.52
Fun	Culture	291	2.53	2	.28

4.6 Influence of Gender and Culture on "boys play more games than girls"

Table 8 shows the results of the influence of gender and culture on the statement "boys play more computer games than girls". No significant statistical differences between gender and culture could be found.

Table 8. The influence of gender and culture on the statement "boys play more computer games than girls"

Dependent	Independent	N	χ^2	DF	p
Boys play more than girls	Gender	360	2.13	2	.35
Boys play more than girls	Culture	350	5.22	2	.07

Table 9. Top 5 favorite games for Males and Female pupils

Males	% (N = 224)	Females	% (N = 206)
Family games	17.41	Racing games	20.87
Racing games	14.29	Sports games	13.59
Action games	12.95	Windows games	11.65
Sports games	11.61	Family games	10.68
Adventure & Classic games	10.27	Action games	10.68

Table 10. Top 5 favorite games for European and African pupils

European	% (N = 167)	African	% (N = 249)
Family games	14.97	Racing games	20.48
Simulation	13.77	Sports games	14.46
Racing games	13.17	Family games	12.85
Action games	11.98	Action games	11.65
Classic & Windows games	10.18	Adventure games	10.84

The top 5 favorite games for males and females are shown in Table 9. The top 5 favorite games for European and African pupils are shown in Table 10.

5 Discussion

The results were rather surprising. Initially the expectations were, in terms of gender, that boys would play more computer games than girls and that they would prefer more violent, action-oriented games in comparison to girls (who would prefer more "girlish" kind of games). In terms of culture, the initial expectations were that the European pupils would play more computer games than the African pupils and that the two groups would prefer different types of games.

The first initial expectation was not supported; no differences were found in how often male and female pupils played computer games. This was somewhat contradicted by the statement "boys play more computer games than girls" where differences in opinions between the genders could be found. Most of the boys (69%)

and girls (65%) agreed that boys played more than girls. Only 31% of the boys and 28% of the girls felt that boys and girls played games in an equal amount. It seems that even among the pupils themselves the perception still holds that boys play more than girls — even though no differences between the genders were found in how often they played computer games. This result is somewhat in accordance with Creative Industries Faculty QUT [8] which found that although fewer girls than boys play computer games, the number of female players is increasing drastically.

The second expectation was also not supported: no differences were found in the game type preferences (including action games) between boys and girls. This could be explained by: "it is perceived to be somewhat more sociable for girls to assume male roles than the other way around; therefore, it has been proposed to encourage girls to play violent and gender-stereotyped games" [11].

The fact that the results of this study are not in accordance with previous theoretical and empirical studies (especially about gender and gaming) can be explained by the technological innovations in both hardware and software in computer gaming that changed the contemporary gaming experience. Technical innovations in hardware, graphics, processor power, as well as the increasing interactivity of games, have changed the dynamics of play and context in which gaming takes place [3].

The similarity found in the types of games that boys and girls played, is supported to a certain extent by the rating of the top 5 categories of games played by males and females; the categories are basically the same, but in different order. It was unexpected that family games came first for boys and action games only came in third place. The fact that racing games was the top scorer for girls was also a surprise, but it corresponds to the results of the Software Games for Girls Workshop [17].

Only in the case of "play for proficiency" and "strategy is an important characteristic of computer games" were significant differences found between male and female. A closer inspection of the data showed that proficiency and strategy were more important for girls than for boys. The strategy characteristic was also found to be important by the Software Games for Girls Workshop [17]. The reason why girls prefer to play for proficiency is more difficult to explain, but based purely on the authors' opinions, it could be because they see that as a way to catch up on the boys.

The initial expectation that the European pupils play more computer games than the African pupils proved to be correct. The difference between the cultures can be attributed to the fact that the European pupils tend to play more on weekends, whereas the African pupils tend to play during the week or do not play at all. This can be explained by the fact that more European pupils will have computers at home than their African counterparts. The African pupils will most probably use the computers at school to play games.

In the case of culture only a few significant differences were found. This is supported to a certain extent by the top 5 choices of games by the two groups. The major difference was "simulation" games that came second for Europeans and did not feature at all for African pupils, and "sports" games that came second for Africans and didn't feature at all for European pupils. The only significant differences in terms of culture were "proficiency" as a reason for playing games and "violence in games puts me off". For "proficiency" it seemed that it was more important for Africans than for Europeans and the reason can be the same as that of girls. The fact that "violence" was significant can be attributed to a small sample size and low numbers in some of the cells.

It is interesting to note that though action and adventure games did not feature in the top two favorites, most pupils suggested those kinds of games as new computer games. One gets the impression that because violence is so prominent in the media, it is the "in" thing. When the pupils are alone in front of the computer they, however, prefer more "tamer" games.

6 Conclusion

Other than expected, our research results indicated that there are no major differences in computer game playing between genders and cultures for grade 10 pupils in this particular city in South Africa. Especially with regard to gender, these results indicated that the situation changed quite a bit over the past few years in comparison to research results found in the literature. Where the literature indicated a major gender inequality in computer game playing, our results indicate that girls (in this case grade 10 scholars) bridged the gender gap as far as frequency, as well as type of computer games, that they played. This is in line with what Funk [11] suggested: Children (including girls) will continue to play computers games as long as it is enjoyable. They are quite likely to ignore offensive content, although perhaps they will still be affected by it, and will focus on action and the challenge. As Funk [11] puts it: "… sometimes girls (and boys) just want to have fun".

The same can be said of cultural inequalities in computer games in South Africa. Much effort has already been channeled into cultural inequalities since 1994, and although many things still need to be done, our research results are indicative of the fact that the "digital-divide" among the different cultural groups in the country seems to be gradually diminishing.

However, this research effort has only touched the proverbial "tip of the ice berg" with regard to the question of cultural differences in computer game playing. Although the results indicate that there are no major differences between cultures in playing computer games, this was tested on grade 10 pupils in one city in South Africa only. One is always careful of generalizing, and it would be wrong to fall into the pitfall of blunt generalization. However, we view the purpose of scientific research studies as the reaching of "generalizable" conclusions when trends seem significant. The fact that the pupils were representative (as far as gender, as well as Euro-centric and Afro-centric culture is concerned) gave us the confidence to extend the conclusions to grade 10 pupils in South Africa as a whole. As more research in this area is still necessary, the research findings should be viewed as indicative rather than authoritive.

References

1. Ahuja, M.K.: Information Technology and the Gender Factor. In: Proceedings of the SIGCPR, Nashville, TN, USA, p. 156 (1995)
2. American Association of University Women: Educating girls in the New Computer Age. American Association of University Women Educational Foundation Press (2000)

3. Bryce, J., Rutter, J.: The Gendering of Computer Gaming: Experience and Space. In: Fleming, S., Jones, I. (eds.) Leisure Cultures: Investigations in Sport, Media and Technology, pp. 3–22. Leisure Studies Association (2003)

4. Cassell, J., Jenkins, H.: Chess for Girls? Feminism and Computer Games. In: Cassell, J., Jenkins, H. (eds.) From Barbie to Mortal Kombat: Gender and Computer Games, pp. 2–45. MIT Press, Cambridge, MA (1998)

5. Cassell, J.: Storytelling as a Nexus of Change in the Relationship between Gender and Technology: A Feminist Approach to Software Design. In: Cassell, J., Jenkins, H. (eds.) From Barbie to Mortal Kombat: Gender and computer games, pp. 298–327. MIT Press, Cambridge, MA (1998)

6. Children NOW. Girls and Gaming: Gender and Video Game Marketing (2000), http://www.childrennow.org/

7. Chu, K.C., Heeter, C., Egidio, R., Mishra, P.; Girls and Games Literature Review (2004), http://spacepioneers.msu.edu/girls_and_games_lit_review.htm.

8. Creative Industries Faculty QUT: Computer Games – Female Users: Increasing Numbers of Women Playing Video Games (2004), http://wiki.media-culture.org.au/index.php/Computer_Games_-_Female_Users

9. ELSPA: GameVision Report (Autumn 2003)

10. Funk, J.B., Jenks, J., Bechtoldt, H.: Children's Experience of Video Games. Unpublished data (2001)

11. Funk, J.B.: Girls Just Want To Have Fun (2001), http://culturalpolicy.uchicago.edu/conf2001/papers/funk2.html

12. Haines, L.: Why Are There So Few Women in Games? Research for Media Training NorthWest (September 2004)

13. Hart, S.N.: Gender and Racial Inequality in Video Games (1997), http://www.geekcomix.com/vgh/genracinequal.shtml

14. Kafai, Y.: Gender Differences in Children's Constructions of Video Games. In: Greenfield, P.M., Cocking, R.R. (eds.) Interacting with video games, Erlbaum, Norwood, NJ (1996)

15. Martin, C.K.: Girls, Video Games, and the Traditional Stereotype of Female Characters (1999), http://ldt.stanford.edu/ldt1999/Students/ckmartin/pdf/videoGames.pdf

16. Meunier, L.: Gender Differences in Computer Use (1996), http://bureau.philo.at/mii/gpmc.dir9606/msg00014.html

17. In: Software Games for Girls Workshop. Focus groups conducted by Electronic Arts and Mattel (2005), http://www.Castilleja.org (Last visited 31/10/2005)

Author Index

Lecture Notes in Computer Science

Sublibrary 1: Theoretical Computer Science and General Issues

Vol. 4638: T. Stützle, M. Birattari, H. H. Hoos (Eds.), Engineering Stochastic Local Search Algorithms. X, 223 pages. 2007.

Vol. 4630: H.J. van den Herik, P. Ciancarini, H.H.L.M. (Jeroen) Donkers(Eds.), Computers and Games. XII, 283 pages. 2007.

Vol. 4628: L.N. de Castro, F.J. Von Zuben, H. Knidel (Eds.), Artificial Immune Systems. XII, 438 pages. 2007.

Vol. 4627: M. Charikar, K. Jansen, O. Reingold, J.D.P. Rolim (Eds.), Approximation, Randomization, and Combinatorial Optimization. XII, 626 pages. 2007.

Vol. 4624: T. Mossakowski, U. Montanari, M. Haveraaen (Eds.), Algebra and Coalgebra in Computer Science. XI, 463 pages. 2007.

Vol. 4623: M. Collard (Ed.), Ontologies-based DataBases and Information Systems. X, 153 pages. 2007.

Vol. 4621: D. Wagner, R. Wattenhofer (Eds.), Algorithms for Sensor and Ad Hoc Networks. XIII, 415 pages. 2007.

Vol. 4619: F. Dehne, J.-R. Sack, N. Zeh (Eds.), Algorithms and Data Structures. XVI, 662 pages. 2007.

Vol. 4618: S.G. Akl, C.S. Calude, M.J. Dinneen, G. Rozenberg, H.T. Wareham (Eds.), Unconventional Computation. X, 243 pages. 2007.

Vol. 4616: A. Dress, Y. Xu, B. Zhu (Eds.), Combinatorial Optimization and Applications. XI, 390 pages. 2007.

Vol. 4614: B. Chen, M.S. Paterson, G. Zhang (Eds.), Combinatorics, Algorithms, Probabilistic and Experimental Methodologies. XII, 530 pages. 2007.

Vol. 4613: F.P. Preparata, Q. Fang (Eds.), Frontiers in Algorithmics. XI, 348 pages. 2007.

Vol. 4600: H. Comon-Lundh, C. Kirchner, H. Kirchner (Eds.), Rewriting, Computation and Proof. XVI, 273 pages. 2007.

Vol. 4599: S. Vassiliadis, M. Berekovic, T.D. Hämäläinen (Eds.), Embedded Computer Systems: Architectures, Modeling, and Simulation. XVIII, 466 pages. 2007.

Vol. 4598: G. Lin (Ed.), Computing and Combinatorics. XII, 570 pages. 2007.

Vol. 4596: L. Arge, C. Cachin, T. Jurdziński, A. Tarlecki (Eds.), Automata, Languages and Programming. XVII, 953 pages. 2007.

Vol. 4595: D. Bošnački, S. Edelkamp (Eds.), Model Checking Software. X, 285 pages. 2007.

Vol. 4590: W. Damm, H. Hermanns (Eds.), Computer Aided Verification. XV, 562 pages. 2007.

Vol. 4588: T. Harju, J. Karhumäki, A. Lepistö (Eds.), Developments in Language Theory. XI, 423 pages. 2007.

Vol. 4583: S.R. Della Rocca (Ed.), Typed Lambda Calculi and Applications. X, 397 pages. 2007.

Vol. 4580: B. Ma, K. Zhang (Eds.), Combinatorial Pattern Matching. XII, 366 pages. 2007.

Vol. 4576: D. Leivant, R. de Queiroz (Eds.), Logic, Language, Information and Computation. X, 363 pages. 2007.

Vol. 4547: C. Carlet, B. Sunar (Eds.), Arithmetic of Finite Fields. XI, 355 pages. 2007.

Vol. 4546: J. Kleijn, A. Yakovlev (Eds.), Petri Nets and Other Models of Concurrency – ICATPN 2007. XI, 515 pages. 2007.

Vol. 4545: H. Anai, K. Horimoto, T. Kutsia (Eds.), Algebraic Biology. XIII, 379 pages. 2007.

Vol. 4533: F. Baader (Ed.), Term Rewriting and Applications. XII, 419 pages. 2007.

Vol. 4528: J. Mira, J.R. Álvarez (Eds.), Nature Inspired Problem-Solving Methods in Knowledge Engineering, Part II. XXII, 650 pages. 2007.

Vol. 4527: J. Mira, J.R. Álvarez (Eds.), Bio-inspired Modeling of Cognitive Tasks, Part I. XXII, 630 pages. 2007.

Vol. 4525: C. Demetrescu (Ed.), Experimental Algorithms. XIII, 448 pages. 2007.

Vol. 4514: S.N. Artemov, A. Nerode (Eds.), Logical Foundations of Computer Science. XI, 513 pages. 2007.

Vol. 4513: M. Fischetti, D.P. Williamson (Eds.), Integer Programming and Combinatorial Optimization. IX, 500 pages. 2007.

Vol. 4510: P. Van Hentenryck, L.A. Wolsey (Eds.), Integration of AI and OR Techniques in Constraint Programming for Combinatorial Optimization Problems. X, 391 pages. 2007.

Vol. 4507: F. Sandoval, A.G. Prieto, J. Cabestany, M. Graña (Eds.), Computational and Ambient Intelligence. XXVI, 1167 pages. 2007.

Vol. 4502: T. Altenkirch, C. McBride (Eds.), Types for Proofs and Programs. VIII, 269 pages. 2007.

Vol. 4501: J. Marques-Silva, K.A. Sakallah (Eds.), Theory and Applications of Satisfiability Testing – SAT 2007. XI, 384 pages. 2007.

Vol. 4497: S.B. Cooper, B. Löwe, A. Sorbi (Eds.), Computation and Logic in the Real World. XVIII, 826 pages. 2007.

Vol. 4494: H. Jin, O.F. Rana, Y. Pan, V.K. Prasanna (Eds.), Algorithms and Architectures for Parallel Processing. XIV, 508 pages. 2007.

Vol. 4493: D. Liu, S. Fei, Z. Hou, H. Zhang, C. Sun (Eds.), Advances in Neural Networks – ISNN 2007, Part III. XXVI, 1215 pages. 2007.

Vol. 4492: D. Liu, S. Fei, Z. Hou, H. Zhang, C. Sun (Eds.), Advances in Neural Networks – ISNN 2007, Part II. XXVII, 1321 pages. 2007.

Vol. 4491: D. Liu, S. Fei, Z.-G. Hou, H. Zhang, C. Sun (Eds.), Advances in Neural Networks – ISNN 2007, Part I. LIV, 1365 pages. 2007.

Vol. 4490: Y. Shi, G.D. van Albada, J.J. Dongarra, P.M.A. Sloot (Eds.), Computational Science – ICCS 2007, Part IV. XXXVII, 1211 pages. 2007.

Vol. 4489: Y. Shi, G.D. van Albada, J.J. Dongarra, P.M.A. Sloot (Eds.), Computational Science – ICCS 2007, Part III. XXXVII, 1257 pages. 2007.

Vol. 4488: Y. Shi, G.D. van Albada, J.J. Dongarra, P.M.A. Sloot (Eds.), Computational Science – ICCS 2007, Part II. XXXV, 1251 pages. 2007.

Vol. 4487: Y. Shi, G.D. van Albada, J.J. Dongarra, P.M.A. Sloot (Eds.), Computational Science – ICCS 2007, Part I. LXXXI, 1275 pages. 2007.